Springer Series in **Materials Science** 11

Edited by K. Alex Müller

Springer Series in **Materials Science**

Volume 1 **Chemical Processing with Lasers**
By D. Bäuerle

Volume 2 **Laser-Beam Interactions with Materials**
Physical Principles and Applications
By M. von Allmen

Volume 3 **Laser Processing of Thin Films and Microstructures**
Oxidation, Deposition and Etching of Insulators
By I. W. Boyd

Volume 4 **Microclusters**
Editors: S. Sugano, Y. Nishina, and S. Ohnishi

Volume 5 **Graphite Fibers and Filaments**
By M. S. Dresselhaus, G. Dresselhaus, K. Sugihara, I. L. Spain, and H. A. Goldberg

Volume 6 **Elemental and Molecular Clusters**
Editors: G. Benedek, T. P. Martin, and G. Pacchioni

Volume 7 **Molecular Beam Epitaxy** Fundamentals and Current Status
By M. A. Herman and H. Sitter

Volume 8 **Physical Chemistry of, in and on Silicon**
By G. F. Cerofolini and L. Meda

Volume 9 **Tritium and Helium-3 in Metals**
By R. Lässer

Volume 10 **Computer Simulation of Ion – Solid Interactions**
By W. Eckstein

Volume 11 **Mechanisms of High Temperature Superconductivity**
Editors: H. Kamimura and A. Oshiyama

Volume 12 **Laser Technology in Microelectronics**
Editors: S. Metev and V. P. Veiko

Volume 13 **Semiconductor Silicon**
Materials Science and Technology
Editors: G. C. Harbeke and M. J. Schulz

H. Kamimura A. Oshiyama (Eds.)

Mechanisms of High Temperature Superconductivity

Proceedings of the 2nd NEC Symposium, Hakone, Japan, October 24–27, 1988

With 203 Figures

Springer-Verlag Berlin Heidelberg New York
London Paris Tokyo

Professor Dr. Hiroshi Kamimura
Department of Physics, University of Tokyo,
7-3-1 Hongo, Bunkyo-ku, Tokyo, 113, Japan

Dr. Atsushi Oshiyama
Fundamental Research Laboratories, NEC Corporation,
4-1-1 Miyazaki, Miyamae, Kawasaki 213, Japan

ISBN-13: 978-3-642-74409-9 e-ISBN-13: 978-3-642-74407-5
DOI: 10.1007/978-3-642-74407-5

Softcover reprint of the hardcover 1st edition 1989

2154/3150-543210 – Printed on acid-free paper

Preface

This volume contains the proceedings of the second in a series of biennial NEC Symposia on Fundamental Approaches to New Material Phases sponsored by the NEC Corporation, Japan. The subject of the second NEC Symposium was "Mechanisms of High Temperature Superconductivity". It was held October 24 to 27, 1988, at Hakone Kanko Hotel in Hakone near Mt. Fuji, in the style of a closed meeting with invited participants only.

Fourty-eight leading theorists and experimentalists invited from Japan, USA and Europe stayed together at the symposium site during the four days of the symposium, and were immersed in attempts to clarify the mechanism of high temperature superconductivity. The venue was a conference room facing Mt. Fuji and the beautiful lake Ashinoko extending from the foot of the slope in the old crater. At this symposium one plenary lecture was presentd by Dr. K. Alex Müller, 1987 Nobel Laureate, together with 32 invited theoretical and experimental lectures.

The aim of this symposium was to deepen our understanding of the microscopic mechanisms of high temperature superconductivity, by discussing the latest experimental results on magnetic, optical, electrical, thermal and mechanical properties of the Cu-O and Bi-O superconductors as well as proposed theoretical models of the mechanisms. The focal points of the symposium were the following:

(I) For the high temperature copper oxide superconductors, the general consensus of the participants of this symposium was that electron correlation effects are of essential importance. Based on this understanding a number of discussions took place: (1) on how to include the correlation effects and on the relative importance of various interactions; (2) on whether the ground state of a normal state is Fermi-liquid-like or not; (3) on where the doped holes mainly exist and on the nature of these holes; (4) on the value of the ratio of the energy gap to the critical temperature; (5) on whether the mechanisms are magnetic, or charge fluctuation, or excitonic, or very unusual ones quite different from the Fermi-liquid picture; (6) on whether a "normal" phase is normal or abnormal, etc.

(II) For the Bi-O superconductors, such as the BaKBiO system, the discussion concentrated on whether or not the mechanism of superconductivity is the same as that of the high temperature copper oxide superconductors.

The discussions at the symposium were very intense and stimulating. In particular, the highlight of the symposium was an extraordinary discussion session on the night of 26 October. From the summary of this discussion session prepared by Professor George Sawatzky who chaired that session very effectively, we learn what was and was not resolved in this symposium. Thus we have included Professor Sawatzky's summary at the end of this volume and readers may regard this as a summary of the whole symposium.

The arrangement of the papers in this volume is not in chronological order of the symposium program, but has been made according to subject matter. The grouping which we have made is subjective, but we hope that it will facilitate reference to the subjects of interest.

On this occasion we would like to express our sincere gratitude to all the participants for providing such excitement and contributing to the great success of this symposium. We are also indebted to the following members of the Steering Committee of the NEC Symposium for establishing the symposium and choosing "mechanisms of high temperature superconductivity" as the key subject of the second symposium: Dr. Satoru Sugano, Professor of the University of Tokyo and Chairman of the Steering Committee; Dr. Michiyuki Uenohara, Senior Executive Vice President and Director of NEC Corporation; Dr. Daizaburo Shinoda, Executive Vice President of NEC Research Institute, Inc. in USA; Dr. Fujio Saito, Vice President of NEC Corporation; Dr. Teruya Shinjo, Professor of Kyoto University; Dr. Akio Yoshimori, Professor of Osaka University; and Dr. Yuichiro Nishina, Professor of Tohoku University. We are also grateful to the following advisory members of the Organizing Committee of the second NEC Symposium for their valuable advice: Dr. Elius Burstein, Professor of Physics at the University of Pennsylvania; Dr. Sadao Nakajima, Professor at Tokai University and Dr. Shoji Tanaka, Director of the International Superconductivity Technology Center, Tokyo. Finally we thank the following members of the Organizing Committee for organizing the second NEC Symposium: Professor Teruya Shinjo who is the co-chairman of the Organizing Committee, Professor Koichi Kitazawa, Professor Sadamichi Maekawa, Professor Shin-ichi Uchida, and Dr. Hiroyuki Abe.

Tokyo, December 1988

Hiroshi Kamimura
Atsushi Oshiyama

Contents

Part I

Plenary Lecture

The Development of the High-Temperature Superconductivity Field

K. Alex Müller

IBM Zürich Research Laboratory, CH-8803 Rüschlikon, Switzerland

To open the 2nd NEC Symposium devoted to "Mechanisms of High-Temperature Superconductivity," the progress that occurred since the discovery of the layered cuprate was reviewed. This was done by recalling first the knowledge gained by the end of 1987, and then referring to the substantial results achieved by fall of 1988. The emphasis was put on the experiments and the structural aspects, mainly because the author was more directly following these, but also as these are really relevant regarding the theoretical models. One may even say that so far the main advance took place on the empirical side.

The status of the field at the end of 1987 has been summarized in two parts of a presentation called "Perovskite-type oxides – The new approach to high-T_c superconductivity" [1] and the reader is referred to this review for details and references. Part I summarized the early work in Rüschlikon on La_2CuO_4 as well as the subsequent work on $YBa_2Cu_2O_7$, which both exhibit Cu-O sheets, and part II then reviewed the electronic aspects. We shall now discuss some highlights of the latter. By spring of 1987, the question "does one again have Cooper pairing in the cuprates?" could be answered in the affirmative: Two different experiments, both microwave-induced Shapiro steps in Josephson junctions and flux quantization $\phi_0 = hc/q$, yielded $q = |2e|$ for the charge of the particles. Also the nature of the carrier charge present was resolved, in that the compounds had to contain holes. Charge-neutrality requirements, Hall-effect measurements and chemical analysis all gave the same answer. By summer of the same year, photoemission experiments indicated that the mobile holes were predominantly located in the oxygen-derived bands $\underline{L}$, i.e. the predominant configuration is $3d^9\,\underline{L}$. The most direct confirmation was through data on core-level excitations from oxygen 1s electrons into empty 2p states occurring at 528 eV.

Investigating the properties of the hole-concentration dependence of T_c in the Sr^{2+}-doped La_2CuO_4 compound, $La_{2-x}Sr_xCuO_4$, a maximum of $T_c = 38$ K at $x = 0.15$ was found and a threshold of 0.05% for the superconductivity discovered. Below this concentration, the carriers are localized, and antiferromagnetic ordering was observed for the low hole concentrations, up to 240 K at $x = 0$. Similar results were found in $YBa_2Cu_3O_7$, which has a maximum T_c of 92 K. In both La_2CuO_4 and $YBa_2Cu_3O_7$ (1 2 3) compounds an isotope effect was reported; in the latter, however, it is small.

Springer Series in Materials Science, Vol. 11
Mechanisms of High Temperature Superconductivity
Editors: H. Kamimura and A. Oshiyama

From the existence of Cooper pairing, a superconducting gap Δ should exist and was indeed found in three different ways: by infrared reflectivity, NMR relaxation, and tunneling. The first two, carried out on $YBa_2Cu_3O_7$, yielded $9 > 2\Delta/kT_c \geq 7$, indicative of strong coupling at least in the *a-b* plane. Tunneling experiments were less clear for two reasons: (i) owing to chemical contamination of the surface, and (ii) because the short coherence length ξ of the Cooper pairs intrinsically introduces considerable weakening of the pair potential near the gap. That short coherence lengths were present was evident from the extremely large critical fields $H_{c_2} \equiv \phi_0/2\pi\xi^2$ in the mega-gauss range. In fact, they are $\xi_c = 2-4$ Å parallel to the *c*-axis and $20-30$ Å perpendicular to it. In the 1 2 3 compound, the short coherence lengths in the layered superconductors are also important theoretically. Along the *c*-axis, a more boson-like property is implied by ξ_c.

The substantial advance in experimental analysis over the past one and a half years was possible owing to the progress in the materials sector: the characterization of the structures, their defects as intergrowths, twins, oxygen over- and understochiometry, etc. became a domain of research in its own right, with a variety of techniques. Furthermore, small single crystals became available both for the La_2CuO_4 and $YBa_2Cu_3O_7$ compounds [2]. As is well known, the former contains sheets of corner-sharing CuO_6 octahedra, the latter corner-sharing CuO_5 pyramids. In 1 2 3 compounds, two of these sheets are arranged mirror symmetrically; the mirror plane being formed by Y^{3+} ions. Between these sheets, at the apex of the pyramids, Cu-O chains are located. By late 1987 and early 1988, two new families of cuprates were discovered which contain bismuth and thallium ions instead of the rare earths. These compounds have the general formula $Bi_2Sr_2Ca_nCu_{1+n}O_{6+2n}$ or Bi replaced by Tl, with near-identical structure [3]. For $n = 0$ one again finds the corner-sharing octahedra, and for $n = 1$ the mirror-symmetric corner-sharing pyramids, as in $YBa_2Cu_3O_7$ in which the mirror plane is occupied by Ca^{2+} or Sr^{2+} instead of Y ions. For $n = 2$ in the mirror plane, a square planar Cu-O sheet is located. The thallium variety of this compound shows the highest $T_c = 125$ K found so far. The $n = 3$ has also been synthesized: There two square planar Cu-O sheets are present between the CuO_5 pyramidal sheets. However, the T_c is lower than that of the $n = 2$ compound. It has to be added that between these sheets two layers of either Bi-O or Tl-O are present. For the Tl cuprates, the series with only one Tl-O layer has been found and the $n = 0$ to $n = 3$ structures synthesized (see R.B. BEYERS, S.S.P. PARKIN in [3]).

The $BaBi_{1-x}Pb_xO_3$ perovskite electron superconductor with a $T_c = 13$ K was of importance in the discovery of the layered cuprates [1]. The latter, right up to this symposium, are doped to be hole superconductors. Thus doping the parent insulator perovskite $BaBiO_3$ to contain holes was a materials challenge, and a major breakthrough could be achieved at AT&T in synthesizing $Ba_{1-x}K_xBiO_3$. This perovskite is a 3-D hole superconductor with a remarkable $T_c = 29.6$ K at $x = 0.4$; the details are reviewed in B. BATLOGG's articles [3, 4]. As the T_c of hole-containing $BaBiO_3$ is more than twice as high as that of the electron-containing compound, one might expect an enhancement of T_c for hole superconductivity over electron superconductivity in the cuprates if the latter are found. In saying this, we have presupposed that the mechanism in $BaBiO_3$ and the cuprates is similar. This is what a plot of T_c versus the Sommerfeld constant $\gamma \propto N(E)$ indicates [3]. A common feature was mentioned by the present author: Bi is a valence skipper, i.e., it has a tendency to disproportionate into Bi^{3+} and Bi^{5+}. Thus, nominal Bi^{4+} has a "soft" outer shell. The latter property also pertains to the Cu^{2+} with its hole in the *d*-shell, which can easily be deformed.

The $Ba_{1-x}K_xBiO_3$ is diamagnetic above T_c. Therefore any magnetic type of mechanism is excluded. Presupposing a similarity between BaBi and the cuprates makes magnetic mechanisms for superconductivity less likely for the latter. In saying this, one should note that at the beginning of 1988, a magnetic mechanism could *quantitatively* explain the $La_{2-x}Sr_xCuO_4$ phase diagram for concentrations x of holes up to the superconducting range. The reader is referred to B. BIRGENAU and Y. ENDOH's reviews in this issue [4] and to A. AHARONY's in an earlier workshop [3]. Introducing a hole on an oxygen, i.e. creating an O^- ion, leads to ferromagnetic interactions between the adjacent Cu^{2+} ions in a Cu-O plane. Because the Cu^{2+} in the La_2CuO_4 insulator are ordered antiferromagnetically below 240 K, frustration is present. This leads to a rapid depression of the antiferromagnetic Néel temperature T_N and, for higher values of x, to a spin glass. The O^- ions responsible for this spin glass are *still localized* in the fuzzy band edge even beyond the metal-insulator transition, M-I, where the additional ones become *itinerant*. The itinerant ones are those which lead to superconductivity. The interaction between the localized "spin glass" and the delocalized "superconducting Cooper pair" ones is not clear, but NMR, muon rotation and neutron data may help to clarify this. Especially NMR was first to establish the above phase diagram.

The existence of localized and itinerant holes on oxygens implies the presence of a Fermi-surface liquid state for the higher doping levels. This is what the more fundamental theories based on magnetic interactions dispute vigorously, invoking sample inhomogeneity etc. There are now two closely related ones: the RVB and that using fractional quantum states (FQS). The latter is presented in this issue by R.B. LAUGHLIN [4]. Basically, it is a transposition of the $m = 3$ basis function of the fractional quantum Hall effect to an $m = 4$ state. By this, the fermionic character appropriate for the particles in the Hall effect with half integral spin $3 \times \frac{1}{2} = \frac{3}{2}$ is changed to integral $4 \times \frac{1}{2} = 2$ character of bosonic superconducting particles. This theory leads to a gap $\Delta(k)$ in all directions of the wave vector $\vec{k}$ as also found by tunneling experiments (see H. TAKAGI's report [4]), whereas RVB warrants closing of the gap in certain directions of $\vec{k}$. A closing is also at variance with specific-heat data in Bi-Sr-Ca-Cu-O (A.K. ZETTEL [4]).

The issue of whether RVB and FQS are just beautiful theories *per se* or also applicable to the new superconductors resides on the existence or not of a Fermi liquid. In this respect, experimental facts are accumulating that such a Fermi surface indeed exists and is of *s*-wave character. Apart from the above-mentioned simultaneous presence of spin glass and superconductivity, there are angle-resolved photoemission and positron annihilation experiments (M. PETER [3]), the London penetration depth $\lambda(T)$, and nuclear quadrupolar and magnetic resonance on O^{17} ($I = 5/2$) nuclei. Futhermore, Andreev reflections have been observed (H. VAN KEMPEN [3]), possible only in the presence of *s*-wave BCS-type pairing with a Fermi surface. In angle-resolved resonant photoemission in $Bi_2Sr_2CaCu_2O_8$, a flat band *A* has been detected that crosses the Fermi level along the $X - \Gamma$ line in the Brillouin zone (T. KATAYAMA-YOSHIDA *et al.* [4]). However, this method is quite surface sensitive. Of importance is that the London penetration depth $\lambda(T)$ follows *very accurately* a $\left[1-(T/T_c)^4\right]^{-1/2}$ behavior from muon rotation experiments (see B. BATLOGG's summary [4]). This law is what is found and calculated for *s*-wave pairing of BCS type. The nuclear magnetic relaxation data $1/T_1$ on O^{17} indicate the same for several cuprate superconductors (see the reports of Y. KITAOKA *et al.* [3, 4]). On cooling, $1/T_1$ decreases towards T_c almost linearly, reminiscent of Korringa relaxation in metals. Just below T_c an enhancement is observed as has been found in the classical BCS superconductor Al. Then, $1/T_1$ is strongly depressed as often measured in metallic superconductors. On the other hand, the temperature dependence of $1/T_1$ of Cu^{63}, especially above T_c, shows an entirely different behavior. It is larger and then saturates on heating. This temperature dependence has been ascribed to AFM fluctuations at Cu sites (see also R.E. WALDSTEDT's report [4]).

The very different temperature dependences of $1/T_1$ on Cu^{63} and O^{17} force one to think that the holes on Cu^{2+} and O^- may be in different electronic bands. Otherwise cross relaxation should lead to the same temperature dependence. The most likely situation is that the holes on Cu are in a narrow antibonding band and those due to doping in a broad nonbonding one. Such a scenario has been emphasized by C.M. VARMA [3] and collaborators. Considerable theoretical work and discussion as to the symmetry of the bands in which the holes are sited has taken and is taking place at this conference. BIRGENAU [4] and AHARONY [3] ascribe the $\underline{L}$ carriers to *p*-orbitals in-plane *perpendicular* to the Cu-O axis, SAWATZKY [4] *along* this axis, and KAMIMURA [4] *out-of-plane* perpendicular to it. Most recent angle-resolved 1*s*-2*p* core level spectroscopy in $Bi_2Sr_2Ca_1CuO_8$ by FINK *et al.* [3] indicate that the orbitals are in-plane. A clear-cut experiment to distinguish between the remaining two possibilities is needed.

In two band models the relative distance between the *d* and *p* levels $\varepsilon = E_d - E_p$ is of importance. Cluster calculations depend on initial conditions and yield positive or negative values of ε within 0.2 eV. From this result and independently of the sign of ε, the charge transfer energy has to be of this magnitude. Thus excitonic mechanisms have been given serious consideration. Among them, C.M. VARMA [3] and J.E. HIRSCH [4] consider in-plane Cu-O transfer, and most recently A. BISHOP *et al.* [5] out-of-plane transfer. The latter would thus become related to earlier propositions of Ginzburg and Little. To make progress, experiments to distinguish between them are of great interest. Excitonic mechanisms also imply a high polarizability, especially of the oxygens. The support for such high dielectric constants is deferred to another paper by the author in the journal *Ferroelectrics*.

Acknowledgements

The author should like to thank all colleagues mentioned in this article and many others, especially I. Morgenstern and T. Schneider.

References

1. J. Georg Bednorz and K. Alex Müller: Nobel Lecture 1987, reprinted in *Rev. Mod. Phys.* **60**, 585 (1988).
2. Many important contributions are contained in the Proceedings of the Interlaken Conference on High Temperature Superconductors, Materials and Mechanisms of Superconductivity, HTSC-M²S, edited by J. Müller and J. L. Olsen, special issue of *Physica C*, vol. 153-155 (North-Holland, Amsterdam, 1988).
3. For reviews, see the Proceedings of the Oberlech Workshop on High T_c Superconductivity, Oberlech, Austria, August 8-14, 1988, edited by J. G. Bednorz and K. A. Müller, to appear in the May 1989 issue of *IBM Journal of Research and Development*.
4. See the respective contribution in these proceedings of the 2nd NEC Symposium on "Mechanisms of High Temperature Superconductivity," Hakone, Japan, October 24-27, 1988.
5. A.R. Bishop, R.L. Martin, K.A. Müller, Z. Tešanović: to be submitted to *Z. Phys. B*.

Part II

Theoretical Approach

Spin-Polaron Pairing Mechanism in the High T_c Copper Oxides

Hiroshi Kamimura, Shunichi Matsuno, and Riichiro Saito

Department of Physics, University of Tokyo, Bunkyo-ku, Tokyo 113, Japan

On the basis of the spin-polaron pair model with s-wave pairing recently developed by Kamimura, Matsuno and Saito, the origin of the x dependence of T_c and that of a high value of T_c in $La_{2-x}Sr_xCuO_4$ are clarified. Predictions are also made on (i) the coexistence of the superconducting phase and spin fluid phase and (ii) the simultaneous occurrence of the superconducting phase and the metal-insulator transition. Anomalous behaviour of NMR and Hall effects in high T_c copper oxides are discussed in terms of the spin-polaron pair model.

1. INTRODUCTION AND SUMMARY

Recently KAMIMURA [1], and KAMIMURA, MATSUNO, and SAITO [2] have developed *spin-polaron pairing mechanism* based on a so-called *two-floor house model* as a possible mechanism of high temperature superconductivity in the copper oxides. The purpose of the present paper is two-fold:(i) to calculate the doping concentration dependence of T_c in $La_{2-x}Sr_xCuO_4$ quantitatively and to clarify its microscopic origin and the origin of a high value of T_c and (ii) to predict some of the characteristic features of the magnetic and transport properties, based on the spin-polaron pair model. Before doing so, the important features of spin-polaron pairing mechanism are summarized in §2. Then in §3 the calculated result of T_c in $La_{2-x}Sr_xCuO_4$ is presented. It is shown that, if we choose the effective coupling between spin-polarons to be 1500K, a theoretical result reproduces recent experimental results satisfactorily. In §4 the origin for the x dependence of T_c and a high value of T_c are clarified as follows. Since the x dependence of T_c is theoretically given as the convolution of s-wave pairing and the density of states of the second floor band in the energy space in our theory, it is clarified that a high value of T_c is due to the interplay of a large value of the coupling constant and a high density of states near the top of the second floor band. On the other hand, the decrease of T_c with increasing x is due to the energy dependence of the s-wave pairing. In §5 it is shown from the present theory that both superconducting and spin-fluid phases may coexist in $La_{2-x}Sr_xCuO_4$. In §6 anomalous behaviors of NMR and Hall effects observed in the high T_c copper oxides are discussed and some of observed features are shown to be explained qualitatively by the spin-polaron pair model.

Springer Series in Materials Science, Vol. 11
Mechanisms of High Temperature Superconductivity
Editors: H. Kamimura and A. Oshiyama
Springer-Verlag Berlin Heidelberg © 1989

2. IMPORTANT FEATURES OF SPIN-POLARON PAIRING MECHANISM

In this section we first summarize the important characteristics of the spin-polaron pairing mechanism developed by us. Since in our mechanism the existence of CuO_2 layer is essentially important, our mechanism is applicable to any of La-(Ba,Sr)-Cu-O, Y-Ba-Cu-O, Bi-Sr-Ca-Cu-O and Tl-Sr-Ca-Cu-O systems all of which include CuO_2 layers. Thus we describe the essential points of our mechanism for the simplest system of $La_{2-x}Sr_xCuO_4$ by way of itemizing them.

(i) *Existence of two bands* (*two-floor house*). In our mechanism the transfer interactions between Cu d and O p orbitals are first diagonalized. Thus in pure La_2CuO_4 the first floor of the two-floor house model as shown in Fig.1 corresponds to the hybridized orbital of Cu dx^2-y^2 and O $p\sigma$ orbitals extended along a CuO_2 layer while the second-floor state corresponds to the hybridized orbital of Cu dz^2, O $p\sigma$ in CuO_2 layer and Op_z above and below Cu ions. In Fig.1 the energy scale is taken for a hole. If we consider only a single CuO_2 layer consisting of CuO_6 octahedra, we can classify the first- and second-floor orbitals for wave vector $\mathbf{k}=\mathbf{0}$ by the irreducible representation of D_{4h} group as b_{1g} and a_{1g} orbitals respectively, following **SAWATZKY**[3]. The atomic basis functions which contribute to a_{1g} orbital are shown in Fig.2. In the present paper we call the first- and second-floor states b_{1g} and a_{1g} bands, respectively.

(ii) *Strong correlation in the first floor*. By the strong correlation at Cu sites for b_{1g} band which is expressed as

Fig.1 Two-floor house model corresponding to two orthogonal a_{1g} and b_{1g} bands, and the occupation of dx^2-y^2 holes in pure La_2 CuO_4 together with the representation of up and down spin states in a two-dimensional quantum spin system

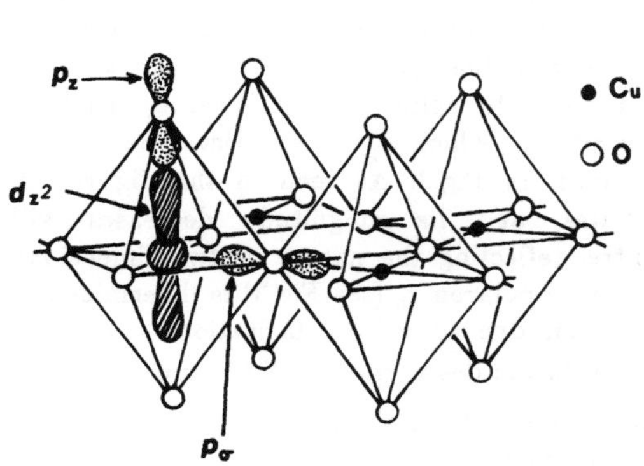

Fig.2 Atomic basis functions, Cu dz^2, O p_σ in CuO_2 layer and O p_z above and below Cu, which contribute to the construction of a_{1g} orbital in the second floor

$$H_{corr}=U\Sigma_m \hat{n}_\uparrow(\text{Cu m}, dx^2\text{-}y^2)\hat{n}_\downarrow(\text{Cu m}, dx^2\text{-}y^2), \quad (1)$$

with $\hat{n}_\sigma$(Cu m, dx^2-y^2) being the number operator in dx^2-y^2 orbital at Cu m site with spin σ, the holes in the first floor are localized at Cu sites. Hereafter we call them the Cu dx^2-y^2 holes, each of which has spin 1/2. The so-called Hubbard U interaction for the Cu dx^2-y^2 orbital in (1) is very strong and of the order of 6 to 8 eV. Then the localized Cu dx^2-y^2 spins interact each other by the superexchange interaction through intervening oxygen 2s and 2p orbitals. Its Néel temperature for a three-dimensional antiferromagnetic ordering of dx^2-y^2 spins is of the order of 200K.

(iii) ***Hund's coupling.*** By doping divalent ions such as Sr in pure La_2CuO_4, extra holes are created in the second floor (a_{1g} band) but not in the first floor (b_{1g} band) by the ***intra-atomic exchange interaction***, because the interaction energy for two holes in the first and the second floors is larger than the energy difference between a_{1g} and b_{1g} band which is about 0.5 eV from the band structure calculation [4]. From the ligand field theory [5] the intra-atomic exchange interaction in a single $Ni^{2+}(3d^8)$ ion in the environment of octahedral symmetry is of the order of 1 eV, but it is expected to be much reduced in a CuO_2 layer system by strong covalency. The essentially important role of Hund's coupling is the key point of our mechanism which was first proposed by KAMIMURA [6] in early 1987.

(iv) ***RVB type spin state in the first floor.*** When extra holes are created in the second floor, the weak superexchange interaction between Cu dx^2-y^2 spins in different CuO_2 layers is destroyed at the certain Sr doping concentration x_c. Then the Cu dx^2-y^2 spins in the first floor act as a two-dimensional quantum spin system. We assume that the Cu dx^2-y^2 spins form a ***RVB type spin-singlet ground state*** for $x > x_c$ in $La_{2-x}Sr_xCuO_4$. This situation is schematically shown in Fig.1, where persons standing upright and on their heads represent spin-up and spin-down states of Cu dx^2-y^2 holes, respectively.

(v) ***Spin-polaron.*** Then, for $x > x_c$ spin-polarons are created by intra-atomic exchange interaction between a_{1g} holes in the second floor and Cu dx^2-y^2 holes in the first floor at Cu sites. Namely, the intra-atomic exchange interaction always favours ferromagnetic spin alignment so that the spins of Cu dx^2-y^2 holes in the first floor tend to be polarized parallel with spins of a_{1g} holes, by partially destroying the RVB state around Cu sites at which a_{1g} holes exist. The creation of a spin-polaron is schematically shown in Fig.3. As seen in this figure, the direction of spin-polarization oscillates, and its magnitude decreases with increasing the distance from the centre, reflecting the nature of two-dimensional quantum spin system. The radius of a spin-polaron r_0 (see Fig.3) is determined by the competition between the energy gain due to spin polarizations caused by Hund's coupling and the energy loss due to the destruction of RVB type state.

(vi) ***Spin-polaron pairing.*** When a number of spin-polarons are created, an attractive antiferromagnetic exchange interaction appears between spin-polarons,

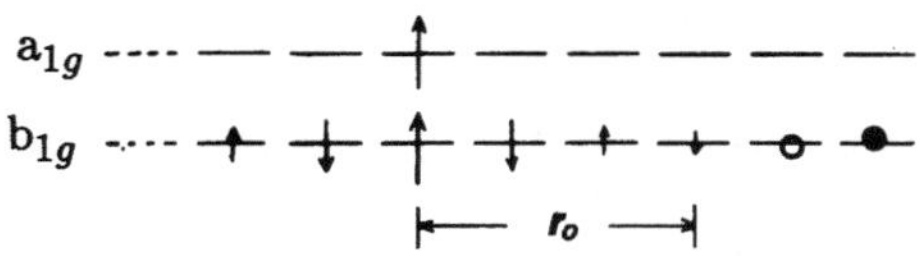

Fig.3 Creation of a spin-polaron. o—● outside the radius of a spin-polaron represents spin-singlet state in the RVB type state

so that they form spin-singlet pairs for a certain value of the strength of the intra-atomic exchange interaction which is of the order of 0.1 eV. The accumulation of the energy gain due to the Hund's coupling at each Cu site in the range of a spin-polaron r_0 leads to a much larger spin-polaron pairing energy.

(vii) *The motion of a spin-polaron pair.* The effective mass of a single spin-polaron is very large, because the transfer interaction of an a_{1g} hole in the second floor is reduced by the overlap integral between different spin states before and after the transfer of a spin-polaron, unless spin-polarons move only along the oxygen sites at which the magnetic interactions due to Cu dx^2-y^2 spins cancel. However, when a spin-polaron pair with spin-singlet is formed, the spin-polaron pair can move by the transfer interaction of the a_{1g} holes in the second floor, because both the spin-polaron pair and its surrounding RVB type state are the spin-singlet so that the reduction of the transfer integral by the overlap integral between spin states is negligibly small. This has been also shown by the calculation of one dimensional spin system by ISHIDA and KAMIMURA [7].

Taking account of the above seven essential points of our mechanism, KAMIMURA *et al.* [1,2] proposed the following Kondo-lattice type Hamiltonian to describe the formation and motion of spin-polaron pairs.

$$H = H_{tr} + H_{super} + H_{intra}$$

$$= \Sigma_{nm\sigma} t_{nm}(a_{n\sigma}^{+} a_{m\sigma} + h.c.) + J\Sigma_{nm} \mathbf{S}_n \cdot \mathbf{S}_m - K\Sigma_n \boldsymbol{\sigma}_n \cdot \mathbf{S}_m, \qquad (2)$$

where t_{nm} represents the transfer of a_{1g} hole in the second floor between Cu n and m sites through p orbitals of intervening oxygen sites, $a_{n\sigma}^{+}$ ($a_{n\sigma}$) the creation (annihilation) operator of the a_{1g} hole at Cu n site, J the superexchange coupling between the spins $\mathbf{S}_n$ and $\mathbf{S}_m$ of dx^2-y^2 holes at nearest neighbour Cu n and m sites (J> 0, antiferromagnetic) and K the intra-atomic exchange integral between the spin of the a_{1g} hole $\boldsymbol{\sigma}_n$ and that of dx^2-y^2 hole at Cu n site (K> 0, ferromagnetic). Assuming that the double occupancy in dx^2-y^2 and a_{1g} orbitals are prohibited, by the strong electron correlation, the above Hamiltonian can be deduced from a general Hubbard Hamiltonian for two bands including lattice distortions introduced by AOKI and KAMIMURA [8], in which the Wannier orbitals are taken as a basis set and as a result the oxygen p orbitals are apparently eliminated.

3. CALCULATED RESULTS OF T_c FOR $La_{2-x}Sr_xCuO_4$

In order to calculate T_c for $La_{2-x}Sr_xCuO_4$ based on the spin-polaron pairing mechanism, we introduce the following effective spin Hamiltonian, noting that the magnitude of the spin of a spin-polaron is 1/2 and the spin-polarons whose motion is governed by the a_{1g} holes in the second floor obey Fermi statistics;

$$H_{eff} = \Sigma_{nm\sigma}\tilde{t}_{nm}(C_{n\sigma}{}^{+}C_{m\sigma}+h.c.) + J_{eff}\Sigma_{nm}\mathbf{s}_n \cdot \mathbf{s}_m, \tag{3}$$

where $\tilde{t}_{nm}$ is the effective transfer integral of a spin-polaron between Cu n and m sites in a crystal, J_{eff} the exchange coupling between spin-polarons ($J_{eff}> 0$, antiferromagnetic), and $C_{n\sigma}{}^{+}$ and $C_{n\sigma}$ represent creation and annihilation operators of a spin-polaron at Cu n site, respectively. Further the spin of a spin-polaron at n site, $\mathbf{s}_n$, has been introduced by a relation,

$$\mathbf{s}_{nz} = 1/2(C_{n\uparrow}{}^{+}C_{n\uparrow} - C_{n\downarrow}{}^{+}C_{n\downarrow})$$

$$\mathbf{s}_{n+} = C_{n\uparrow}{}^{+}C_{n\downarrow}$$

$$\mathbf{s}_{n-} = C_{n\downarrow}{}^{+}C_{n\uparrow}. \tag{4}$$

It should be mentioned that (3) is the effective Hamiltonian which describes the motion of spin-polaron pair but not that of an isolated spin-polaron. Thus $\tilde{t}_{nm}$ in (3) should be understood as the transfer integral of a spin-polaron pair but not that of an isolated spin-polaron. In the mean field approximation the gap equation corresponding to the effective Hamiltonian (3) is obtained in the wave vector representation as follows;

$$\Delta(\mathbf{k},T) = -(1/2N)\Sigma_q V(\mathbf{k},\mathbf{q})\Delta(\mathbf{q},T)\tanh[E(\mathbf{q},T)/2k_BT]/E(\mathbf{q},T), \tag{5}$$

where

$$V(\mathbf{k},\mathbf{q}) = -J_{eff}[(1/2)\gamma(\mathbf{k}-\mathbf{q}) + \gamma(\mathbf{k}+\mathbf{q})] \tag{6a}$$

$$E(\mathbf{k},T) = [(\varepsilon(\mathbf{k})-\mu)^2 + \Delta(\mathbf{k},T)^2]^{1/2} \tag{6b}$$

$$\gamma(\mathbf{k}) = \cos(k_x a) + \cos(k_y a), \tag{6c}$$

with μ being the chemical potential, a the lattice constant between Cu sites, and $\varepsilon(\mathbf{k})$ the energy dispersion of an a_{1g} hole.

Since the doping concentration of divalent ions is very low, that is of the order of $10^{21}cm^{-3}$, μ for a_{1g} holes (in the second floor) is not so large. Therefore, the Fermi level is located near the top of a_{1g} band. Because of a small wave vector $\mathbf{k}$ corresponding to a small Fermi surface for a_{1g} holes, one can approximate the order parameter for spin-polaron pairing $\Delta(\mathbf{k},T)$ as the form of s-wave pairing;

$$\Delta_s(\mathbf{k}) = \Delta(T)\gamma(\mathbf{k}). \tag{7}$$

Then T_c is calculated from the following equations;

$$2 = (3/4N)J_{eff}\Sigma_{\mathbf{k}}\gamma^2(\mathbf{k})\tanh[(\varepsilon_{\mathbf{k}}-\mu)/2k_BT_c]. \tag{8}$$

In order to calculate T_c of $La_{2-x}Sr_xCuO_4$ as a function of x (that is μ), one needs information on $\varepsilon(\mathbf{k})$ and J_{eff}. As mentioned in the essential point (vii) of the spin-polaron pair model in §2, once a spin-polaron pair is created, it can move according to the transfer interaction of the a_{1g} hole in the second floor. In this context the transfer integral of a spin-polaron $\tilde{t}_{nm}$ in (3) may be taken as that of a_{1g} hole, that is t_{nm} in Hamiltonian (2); $\tilde{t}_{nm}=t_{nm}$, though the transfer interaction of a single spin-polaron is very small. Then one can use the result of a real band structure calculation for the Cu-O hybridized a_{1g} band in $La_{2-x}Sr_xCuO_4$ for $\varepsilon(\mathbf{k})$ in (8). KAMIMURA *et al.* [1,2]. have expressed the numerical result of band structure calculation for the a_{1g} band by SHIRAISHI, OSHIYAMA, SHIMA, NAKAYAMA, and KAMIMURA [9,10] by the following analytical expression

$$\varepsilon(\mathbf{k}) = A[\cos(k_xa) + \cos(k_ya)] + B\cos(k_xa)\cos(k_ya)$$
$$+C[\cos(2k_xa)+\cos(2k_ya)] + D\cos(k_xa)\cos(k_ya)\cos(k_zc), \tag{9}$$

with A=179 meV, B=−71 meV, C=−40 meV, and D=−22 meV.

In this case a remaining parameter in the gap equation is J_{eff}. KAMIMURA *et al.* [1,2] have varied J_{eff} as a parameter and found that the experimental results of the x dependence of T_c in $La_{2-x}Sr_xCuO_4$ reported by TORRANCE *et al.* [11,12] is reproduced satisfactorily by taking J_{eff} as 1500K for the case of s-wave pairing. Both the calculated results of s-wave pairing with J_{eff}=1500K and experimental result concerning the x dependence of T_c for $La_{2-x}Sr_xCuO_4$ are shown in Fig.4. In so doing we have assumed that in the localized region of $x < x_c \sim 0.06$ the RVB type state does not yet appear in the first floor on account of that the interlayer magnetic interactions or three-dimensional spin-glass type interactions still exist. Or, even though spin-polarons are created in $x < x_c$, they cannot move because the Néel state still exists in the first floor. In this context we have shifted a theoretical onset point of superconductivity at x=0 to the experimental onset point x_c in Fig.4, although this conventional treatment is not correct rigorously. According to this argument, the doped holes cannot move for $x < x_c$. Therefore, the number of spin-polaron pairs which contribute to superconductivity is expected to increase discontinuously from zero to a finite value at $x=x_c$. As a result T_c will increase more sharply from zero at $x=x_c$, compared with the calculated result. Nevertheless, as seen in Fig4., the agreement between the theoretical calculation and experimental result is remarkably good.

We conclude from the argument of this section that $x=x_c$ should correspond to a critical point at which the transition from a three-dimensional antiferromagnetic

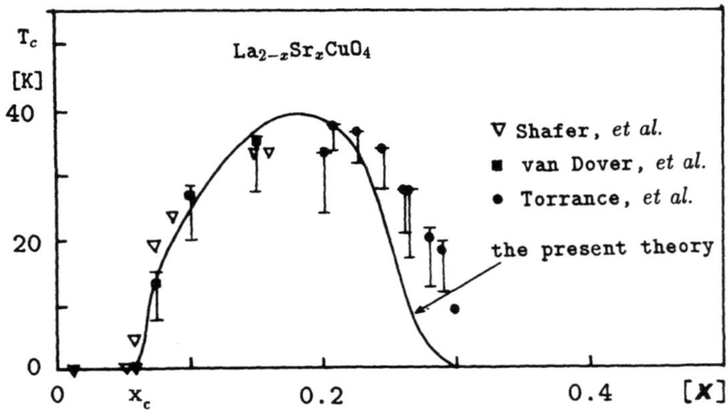

Fig.4 Calculated and observed x dependence of Tc in $La_{2-x}Sr_xCuO_4$. Experimental data are taken from ref.[11]

state (or Néel state) to RVB type state in the first floor, the insulator-metal transition and the transition to a superconducting phase take place simultaneously.

4. ORIGIN FOR THE DOPING CONCENTRATION DEPENDENCE

In this section we show theoretically that a high value of T_c in $La_{2-x}Sr_xCuO_4$ comes partly from a sharp peak in the density of states of the a_{1g} band corresponding to the second floor. For this purpose we have calculated the density of states of the a_{1g} band corresponding to (9). The calculated result is shown in Fig.5, where only the density of states near the top of a_{1g} band is shown.

In (9) the first to third terms are due to the intra-layer transfer interactions and the fourth term is due to the interlayer transfer interaction. If we consider

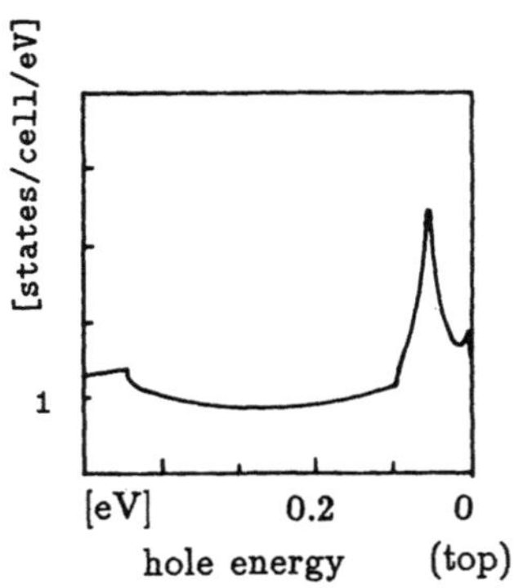

Fig.5 Density of states of a_{1g} band near the top of the band, calculated by fitting to the energy band structure of $La_{2-x}Sr_xCuO_4$ by SHIRAISHI *et al.* [4,9,10]

only the first term the density of states has a logarithmic singularity at the middle of the band which is characteristic of a two-dimensional saddle point. Because of the existence of the second term in (9) the peak in the density of states shifts to near the top of the band and the height of the peak becomes a little lower by the effect of the fourth term, i.e. the three-dimensional dispersion. Thus the peak in the density of states in Fig.5 is due to the contribution from a saddle point in the a_{1g} band at which the mass along the CuO_2 layer and that along the c-axis have different signs.

When holes are doped from the top of the a_{1g} band, the Fermi energy shifts from the top of the band to lower energy. Then T_c increases with increasing the density of states and takes a maximum value at a certain value of x, x_1, which approximately corresponds to a sharp peak in the density of states. Since T_c in (8) is calculated as the convolution of the density of states of the a_{1g} band and the order parameter of the s-wave pairing (7) in the energy space, T_c decreases from its maximum with further increasing x, reflecting the decrease of $\Delta(\mathbf{k})$ in (7) with increasing the energy. Thus a high value of T_c at $x=x_1$ is due to (i) a large value of the two-dimensional antiferromagnetic exchange between spin-polarons J_{eff} and (ii) the appearance of a sharp peak in the density of states of a_{1g} band associated with the motion of spin-polaron pairs. Further the decrease of T_c in $x > x_1$ may be considered as one of the evidences of the s-wave pairing.

5. COEXISTENCE OF THE SUPERCONDUCTING PHASE AND A SPIN FLUID PHASE

In the present spin-polaron pairing mechanism the spin-polaron pairs are created at the critical doping concentration x_c at which the three-dimensional antiferromagnetic phase disappears and consequently the RVB type state is produced in the first floors. As a result the spin-polaron pairs can move and contribute to the superconductivity. Thus in the superconducting phase the spin correlation still exists in the region of RVB type state in the first floor between spin-polaron pairs. Thus the spin correlation length in the superconducting phase corresponds to that in the RVB region which is equal to the average distance between spin-polaron pairs in a CuO_2 layer, that is proportional to $1/\sqrt{x}$, which is consistent with experimental results [13,14]. Since such a RVB region moves with the motion of spin-polaron pairs, we call such a phase of the spin system the spin fluid phase. Thus we conclude that the coexistence of the superconducting phase and the spin fluid phase is one of the characteristic features of the present spin-polaron pair model. This feature is schematically shown in Fig.6.

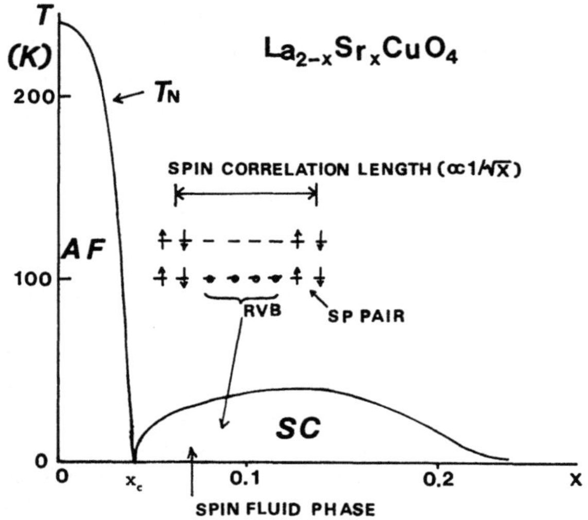

Fig.6 Phase diagram predicting the coexistence of superconducting phase and spin fluid phase, together with a critical concentration x_c at which the superconductivity and the insulator-metal transition occur simultaneously

6. REMARKS ON CHARACTERISTICS PREDICTED FROM THE SPIN-POLARON PAIR MODEL

(1) <u>NMR results</u>

First we would like to discuss NMR results at Cu and O sites at which the observed temperature dependence of spin-lattice relaxation rate $1/T_1$ is remarkably different for Cu and O sites both in superconductivity and normal phases [15,16,17]. In the present model the spin-polaron pairs in the superconducting phase can move according to the transfer motion of a_{1g} holes in the second floor. Since the magnetic interaction effect due to Cu dx^2-y^2 spins cancels at O sites because of the spin-singlet nature of Cu dx^2-y^2 spins, we may expect that mechanisms contributing to spin lattice relaxation should be different at Cu and O sites. Further in the normal state the spin-polaron pairs are dissociated into spin-polarons. As mentioned in the essential point (vii) in §2, spin-polarons cannot move from Cu to Cu sites because of small overlap between different spin states. Thus the spin-polarons in the normal state might move only along the oxygen sites in order to avoid the spin scattering due to Cu dx^2-y^2 spins. In that case the a_{1g} holes at O sites behave just like Fermi liquid particles. Thus in the normal state we may expect that $1/T_1$ at O sites obeys the Koringa relation. On the other hand, at Cu sites the existence of localized Cu dx^2-y^2 spins influence critically the temperature dependence of $1/T_1$ and thus it is expected to deviate substantially from the Koringa relation in the normal state.

(2) Polarized X-ray absorption at Cu sites

BIANCONI *et al.* [18] recently measured the polarized X-ray absorption spectra from Cu 2p deep core levels to empty states just above the Fermi level and observed the existence of Cu dz^2 character as well as dx^2-y^2 character near the Fermi level. This experimental result is consistent with the present two-floor house model.

(3) Hall effect in the normal state

IYE [19] reported the anisotropic behaviour of the Hall effect; the hole-like behaviour for a magnetic field applied along the c-axis while the electron-like behaviour for a magnetic field applied in the a-b plane, where the current flows in the a-b plane. This anomalous behaviour of the Hall effect in the normal state may be explained by the saddle point characteristic of the Fermi surface near the sharp peak in the density of states shown in Fig.5.

(4) Effect of coulomb repulsive interaction between dx^2-y^2 and a_{1g} holes at Cu sites

In the present paper we have not considered the effect of off-diagonal coulomb repulsive interaction between dx^2-y^2 and a_{1g} holes at Cu sites, which is expressed as

$$H_{coul} = U_{12}\Sigma_m \hat{n}(\mathrm{Cu}\ m,\ dx^2\text{-}y^2)\hat{n}(\mathrm{Cu}\ m,\ dz^2). \qquad (10)$$

When we take account of this effect, we have to perform the configuration interaction calculation for the result of the band structure calculation done by SHIRAISHI *et al.*. Then it is expected that the values of A,B,C, and D in (9) will change. In particular the value of A will be much reduced from 179 meV obtained from the band structure calculation. Thus we may expect that the spin-polaron pairs stay in a longer time at O sites, compared with the present result.

Furthermore, when the spin-polaron pairs are dissociated into spin-polarons in the normal state, spin-polarons hardly move from Cu sites to Cu sites, because the spin states at Cu sites before and after transfer are different, as mentioned above. If we take account of the configuration interaction effect due to the interaction (10), spin-polarons become able to move mainly from O to O sites. As a result the spin scattering with Cu dx^2-y^2 spins are avoided and the effective mass of a single spin-polaron becomes lighter. This argument supports various experimental evidence that doped holes stay mainly at O sites [14,15,16,17,20].

(5) The d-wave pairing

Based on the present spin-polaron pairing mechanism, we can also calculate T_c for d-wave pairing, for which the order parameter is expressed as

$$\Delta_d(\mathbf{k},T) = \Delta(T)[\cos(k_x a)-\cos(k_y a)]. \qquad (11)$$

In that case the equation to determine T_c is given as

$$2 = (3/4N)J_{eff}\Sigma_k[\cos(k_xa)-\cos(k_ya)]^2\tanh[(\varepsilon_k-\mu)/2k_BT_c]. \qquad (12)$$

In this case we find from a numerical result that T_c of $La_{2-x}Sr_xCuO_4$ is zero for $0 < x < 0.3$ and then increases monotonously with increasing x up to 1.0. We can conclude from this result that the d-wave pairing does not occur in the copper oxide superconductors as far as the spin-polaron pairing mechanism is adopted.

In summary we have first given a brief review on the spin-polaron pairing mechanism, summarizing the important features of this mechanism. Then, based on this mechanism the origin of the doping concentration dependence of T_c and that of a high value of T_c in $La_{2-x}Sr_xCuO_4$ are clarified. Further predictions have been made on the coexistence of superconducting and spin fluid phases, the simultaneous occurrence of superconductivity and the metal-insulator transition and anomalous behaviour of NMR and Hall effects.

References

1. H. Kamimura: In Proc. Adriatico Conf. on Theoretical Understanding of High Temperature Superconductivity , ed. by Yu Lu, E. Tosatti (World Scientific, Singapore, 1988), to be published
2. H. Kamimura, S. Matsuno, R. Saito: Solid State Commun. 67, 363 (1988)
3. G.A. Sawatzky: In Proc. Adriatico Conf. on Theoretical Understanding of High Temperature Superconductivity , ed. by Yu Lu, E. Tosatti (World Scientific, Singapore, 1988), to be published. Related references therein and in this conference proceedings
4. H. Kamimura, T. Nakayama, A. Oshiyama, N. Shima, K. Shiraishi: In Proc. 19th Int. Conf. on Physics of Semiconductors, ed. by W. Zawadzki, J.M. Langer ,Institute of Polish Academy, Warsaw, 1989), to be published
5. S. Sugano, Y. Tanabe, H. Kamimura: Multiplets of Transition Metal Ions in Crystals (Academic Press. New York, 1970)
6. H. Kamimura: Jap. J. Appl. Phys. 26, L627 (1987);
   H. Kamimura: In Proc. Adriatico Conf. on High Temperature Superconductors, ed. by S. Lundgirst, E. Tosatti, M.P. Tosi,Yu Lu (World Scientific, Singapore, 1987) p.873
7. K. Ishida, H. Kamimura: private communication
8. H. Aoki, H. Kamimura: Solid state commun. 63, 665 (1987)
9. K. Shiraishi, A. Oshiyama, N. Shima, T. Nakayama, H. Kamimura: Solid state Commun. 66, 629 (1988)
10. N. Shima, K. Shiraishi, T. Nakayama, A. Oshiyama, H. Kamimura: In Proc. Int. Conf. on Electronic Materials, ed. by T. Kamiya, K. Mizushima, R. Chang (MRS, Pittsburgh, 1988), to be published
11. J.B. Torrance, Y. Tokura, A.I. Nazzal, A. Bezinge, S.S.P. Parkin: Phys. Rev. Lett. 61, 1127 (1988)
12. Y. Tokura: in this conference proceedings

13. R.J. Birgenau: in this conference proceedings. Related references therein
14. Y. Endoh: in this conference proceedings. Related references therein
15. Y. Kitaoka, K. Ishida, K. Fujioka, Y. Kohori, K. Asayama: in this conference proceedings. Related references therein
16. H. Yasuoka: in this conference proceedings. Related references therein
17. R.E. Walstedt, W.W. Worren,Jr, R.F. Bell, G.F. Brennert, G.P. Espinosa: in this conference proceedings. Related references therein
18. A. Bianconi, M. de Santis, A.F. Flank, A. Fontaine, P. Lagarde, A. Marcelli, H. Katayama-Yoshida, A. Kotani: In Proc. Adriatico Conf. on Theoretical Understanding of High Temperature Superconductivity, ed. by Yu Lu, E. Tosatti (World Scientific, Singapore, 1988), to be published. Related references therein
19. Y. Iye: in this conference proceedings. Related references therein
20. A. Fujimori: in this conference proceedings, and related references therein

The Importance of Correlation, Multiplet Structure, Hybridization and Symmetry in the Electronic Structure of High T_c Cu Compounds

H. Eskes, H. Tjeng, and G.A. Sawatzky

Department of Applied and Solid State Physics, Materials Science Center, University of Groningen, Nijenborgh 18, 9717 AG Groningen, The Netherlands

Combining experimental and theoretical results on Cu oxide model compounds with those on high T_c materials we draw conclusions concerning the magnitudes of the various interactions which are of interest in selecting appropriate models to describe the electronic structure of high T_c materials. We discuss in some detail the importance of correlation, symmetry, hybridization and multiplet structure in understanding the properties of high T_c materials. We will find that the holes induced by doping are of primarily O_{2p} character which together with the already present Cu 3d holes probably form strongly exchange coupled local singlets. We discuss the influence of these quasi particles on the magnetic long range order and suggest that magnetic mechanisms are responsible for an attractive interaction between these quasi particles.

1. Introduction

In this paper we discuss the role of the atomic Cu and oxygen Coulomb (U_{dd} and U_{pp}) and exchange interactions, the O-Cu charge transfer energy (Δ), the O2p-Cu3d hybridization (T) and the Cu local point group symmetry on the electronic structure of high T_c compounds. We will assume that the most important action is on the CuO_2 planes although the possible role of the out of plane oxygens will also be discussed briefly. As far as the CuO_2 planes in the high T_c's are concerned the Cu-O distances and the Cu coordination (4-fold) are very close to those encountered in CuO [1] in which the Cu is in a nearly square planar O coordination. CuO is therefore a good model compound for studying U, Δ and T [2] although the oxygen is coordinated in a different way so that the O-2p band width is somewhat smaller (4 eV) in CuO [2] than the high T_c's [3] ($\approx$5 eV). The advantage of using compounds like CuO and Cu_2O is that these are well defined, stable compounds also as far as the surface is concerned so that the whole arsenal of very powerful electron spectroscopy techniques can be used in studies similar to those done previously for other Cu and Ni compounds [4].

Our main objective will be to determine the character of the first ionization states in the insulating, non-substituted compounds like La_2CuO_4, which generally are considered to be the states responsible for the metallic and superconducting behaviour in the substituted or non-stoichiometric compounds. A classification of these states for transition metal compounds including Cu oxides was already made in 1985 in the ZSA scheme proposed by Zaanen et al. [5,6]. ZSA suggested that for the late divalent TM oxides $U_{dd}>\Delta$ which would place CuO in the charge transfer type of band gap class for which the first ionization state is of primarily O-2p character. A somewhat simplified version of the ZSA scheme has since then been used in numerous recent theoretical discussions of the high T_c by, for example, Emery [6], Varma et al. [7], and Zang and Rice [8] among many

Springer Series in Materials Science, Vol. 11
Mechanisms of High Temperature Superconductivity
Editors: H. Kamimura and A. Oshiyama

others. The general consensus now is that indeed $U_{dd}>\Delta$ and the first ionization state of the insulators or hole states in the superconductors is of primarily O-2p character. The most direct experimental evidence of this came from O_{1s} X-ray absorption edge studies [9] and electron energy loss spectroscopy [10] both of which showed an absorption edge pre-peak of O 2p character in doped La_2CuO_4 and in $YBa_2Cu_3O_{7-y}$ for $y<0.5$. A recent systematic study shows a linear relationship between the intensity of this pre-peak and y [11].

The basic questions which remain are concerned with the very important details of these primarily O-2p hole states like what is their spatial distribution, their symmetry, what is the sign and magnitude of the exchange interactions with the Cu spins, how large is U_{pp} and the interatomic interaction V_{pd} between a hole on the oxygen and one on a nearest neighbor Cu and how do these compare to the dispersional width of these hole states. In short, what is the detailed nature of the quasi particles responsible for the metallic behaviour.

2. An Ansatz

To decide on an appropriate ansatz in describing the electronic structure we consider the relative magnitudes of various interactions. Band theory [3] tells us that the O 2p band width is large ≈4-5 eV. We know that $U_{dd}>\Delta$ and large (≈6-8 eV). Also we know that the O2p-Cu3d hybridization especially for the x^2-y^2 (in plane) orbitals is large [12], and in the insulating antiferromagnetic state Cu is in a d^9 configuration so we have a high density of holes already present on the Cu ions. With small doping a relatively low density of holes is introduced into the O 2p states so at least here it might be valid to neglect U_{pp}. Bearing these in mind a reasonable ansatz could be to consider a Cu^{2+} ion as an impurity in a host consisting of an initially full O 2p band or perhaps a cluster of full shell O 2p ions. In this way the translation symmetry of the Cu ions is neglected with the advantage that we can take U_{dd} into account explicitly. In first instance we also consider the Cu^{2+} impurity to be in a square planar coordination of O ions (D_{4h} point group).

Some persons argue that if we know that the additional holes are in O 2p states it would be more logical to consider a cluster with O at the centre surrounded by 2 Cu ions. Although this would be another possible ansatz we will argue below that this ansatz is very expensive in terms of energy as compared to the D_{4h} ansatz with Cu at the centre.

3. Ground State

We will want to determine the ground state of such an impurity which is the lowest energy state of a one-hole problem with states d^9_m and $d^{10}\underline{L}_{km}$ where m is a spin/symmetry label and $\underline{L}_{km}$ is a hole in the (ligand) O 2p band with energy ε_{km}. In D_{4h} symmetry the irreducible representations spanned by one d hole are $b_1(x^2-y^2)$, $a_1(3z^2-r^2)$, $b_2(xy)$, $e(xz, yz)$ where in brackets are the commonly used labels according to a coordinate system as given in Fig. 1.

The Anderson impurity Hamiltonian ($H=H_0+H_1$) is given by

$$H_0 = \sum_m \int d\varepsilon\ \varepsilon\ c^+_{\varepsilon m} c_{\varepsilon m} + \sum_m \varepsilon_m d^+_m d_m + \sum_m \int d\varepsilon\ (T(m)\ d^+_m c_{\varepsilon m} + h.c.) \quad (1)$$

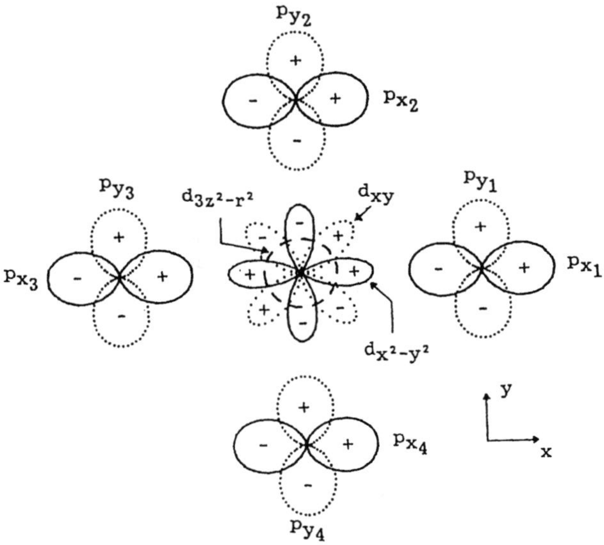

Fig. 1. Shows the orbitals and the coordinate system used in a cluster-like calculation. The z-direction points out of the page. The O-2p_z and Cu-3$d_{xz,yz}$ orbitals are not drawn.

$$H_1 = \sum_{m,m',n,n'} U(m,m',n,n') \; d_m^+ d_{m'} \, d_n^+ d_{n'} \qquad (2)$$

where c^+ creates holes in the ligand (oxygen) band with energy ε and d^+ creates Cu-3d holes. The indices m, m', n and n' denote the spin and orbital quantum numbers. T(m) is the transfer integral which depends strongly on the symmetry m. In square planar, four-fold coordination $T(a_1) = T(b_1)/\sqrt{3}$ and $T(e) = T(b_2)/\sqrt{2}$. Since π bonding is about 1/2 of the σ bonding [13] we take $T(b_2) = T(b_1)/2$. These differences in T(m) yield the ligand field splitting and since T for b_1 is the largest, the ground state has one hole in a b_1 symmetry orbital with predominantly Cu(d^9) character for a large charge transfer energy $\Delta = E(d^{10}\underline{L}) - E(d^9)$. The energy of the d-d transitions and charge transfer optical transitions can be determined from the same calculation.

The same type of calculation can be done using a $(Cu\ O_4)^{6-}$ cluster where now the band is replaced by a linear combination of O 2p orbitals of the appropriate symmetry. The "host" band width now is replaced by O-O transfer integrals (both σ and π type) [14] which split the O 2p states into molecular orbitals of various symmetry. It is of great importance to note that the $\underline{L}$ states of b_1 symmetry in a cluster type calculation are of the form (see Fig. 1)

$$\Psi_{\underline{L}}(b_{1g}) = \frac{1}{\sqrt{4}} (-p_{1x} + p_{2y} + p_{3x} - p_{4y}) \qquad (3)$$

and are the <u>lowest</u> energy ligand hole states [14,15]. In fact delocalizing the ligand hole in this way lowers the kinetic energy by 1/4 W where W is the total O 2p band width relative to localizing the O 2p hole on <u>one</u> O atom. This is a considerable gain in energy since W ≈ 4-5 eV. The ground state wave function in a cluster will be of the form

$$\Psi_G(b_1) = \alpha|d^9_{x^2-y^2}\rangle + \beta|d^{10}\ \Psi_{\underline{L}}(b_{1g})\rangle$$

4. The Electron Removal Spectrum

Since the ground state calculation involves only one hole, H_1 in Eq. 2 is not operative when written in a hole representation. However, the calculation of the first ionization state involves a two-hole problem and the possible configurations $|d^8\rangle$, $|d^9\underline{L}_k\rangle$ and $|d^{10}\underline{\underline{L}}_{kk'}\rangle$ and U_{dd} enters in the energy of the d^8 states. For $U_{dd}>\Delta$ the energy level scheme is inverted relative to that used to calculate the ground state (i.e. the d^8 state is higher in energy than $|d^9\underline{L}\rangle$). The situation is in addition complicated by the multiplet splitting within a d^8 configuration. The multiplet splitting is determined by the B and C Racah parameters which tell us that the energy of a d^8 state depends on the spin and orbital states occupied by the two holes. In practice this means that U_{dd} depends on the state we are considering. The matrices determining the d^8 energies for the various possible configurations are given in table 1 [16]. The most important states for us are the $^1A_1$ and $^3B_1$ which have two holes in a b_1 orbital and one in a b_1 and one in an a_1 respectively. Since B and C are both positive the $^3B_1$ state is obviously the lowest energy d^8 state for the free ion (Hund's rule). We know from a large number of previous studies on a variety of 3d, 4d, 5d transition and rare earth metals, alloys and compounds that the B and C Racah parameters are not screened in the solid [17]. We can therefore use free ion values [18] which for Cu^{3+} yield B=0.15 eV C=0.58 eV. For these values the $^1A_1(b_1^2)$ state lies 3.5 eV higher in energy than does the $^3B_1(a_1b_1)$ state. The monopole coulomb interaction given by A is however strongly screened in the solid and will have to be determined experimentally. Upon switching on the hybridization with the $d^9\underline{L}$ states the various d^8 states shift quite differently because of the strong differences in U_{dd}.

Table 1. Displays the irreducible representations spanned by two d holes (d^8) in D_{4h} symmetry and the Coulomb and exchange matrix elements in terms of the Racah A, B and C parameters.

$^3A_2$	b_1b_2	e^2
b_1b_2	A+4B	6B
e^2	6B	A-5B

$^1B_1$	a_1b_1	e^2
a_1b_1	A+2C	-2B√3
e^2	-2B√3	A+B+2C

^3E	eb_1	ea_1	eb_2
eb_1	A-5B	-3B√3	3B
ea_1	-3B√3	A+B	-3B√3
eb_2	3B	-3B√3	A-5B

^1E	eb_1	ea_1	eb_2
eb_1	A+B+2C	-B√3	-3B
ea_1	-B√3	A+3B+2C	-B√3
eb_2	-3B	-B√3	A+B+2C

$^1A_1$	a_1^2	b_1^2	b_2^2	e^2
a_1^2	A+4B+3C	4B+C	4B+C	(B+C)√2
b_1^2	4B+C	A+4B+3C	C	(3B+C)√2
b_2^2	4B+C	C	A+4B+3C	(3B+C)√2
e^2	(B+C)√2	(3B+C)√2	(3B+C)√2	A+7B+4C

a_1b_1 $^3B_1$ = A-8B

b_1b_2 $^1A_2$ = A+4B+2C

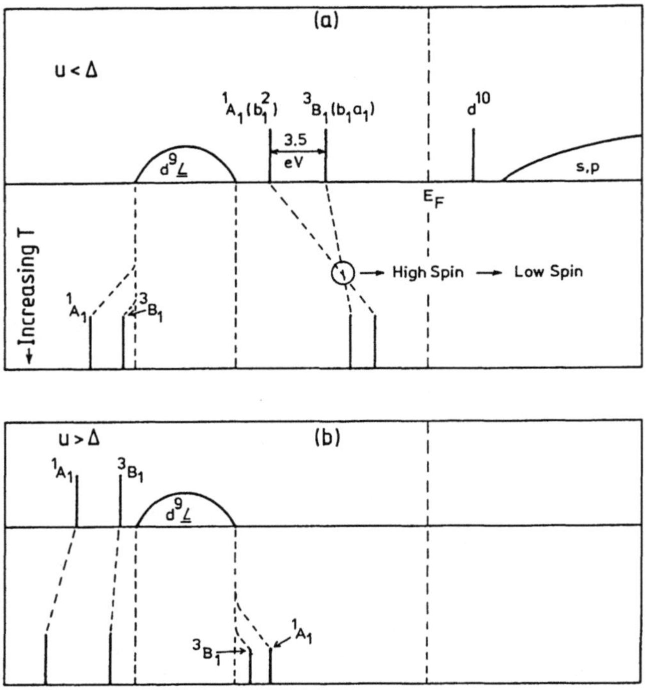

Figure 2. Shows how the d^8 states $^1A_1$ and $^3B_1$ shift upon switching on the hybridization T(m) with the $d^9\underline{L}$ band states. 2a shows the case where U<Δ for which for T(m)=0 the $^3B_1$ d^8 state is lowest in energy but for T(m) larger than a critical value the $^1A_1$ (low spin) state is lowest. In 2b is shown the case for U>Δ for which for T(m) larger than some critical value bound states are split off from the lower energy side of the $d^9\underline{L}$ band and the lowest energy state is $^1A_1$ in D_{4h} symmetry.

This is shown pictorially for two states, the $^1A_1$ and $^3B_1$ in Fig. 2 for two cases $U(^1A_1)$ and $U(^3B_1)$ both smaller than Δ and $U(^1A_1)$ and $U(^3B_1)$ both larger than Δ. For simplicity we have left out the $d^{10}\underline{\underline{L}}$ states which in fact can be quite important. Since in D_{4h} symmetry $T(b_1) = \sqrt{3}\ T(a_1)$ the $^1A_1$ state is affected more by hybridization than is the $^3B_1$ state. We see from Fig. 2 that only for very large $T(b_1)$ will the hybridization be large enough for the $^1A_1$ state to finally end up as the lowest for U<Δ. However, for U>Δ the $^1A_1$ state is the lowest even for quite small $T(b_1)$. We [19] have recently presented this in the form of a phase diagram reproduced in Fig. 3, but now for the complete calculation including the $d^{10}\underline{\underline{L}}$ states. More details of the above concepts are given in references 4, 5, 19, 20.

5. Magnitudes of parameters

We see from the above how important it is to know even roughly the relative magnitudes of the various interactions. As mentioned above the first ionization states are most likely of mainly O 2p character [9-11] so $U_{dd}>\Delta$. We now briefly summarize recent information concerning the magnitudes of

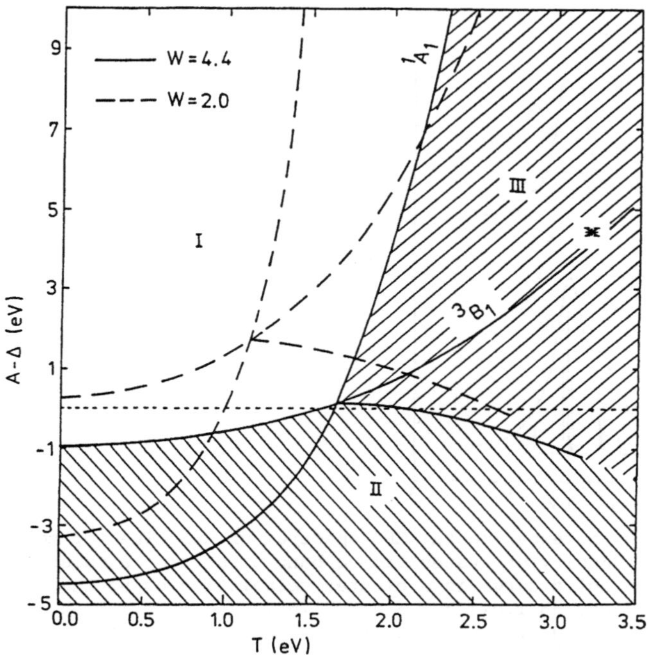

Figure 3. A plot showing the parameters A-Δ and $T(b_1)$ for which the $^1A_1$ and the $^3B_1$ state just form discrete states and a division into regions corresponding to : I - no discrete state, II - $^3B_1$ discrete state, and III - $^1A_1$ discrete state as the lowest energy states. The solid lines are drawn for an oxygen bandwidth of W=4.4 and the dashed lines for W=2. The calculations are down with Δ=2.75. The asterisk is the position obtained using McMahan et al.'s [15] parameters.

the various parameters. From a detailed photoemission, inverse photoemission, resonant photoemission and Auger spectroscopy study of CuO and Cu_2O [2] and a comparison to a cluster, band structure and impurity calculations we find the results given in table 2. A few remarks about these are in place. A fit to the XPS spectrum using a cluster theory of CuO yields the spectra shown in Fig. 4. The position of the satellite (d^8) is determined by A-Δ and the multiplet spread by Racahs B and C parameters in

Table 2. The parameters obtained for Cu_2O and CuO from comparison of a bandstructure (1), a model cluster (2) and a model impurity (3) calculation to various forms of electron spectroscopy. The acceptable ranges of the parameters are strongly correlated. $\Delta-(T_{pp\sigma}-T_{pp\pi}) = 1.8 \pm 4$ eV and $A-\Delta = 4 \pm 0.5$ eV. The values for $T(b_1)$ can be compared to 3.2 eV from ref. 15 who also found $T_{pp\sigma}-T_{pp\pi} = 1.3$ eV.

	A	U_{pp}	V_{pd}	$\Delta=E(d^{10}\underline{L})-E(d^9)$	$T(b_1)$	$T_{pp\sigma}-T_{pp\pi}$	W
$Cu_2O^{(1)}$	7.5 eV	4.6 eV	<1 eV	-	-	-	
$CuO_{(2)}$	6.0-7.0 eV	-	-	2.0-3.5 eV	2.0-3.0 eV	0.5-1.5 eV	-
$CuO^{(3)}$	6.0-7.0 eV	-	-	2.0-3.5 eV	2.5-3.2 eV	-	4-5 eV

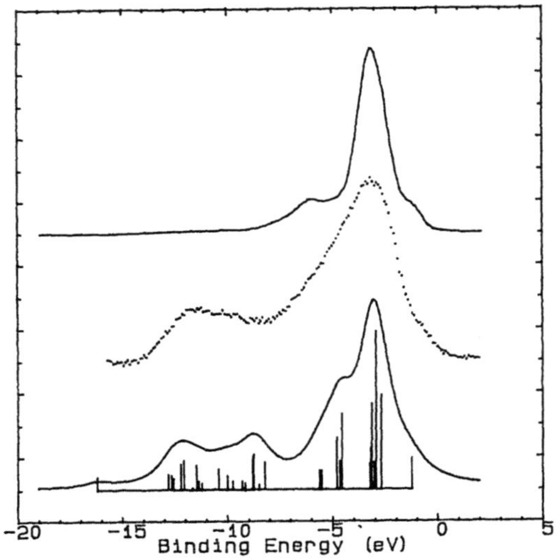

Figure 4. The dots show the experimental XPS spectrum of CuO. The top solid curve is the d partial density of states from a band structure calculation from ref. 2. The bottom solid curve is a cluster calculation using the parameters A=6.5 B=0.15 c=0.58 $T(b_1)$ = 2.5 Δ = 2.75 $T_{pp\sigma}-T_{pp\pi}$ = 1.0 eV where T_{pp} are the σ and π O-O transfer integrals which are close to the band theory values of ref. 15.

addition to the hybridization. Note the small peak at ≈16.2 eV in the theoretical curve which is a $^1A_1$ state derived from the free atom ^1S state at the extreme high energy side of the spectrum. This peak has subsequently been observed at 70-74 eV photon energy as shown in Fig. 5 demonstrating the detail with which theory can describe the results. The Δ value is determined primarily by the band gap and since we are neglecting the Cu d dispersional width in both the cluster or impurity calculation it is difficult to compare the calculated gap to experiment since the latter includes a contribution due to d band dispersion. Note that the band gap

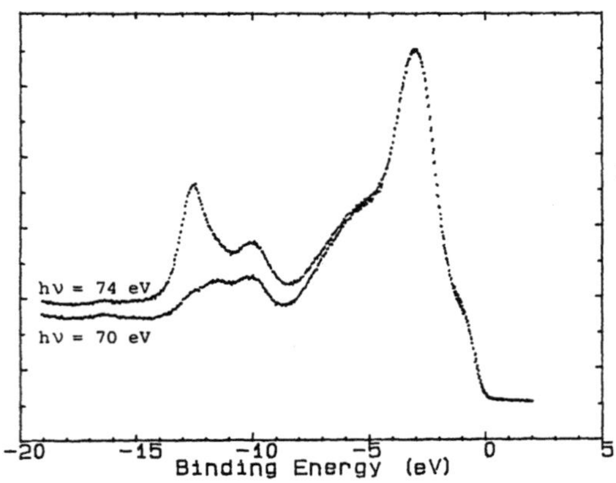

Figure 5. Shows the photoemission spectrum at the 3p-3d resonance photon energy of 74 eV and just off resonance at 70 eV. Here the small predicted peak at 16.2 eV is clearly visible.

calculation involves also the energy of the first electron addition state [5] which in our case is simply a d^{10} state. For $U_{dd}>\Delta$ our calculated value $E_{gap}^{cluster} \approx E_{gap} - w/2$ where w is the dispersional width in the "d^{10}" electron addition spectrum. We estimate w<1 eV.

The values of U_{pp} and the oxygen Cu interatomic interaction V_{pd} are obtained from the Auger spectra of Cu_2O. This is a full band system and therefore the Auger spectra are quite easily analyzed [21,2].

We are of the opinion that the parameters in table 2 are also suitable for the high T_c materials which have very similar CuO distances and coordinates. In fact the d^8 like satellite structure so clearly visible in Figs. 3 and 4 for CuO comes at almost the same energy in $YBa_2Cu_3O_7$ [22]. Also the band gap in La_2CuO_4 is close to that of CuO showing that U and Δ must be similar to those found for CuO.

6. Character of the Hole State

Of most interest is the character of the first ionization state which is the low energy shoulder clearly seen in Fig. 5. Before describing this using the above parameters and theory, we look back at our ansatz.

Our ansatz uses a cluster or host with Cu at the centre. As mentioned above some models for high T_c mechanisms (although this may not be crucial) start with the extra hole localized on an oxygen atom between two Cu ions [23]. For O-O transfer integrals of $T_{pp\sigma}-T_{pp\pi} \approx 1.0$ eV the b_1 symmetry orbital in D_{4h} with Cu at the centre is 1.0 eV lower in energy than an orbital localized on a single O site! In addition the effective Cu-O transfer integral ($T(b_1)$) for a b_1 symmetry orbital is two times that of a $d_{x^2-y^2}$ orbital with only one σ directed O p orbital. This is because of a coherent addition of the contributions from the four oxygen atoms. This factor of 2 in the effective transfer integral causes a factor of 4 (i.e. $T^2/\Delta E$) in the hybridization stabilization of the lowest energy states. This amounts to about an extra 0.5 - 1 eV stabilization for the local singlet state ($^1A_1$) [8,19] as compared to a spin 1/2 state formed by two Cu ions with an intervening σ bonding O hole. The "on one atom localized" ansatz is therefore better only if a different symmetry oxygen orbital like a p_z (π bonding) orbital as suggested by Johnson et al. [24] or the π bonding in plane x or y oxygen orbitals as suggested by Goddard [25] have strongly different on site energies than those orbitals directed to the Cu. The importance of the in plane oxygen p_z orbitals at least for the Bi compound has recently been rejected by polarization dependent O_{1s} x-ray absorption [26] and ELS [27] measurements. The same measurements on $YBa_2Cu_3O_7$ single crystals exhibit some z polarized O 2p holes which could be attributed to either p_z orbitals on the CuO_2 planes or to p_z orbitals on the apex oxygens forming part of the CuO chains [27, 28].

What about the influence of out of plane O 2p orbitals in the apex oxygens? In principle these could be very important. As emphasized above (see Fig. 2) the reason for the $^1A_1$ state with one hole in a Cu $d_{x^2-y^2}$ orbital and the other in an $O(b_1)$ type orbital having the lowest energy in pure square planar symmetry is because $T(b_1)$ is much larger than $T(a_1)$. For example, in O_h symmetry (octahedral) the out of plane oxygen -Cu distance is the same as the in plane distance and in this case obviously $T(b_1)=T(a_1)$ since the $d_{x^2-y^2}$ and $d_{3z^2-r^2}$ orbitals are degenerate in O_h. For this case we find the triplet state to be the lowest energy state even for U>Δ. This is clearly demonstrated in Fig. 6 showing the impurity calculation for O_h

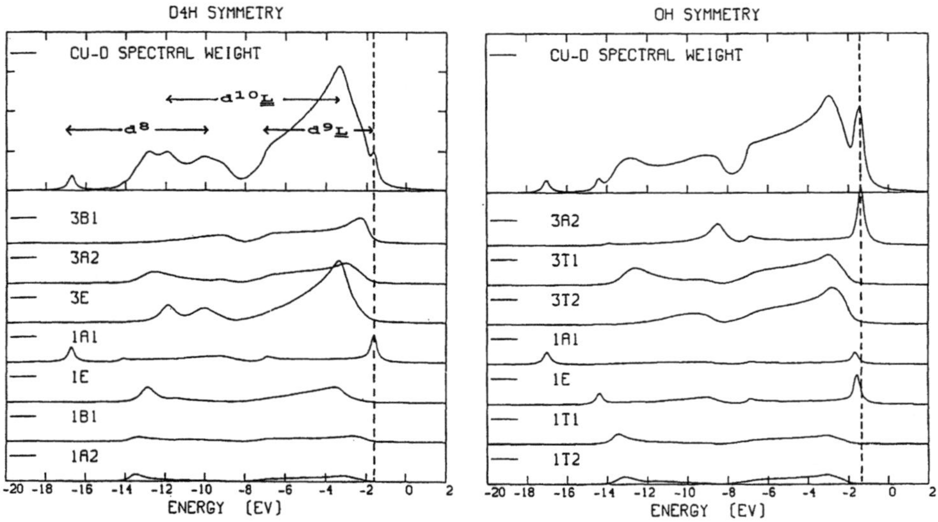

Figure 6. Shows an impurity calculation (see reference 19 and 20) of the d electron removal spectral weight for D_{4h} (left) and O_h (right) symmetry. The parameters used were A-6.5 B=0.15 C=0.58 $T(b_1)=2.65$ $\Delta = 2.75$ and an oxygen band width of 5 eV. Also shown is the decomposition into the various point group irreducible representations showing clearly that the first ionization state in D_{4h} is pure $^1A_1$ and that in O_h is $^3B_1$.

and D_{4h} symmetry for otherwise equal parameters. Here we see that the lowest energy state for D_{4h} is $^1A_1$ for reasonable parameters while it is $^3B_1$ in O_h (or $^3A_2$ in O_h conventions). It is rather interesting to see for which value of the out of plane Cu-O distance the lowest energy state changes from singlet ($^1A_1$) in D_{4h} to triplet ($^3B_1$) for O_h. We find the crossing point for the parameters of Fig. 5 to occur at $T(a_1) \simeq 0.9T(b_1)$. Using Harrison's [29] parameterization scheme $T_{Cu-O} \propto R^{-3.5}$, where R denotes the Cu-O distance, we find that the crossing point occurs at an out of plane Cu-O distance of about 2.0 Å for an in plane separation of 1.9 Å.

In the above we have assumed the out of plane O p_z orbital to have the same on site energy as the oxygen in plane p_x, p_y orbitals. Recently, Fujimori [30] showed that if the on-site out of plane O p_z orbital energy were about 1.7 eV higher than that of the in plane O p_x or p_y orbitals the triplet state ($^3B_1$) would be the lowest in energy even for the much larger Cu-O out of plane distance. As mentioned above there is strong evidence [26,27] that at least for the Bi compound p_z orbitals are not involved.

Going back to the singlet state it is of interest to look in more detail at the wave function composition, spatial extent as well as the effective exchange interaction. For the parameters of Fig. 5 the $^1A_1$ state has 64% $d^9\underline{L}$, 28% $d^{10}\underline{\underline{L}}$ and only 8% d^8 character. In a way this pushed out bound state looks much like a Haldane-Anderson multiple charged impurity state [31]. According to table 1 the $^1A_1$ state can have a_1^2, b_1^2, b_2^2 and e^2 symmetry. We find the lowest energy state to be 99% b_1^2 and only 1% of other character. We have estimated the spatial extent of the additional hole by neglecting the $d^{10}\underline{\underline{L}}$ states and using a tight binding O 2p band structure.

The amount of O 2p hole character in the first oxygen shell is directly related to the separation of the $^1A_1$ bound state from the band. For a separation of 0.5 eV 75% is localized in the first neighboring O shell. The larger the separation the more the wave function is localized in the first shell. The details of these calculations will be published elsewhere [32].

7. Sign and magnitude of the O-Cu exchange interaction

The stabilization of the singlet state can be viewed as a result of the exchange interaction between the O 2p hole and a $d_{x^2-y^2}$ hole. The separation of this state as well as the $^3B_1$ state from the band edge for various parameter choices is shown in Fig. 7. For A-Δ between 4 and 6 eV and $T(b_1)$ between 2.4 and 3 eV the effective stabilization of the singlet state is as large as 0.5 eV which is much larger than the already exceptionally large Cu-Cu superexchange interaction of 0.1 eV [33]. In fact in order for the local singlet to be stable its stabilization energy should be larger than the superexchange interaction lost by the participation of a Cu spin in the singlet.

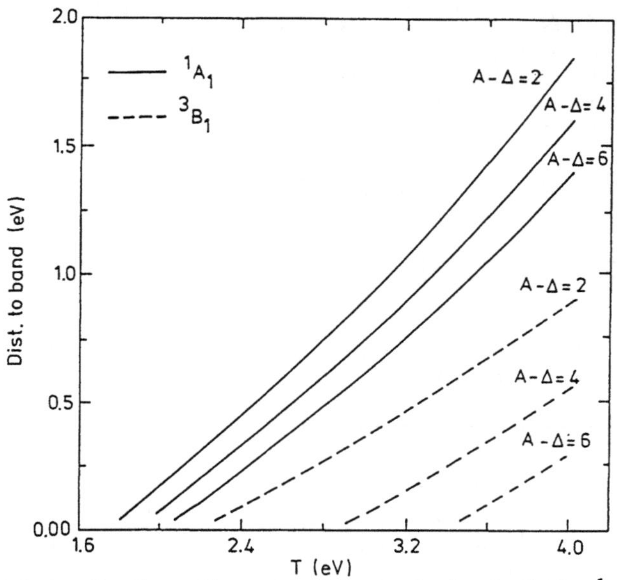

Figure 7. Shows the energy above the band of the $^1A_1$ and $^3B_1$ discrete states as a function of $T(b_1)$ for various values of A-Δ. The separation of the $^1A_1$ state from the band edge can be considered as a measure of the exchange between a x^2-y^2 (b_1) symmetry hole on the oxygens and a central Cu x^2-y^2 hole.

The exchange interaction between the O2p hole and a Cu 3d hole should really be determined from the energy difference of the $^1A_{1g}$ and $^3A_{1g}$ states of which the $^3A_{1g}$ is non bonding as far as d^8 states are concerned. In perturbation theory for a cluster this is given by

$$J = T^2(b_1)\left[\frac{1}{\Delta+U_{pp}} + \frac{1}{U_{dd}({}^1A_{1g})-\Delta}\right]$$

which is larger than 1 eV for reasonable parameters.

It is interesting to look at the dependence of the O-Cu exchange on the symmetry of the O hole state since various models for superconductivity assume various symmetries of O hole states. Consider a simple diatomic cluster compound of a Cu(d^9) ion with a hole in an x^2-y^2 orbital and the O labeled 1 in fig. 1. The O hole could be in a p_x, p_y or p_z orbital. The calculation of the exchange within perturbation theory involves the states $d_{x^2-y^2}p_i$, $d_{x^2-y^2}d_i$, p_ip_j corresponding to $d^9\underline{L}$, d^8 and $d^{10}\underline{L}^2$. These configurations are coupled by a transfer integral which depends on symmetry and their diagonal energy depends on Δ, U_{pp} as well as $U_{dd}(x^2-y^2,i)$. The exchange is then the energy difference between the singlet and triplet in each case. The Hamiltonian matrices are given in table 3 for each of the above cases with the $U_{dd}(x^2-y^2,i)$ taken from table 2 and $T_{x^2-y^2} = T(b_1)/2$ and T_{xz}, T_{xy} the π transfer integrals $T_{xy}=T_{xz} \simeq T_{x^2-y^2}/2$ as discussed above. We see immediately a large difference between a p_x hole and p_y or p_z which is a result of the $d_{x^2-y^2}$ orbital which mixes only with p_x being singly occupied. For the p_x hole only singlet d^8 states are involved so that the total hybridization contributes to the exchange whereas for a p_y or p_z hole both the singlet and triplet d^8 states can be involved so only the difference in the hybridizations contributes to the exchange. The exchange interaction is for these cases

$$J_{x,x^2-y^2} = E_S - E_T \approx -T^2_{x^2-y^2}\left[\frac{1}{\Delta+U_{pp}} + \frac{1}{A+4B+3C-\Delta}\right]$$

$$J_{y,x^2-y^2} \approx +T^2_{xy}\left[\frac{2C}{(A+4B-\Delta)(A+4B+2C+\Delta)}\right]$$

Table 3. Triplet and singlet two hole states and their matrix elements for the oxygen hole in the p_x, p_y and p_z orbital.

	TRIPLETS				SINGLETS				
p_x	$\|d_{x^2-y^2}p_x\rangle$	0	$T_{x^2-y^2}/\sqrt{3}$		$\|d_{x^2-y^2}p_x\rangle$	0	$T_{x^2-y^2}/\sqrt{3}$	$T_{x^2-y^2}$	$T_{x^2-y^2}$
	$\|d_{x^2-y^2}d_{3z^2-r^2}\rangle$	$T_{x^2-y^2}/\sqrt{3}$	A-8B-Δ		$\|d_{x^2-y^2}d_{3z^2-r^2}\rangle$	$T_{x^2-y^2}/\sqrt{3}$	A+2C-Δ	0	0
					$\|d_{x^2-y^2}d_{x^2-y^2}\rangle$	$T_{x^2-y^2}$	0	A+4B+3C-Δ	0
					$\|p_xp_x\rangle$	$T_{x^2-y^2}$	0	0	$\Delta+U_{pp}$
p_y	$\|d_{x^2-y^2}p_y\rangle$	0	T_{xy}	$T_{x^2-y^2}$	$\|d_{x^2-y^2}p_y\rangle$	0	T_{xy}	$T_{x^2-y^2}$	
	$\|d_{x^2-y^2}d_{xy}\rangle$	T_{xy}	A+4B-Δ	0	$\|d_{x^2-y^2}d_{xy}\rangle$	T_{xy}	A+4B+2C+Δ	0	
	$\|p_xp_y\rangle$	$T_{x^2-y^2}$	0	$\Delta+U_{pp}$	$\|p_xp_y\rangle$	$T_{x^2-y^2}$	0	$\Delta+U_{pp}$	
p_z	$\|d_{x^2-y^2}p_z\rangle$	0	T_{xz}	$T_{x^2-y^2}$	$\|d_{x^2-y^2}p_z\rangle$	0	T_{xz}	$T_{x^2-y^2}$	
	$\|d_{x^2-y^2}d_{xz}\rangle$	T_{xz}	A-5B-Δ	0	$\|d_{x^2-y^2}d_{xz}\rangle$	T_{xz}	A+B+2C-Δ	0	
	$\|p_xp_z\rangle$	$T_{x^2-y^2}$	0	$\Delta+U_{pp}$	$\|p_xp_z\rangle$	$T_{x^2-y^2}$	0	$\Delta+U_{pp}$	

$$J_{z,x^2-y^2} \approx +T^2_{xz} \left[\frac{2C+6B}{(A-5B-\Delta)(A+B+2C-\Delta)}\right]$$

showing that only for p_x do we get a large antiferromagnetic exchange. The Cu-Cu superexchange interaction is given by [14, 20]

$$J_{super} \approx \frac{-4T^4_{x^2-y^2}}{\Delta^2} \left[\frac{2}{2\Delta+U_{pp}} + \frac{1}{U(^1A_1)}\right]$$

In table 4 values of J_i for $T(b_1) = 2$ eV, $\Delta = 2.5$ eV, $U_{pp} = 5$ eV, $A = 7$ eV, $B = 0.15$ eV and $C = 0.56$ eV are given. This clearly demonstrates how important the symmetry of the hole state is for the exchange interaction. We see that a b_1 symmetry and to a lesser degree a p_x hole is disastrous for the antiferromagnetic order whereas a p_y or p_z like hole state presents a relative weak perturbation on the magnetic order.

Table 4. The exchange energies for a dx^2-y^2 hole and an oxygen hole in the b_1, p_x, p_y and p_z orbital as well as the d-d superexchange energy, using $T(b_1) = 2$ eV, $\Delta = 2.5$ eV, $U_{pp} = 5$ eV, $A = 7$ eV, $B = 0.15$ eV and $C = 0.58$ eV.

$J(b_1)$	$J_{x,x^2,-y^2}$	J_{y,x^2-y^2}	J_{z,x^2-y^2}	J_{super}
-1.12	-0.28	+0.01	+0.02	-0.20 eV

8. Propagation and Consequences for Anti-Ferromagnetic Order

If as discussed above the O hole + Cu hole anti-ferromagnetic exchange is much larger than the superexchange, local singlets of A_1 symmetry will be formed which will in a way look like a magnetic dilution effect. Such a dilution can have disastrous consequences for a 2D spin 1/2 antiferromagnet. The neighboring Cu spins now only have 3 rather than 4 neighbors coupling them to the opposite sublattice. This will tend to strongly increase the quantum fluctuations in the vicinity of the singlet state. It is interesting in this regard to recall that the energy of a collection of singlets and that of the Néél state are the same for a system with 3 nearest neighbors. In fact the stronger the deviation from the Néél state the easier it is for the bound state singlet to propagate without leaving behind incorrectly oriented spins [34]. This in turn lowers the kinetic energy of the added hole state which again favours a situation of non-Néél like order. So each hole would like to create around it a region of non Néél like order looking perhaps more like RVB. This looks much like Shrieffer's [35] spin bag (or perhaps an RVB bag) except that we are dealing with localized moments rather than spin density waves so a magnetic polaron might be a better term [36]. Such a local spin or RVB bag is of course an attractive place for a second hole. The disappearance of long range magnetic order in $La_{2-x}Sr_xCuO_4$ for $x>0.02$ indicates that the region influenced by one singlet has a radius of about 2 lattice spacings.

What about the optical properties of such a system of local singlet quasi particles? As discussed by Kane et al. [37] using a single band model one would expect to see Drude like behaviour at low energy corresponding to a coherent motion of the quasi particle followed by a structure corresponding to the incoherent motion in which a wake of spin excitations is left behind. At around 0.3 to 0.5 eV we would in addition expect a threshold for optical transitions from the band like to the singlet bound states (see fig. 5) followed at about 1.5 to 2 eV by the usual strong charge transfer transitions. This is qualitatively not unlike what is seen in recent measurements by Thomas et al. [38] on $YBa_2Cu_3O_7$. All of these low energy structures should be x, y polarized if as we suggested the b_1 symmetry orbitals are responsible although the chains in $YBa_2Cu_3O_7$ will contribute also with z polarization.

9. Conclusion

In conclusion we have discussed the role of correlation, multiplet structure, hybridization and especially local symmetry in the electronic structure of the CuO layers in the high T_c materials. We have obtained reliable values for the various interactions usually considered to be of importance. Using these we have discussed the circumstances under which a local singlet or local triplet state of $d^9\underline{L}$ character are the lowest energy ionization states of the insulators and have gone on to describe a possible mechanism for attractive interactions for the case in which the local singlet state is lowest in energy.

10. Acknowledgements

This investigation was supported by the Netherlands Foundation for Chemical Research (SON) and the Foundation for Fundamental Research on Matter (FOM) with financial support from the Netherlands Organization for the Advancement of Pure Research (NWO).

References

1. S. Åsbrink and L.j. Norrby, Acta Cryst. B26, 8 (1970).
2. J. Ghijsen, L.H. Tjeng, J. van Elp, H. Eskes, J. Westerink, G.A. Sawatzky and M.T. Czyzyk, Phys. Rev. B. (in press).
3. A.T. Park, K. Terakura, T. Oguchi, A. Yanase and M. Ikeda, Technical Report ISSN Ser. A. 1960 (1988).
4. G. v.d. Laan, C. Westra, C. Haas and G.A. Sawatzky, Phys. Rev. B23, 4369 (1981);
   A. Fujimori, F. Minami and S. Sugano, Phys. Rev. B29, 5225 (1984);
   G.A. Sawatzky and J.W. Allen, Phys. Rev. Lett. 53, 2239 (1984);
   J. Zaanen, G.A. Sawatzky and J.W. Allen, J. Magn. Magn. Mat. 54-57, 607 (1986);
   For recent reviews see chapters by A. Kotani (page 63), S.-J. Oh, (page 125), A. Fujimori (page 136) and G.A. Sawatzky (page 99) in "Core-Level Spectroscopy in Condensed Systems", edited by J. Kanamori and A. Kotani, Springer 1988.
5. J. Zaanen, G.A. Sawatzky and J.W. Allen, Phys. Rev. Lett. 55, 65 (1985). For a recent review of these ideas we refer to papers by C.O. Almbladh, B. Brandow, G.A. Sawatzky, in the proceedings of the Narrow Band Workshop, June 1987, Staverden, in press Plenum.
6. V.J. Emery, Phys. Rev. Lett. 58, 2794 (1987).
7. C.M. Varma, S. Schmitt-Rink and E. Abrahams, Proceedings of the International Conference on New Mechanisms of Superconductivity, Berkeley CA. 1987, Plenum Press, New York, edited by V. Kresin and S. Wolf.
8. F.C. Zhang and T.M. Rice, Phys. Rev. B37, 3759 (1988).
9. J.A. Yarmoff, D.R. Clarke, W. Drube, U.O. Karlsson, A. Taleb-Ibrahimi and F.J. Himpsel, Phys. Rev. B36, 3967 (1987).

10. N. Nücker, J. Fink, J.C. Fuggle, P.J. Durham, W.N. Timmerman, Phys. Rev. B37, 5158 (1988).
11. P. Kuiper, G. Kruizinga, J. Ghijsen, M. Grioni, P.J.W. Weijs, F.M. F. de Groot, G.A. Sawatzky, H. Verweij, L.F. Feiner and H. Petersen, Phys. Rev. B (in press).
12. L.F. Mattheis, Phys. Rev. Lett. 58, 1028 (1987).
13. L.F. Mattheis, Phys. Rev. B5, 290 (1972).
14. F. Mila (preprint)
H. Eskes, L.H. Tjeng and G.A. Sawatzky (preprint).
15. A.K. McMahan, R.M. Martin and S. Satphathy, Phys. Rev. B. (in press).
16. J. Zaanen, Thesis, University of Groningen, 1986.
J.S. Griffith, "The Theory of Transition Metal Ions", Cambridge University Press, Cambridge, 1961.
17. D. van der Marel and G.A. Sawatzky, Phys. Rev. B (to be published) and references therein.
18. C.E. Moore, "Atomic Energy Levels", NBS circular no. 467 U.S. GPO, Washington, DC, 1958, Vol. 1-3.
19. H. Eskes and G.A. Sawatzky, Phys. Rev. Lett. 61, (1415) 1988.
20. G.A. Sawatzky, Proceedings Adriatico Research Conferences, Trieste, July 1988, in press.
21. M. Cini, Solid State Commun. 24, 681 (1977);
G.A. Sawatzky, Phys. Rev. Lett. 39, 504 (1977).
22. See for example R.L. Kurtz and R.L. Stockbauer, Phys. Rev. B35, 8818 (1987).
23. A. Aharony, R.J. Birgeneau, A. Coniglio, M.A. Kastner and H.E. Stanley, Phys. Rev. Lett. 60, 1330 (1988).
24. K.H. Johnson, M.E. McHenry, C. Counterman, A. Collins, M.M. Donovan, R.C. O'Handley and G. Kalonji, Physica C 153, 1165 (1988).
25. Y. Guo, J.-M. Langlois, W.A. Goddard III, Science 239, 896 (1988).
26. P. Kuiper et al., to be published.
27. N. Nücker, H. Romberg, X.X.XI, J. Fink, B. Gegenheimer, Z.X. Zhao, preprint.
28. N. Nücker, J. Fink, J.C. Fuggle, P.J. Durham and W.M. Temmerman, Physica C 153-155, 119 (1988).
29. W. Harrison, "Electronic Structure and the Properties of Solids" (W.H. Freeman and Co., San Francisco) (1980).
30. A. Fujimori, preprint.
31. F.D.M. Haldane and P.W. Anderson, Phys. Rev. B13, 2553 (1976).
32. H. Eskes and G.A. Sawatzky, to be published.
33. G. Shirane et al., Phys. Rev. Lett. 59, 1613 (1987).
34. G. Baskaran, Z. Zou and P.W. Anderson, Solid State Commun. 63, 973 (1987).
35. J.R. Schrieffer, X.G. Wen and S.C. Zhang, Phys. Rev. Lett. 60, 944 (1988).
36. H. Kamimura, Jpn. J. Appl. Phys. 26, 6627 (1987) and in the proceedings of the Adriatico Research Conference, Trieste, July 1988.

Finite Systems Studies and the Mechanism of High T_c

J.E. Hirsch

Department of Physics, B-019, University of California,
San Diego, La Jolla, CA 92093, USA

I discuss results of exact and Monte Carlo calculations of models for the oxide superconductors. These results show that magnetic mechanisms are highly unlikely to lead to high temperature superconductivity. A Cu-O charge-transfer excitation is found to be a possible mechanism for pairing of oxygen holes, if the parameters are right. However, the body of experimental results and our results from small systems point to another, novel and highly universal mechanism for superconductivity in these and other materials.

I. Introduction

Recent experimental developments [1] have in my view delivered a fatal blow to theories based on magnetic mechanisms as the origin of high T_c superconductivity. The discovery of 30° K superconductivity in $Ba_{1-x}K_xBiO_3$, a material with no traces of magnetism, rules out magnetic mechanisms unless one assumes that the origin of superconductivity in Cu and non-Cu oxides is entirely different. I believe that ascribing entirely different mechanisms to two classes of oxide materials with largely similar properties and both having T_c's substantially higher than what was known before 1986 defies common sense.

Numerical work on model Hamiltonians, however, had already indicated that magnetic mechanisms will not give rise to high T_c [2,3]. Within a model with one orbital per O and one orbital per cation [4], these calculations suggested instead a charge-transfer excitation mechanism [3,5,6]. This mechanism can operate both in Cu and non-Cu based compounds, and thus the recent discovery mentioned above does not invalidate it. An attractive feature of this mechanism is that certain structural and other features that are specific to the oxide materials (both with and without Cu) are required for it to be feasible [3,5]. On the other hand, the parameter regime where this mechanism is found to be effective is somewhat restricted and it may require too large Coulomb interaction between nearest neighbors. In the first part of this paper (Sects. II and III) we review our work and conclusions on these issues.

On the other hand, if we abandon the single-orbital per atom model another, highly universal mechanism involving charge fluctuations suggests itself as compelling [7]. In the last part of this paper we discuss the evidence that points towards this mechanism and some numerical work on an effective Hamiltonian to describe the essential physics.

Springer Series in Materials Science, Vol. 11
Mechanisms of High Temperature Superconductivity
Editors: H. Kamimura and A. Oshiyama

II. Magnetic Mechanisms

Soon after the discovery of high T_c superconductivity, ANDERSON [8] put forward the two-dimensional Hubbard model as the model to describe the essential physics of the phenomenon. A large amount of theoretical work followed that supported this point of view. SCHRIEFFER and coworkers [9] proposed a spin-bag picture to describe the properties of the two dimensional Hubbard model that would also lead to superconductivity in this system. Previous numerical and theoretical work [10] had also suggested superconductivity in the two-dimensional repulsive Hubbard model.

However, numerical work has so far not confirmed any of the above pictures. Detailed numerical work has been performed on the two-dimensional Hubbard model in the past two years. In the half-filled band, it has now been convincingly established that the model exhibits antiferromagnetic order that is not destroyed even by strong charge fluctuations (small Hubbard U) [11]. The magnetic properties are well described by spin-wave theory [11-14], and thus far from being described by a "resonating valence bond" (RVB) insulating state, as originally proposed [8]. Comparison with experiment [14,12] suggests that the model is appropriate to describe the *magnetic* properties of the Cu-O materials.

When the model is doped away from half-filling, results of Monte Carlo simulations [2] indicate that no tendency to superconductivity exists down to temperatures a fraction of J, the antiferromagnetic coupling, where there are very strong antiferromagnetic spin fluctuations. Although the numerical work cannot rule out superconductivity at exponentially smaller temperatures, this is unlikely to occur and would anyway not be relevant for the high T_c phenomenon in oxides. In addition, exact diagonalization results on 8-site clusters [2] showed no tendency to superconductivity down to zero temperature. While boundary effects could play an important role in such a small system, one usually does see at least a tendency to other instabilities in such small systems when expected such as antiferromagnetism, Spin-Peierls, charge-density wave and superconductivity (the latter in an attractive Hubbard model or an electron-phonon model). In addition, we should recall that the coherence length in the high T_c materials is believed to be only a few lattice spacings. The combined evidence of Monte Carlo and exact diagonalization studies of the Hubbard model in my view convincingly establishes that it is not the right model to explain superconductivity in the oxide materials. Furthermore, our simulations of a 3-band model for CuO_2 planes with a Hubbard U on the Cu [3] also failed to show enhanced tendency to pairing, indicating that a Hubbard U by itself cannot induce superconductivity [15].

III. Cu-O Charge Transfer Mechanism

Within a model describing a single orbital on the cation and the anion, the tight-binding Hamiltonian for electrons or holes in a plane is:

$$\begin{aligned} H &= \sum_{\langle i,k \rangle} t(d_{i\sigma}^{+} c_{\ell\sigma} + h.c.) + (\epsilon - \mu) \sum_{\ell\sigma} c_{\ell\sigma}^{+} c_{\ell\sigma} - \mu \sum_{i\sigma} d_{i\sigma}^{+} d_{i\sigma} + t' \sum_{\langle \ell\ell' \rangle} (c_{\ell\sigma}^{+} c_{\ell'\sigma} + h.c.) \\ &+ U \sum_{i} n_{i\uparrow} n_{i\downarrow} + U_p \sum_{\ell} n_{\ell\uparrow} n_{\ell\downarrow} + V \sum_{\langle i\ell \rangle} n_i n_\ell + V_{OO} \sum_{\langle \ell\ell' \rangle} n_\ell n_{\ell'} \end{aligned} \quad (1)$$

where we have included interactions and hoppings up to a distance $\sqrt{2}a$, with a the cation-anion distance, and neglected longer-ranged terms. The operators $c^+_{\ell\sigma}$ create holes in the $2p^6$ band of O^{--}, and $d^+_{i\sigma}$ creates holes in the $3d^{10}$ shell of Cu^+ or $6s^2$ shell of Bi^{3+}. We omit orbital angular moments labels thus restricting the occupation of these orbitals to 0, 1 and 2 holes. ϵ is the energy difference between anion and cation single particle levels. It is determined by the cation and anion as well as the over-all structure.

Consider some bare parameters for the Hamiltonian (1). The on-site repulsion on O sites [16] $U_p = E(O^{--}) + E(O) - 2E(O^-) = 10.2$eV, which is surprisingly large due to the fact that O^{--} is a highly unstable species. The reason O^{--} exists is of course that it is stabilized by electrostatic energy in ionic solids. The on-site repulsion on the cations is not much larger: for Cu [17], $U = E(Cu^{+++}) + E(Cu^+) - 2E(Cu^{++}) = 16.5$eV, and for Bi, $U = E(Bi^{3+}) + E(Bi^{5+}) - 2E(Bi^{4+}) = 10.7$eV. Therefore, it is unrealistic to assume that U_p can be omitted in models where U plays an essential role [18]. U_p is, however, unimportant for magnetic properties: the lattice structure of the planes favors an antiferromagnetic state with the moments centered around the cation sites when there is one hole per unit cell; the strength of the antiferromagnetic exchange is determined mainly by U, with U_p playing a secondary role. The bare Coulomb repulsion between two neighboring holes is $V = e^2/a = 7.4$eV for the nearest-neighbor distance on Cu-O planes, clearly not negligible compared to the on-site interactions.

The basic pairing mechanism in this model in the strong coupling limit is illustrated in Fig. 1 [3]: two added O holes can have a lower energy if they are on the same O atom and polarize their environment:

$$E_2 = 2V + 4\epsilon + U_p \tag{2}$$

rather than far apart from each other, each with energy:

$$E_1 = \epsilon + 2V \tag{3}$$

if the parameters are such that the effective interaction:

$$U_{eff} = E_2 - 2E_1 = -2V + U_p + 2\epsilon \tag{4}$$

is negative. For the bare parameters of CuO_2 given above, U_{eff} is negative for $\epsilon <$ 4.7eV. This argument implicitly assumes that U on the cations is much larger than ϵ; otherwise the added holes go predominantly onto the cations rather than anions and the

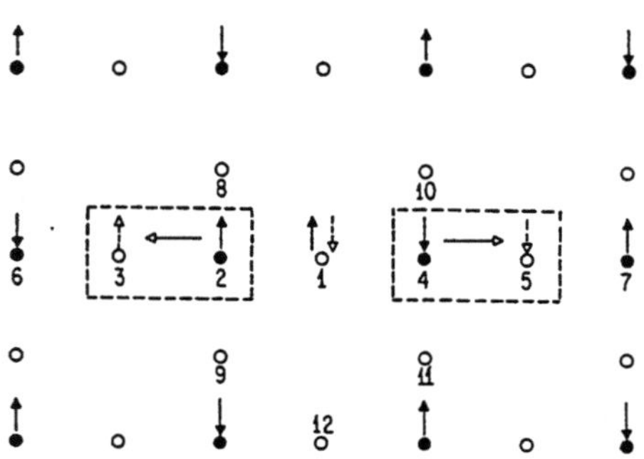

Fig. 1. Schematic illustration of pairing mechanism in 3-band model [3]. As holes are added to the center O ion, holes on neighboring cations are pushed away onto O ions. The dashed lines enclose the two polarizable "side units" to the center O ion that produce the effective attractive interaction on that O

pairing mechanism described becomes ineffective. A similar argument can be used for pairing of O holes on nearest neighbor O atoms [19]. The strong coupling analysis there, however, yields an effective interaction $U_{eff} = \epsilon - V + 2V_{OO}$. If we assume the bare electrostatic value $V_{OO} = V/\sqrt{2}$ we find always a positive U_{eff} under these conditions. Nevertheless, this process does help in reducing the value of the bare nearest-neighbor repulsion substantially.

Note that the lattice structure with doubly coordinated anion and higher coordinated cation is crucial to this argument. Consider instead, for example, a square lattice structure as in the BaO planes in $BaBiO_3$. If we attempt to move charge from Ba to a neighboring O when we add holes to an O (analogously to Fig. 1), this has to overcome the repulsion of 3 rather than 1 nearest-neighbor Ba, and thus it is always energetically unfavorable. Thus, that structure does not allow for the polarization mechanism described above.

The Coulomb interaction parameters in the real material will not be given by these bare values but are going to be screened by processes involving orbitals not included in the Hamiltonian (1). Many calculations suggest that the intra-atomic parameters are reduced by roughly a factor of 2 [20]. To estimate the inter-atomic repulsion is more difficult, and estimates vary.

For finite hopping t, we can obtain semiquantitative estimates for the effective interaction by diagonalizing the two-site units involving the motion of the holes that are pushed away by the added holes (sites 2 and 3, and 4 and 5 in Fig. 1) [21]. Due to the existence of V_{OO}, the motion of the hole on the left is going to be predominantly between sites 2 and 3 rather than including also sites 8 and 9, and similarly on the right. One finds, for example, that there is an optimum value for ϵ which increases with V. For the unit on the left, it is determined approximately by the condition $(V+\epsilon)n_6 \sim V$ which causes the effective energy of sites 2 and 3 to be equal (i.e. maximum resonance) when the first hole is added. The effective frequency of these polarizable side units is $\Delta E \cong 2\sqrt{(\epsilon/2)^2 + t^2}$, which is of the same order of magnitude as the band width, so that for small levels of doping ($\epsilon_F \ll \Delta E$) the effective interaction is essentially instantaneous.

Extensive cluster calculations in one- and two-dimensional geometries give detailed information on the parameter range where pairing will exist [5], and show that pairing can also occur for significantly smaller values of V than the previous analysis would suggest. Figure 2 shows one example of a phase diagram obtained from diagonalization of

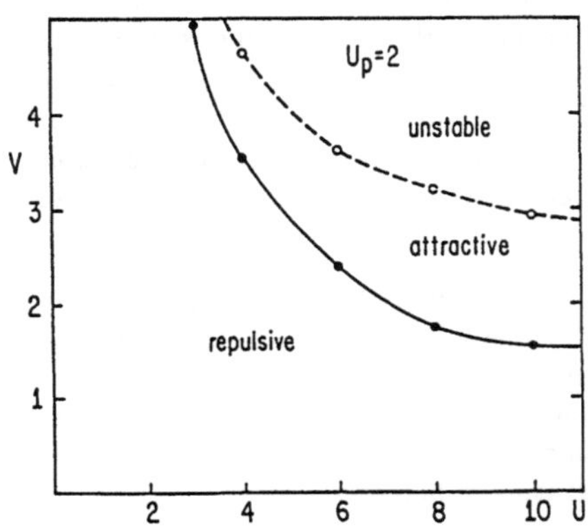

Fig. 2. Phase diagram obtained from diagonalization of a (12)-site Cu-O cluster with Hamiltonian (1). $t = 1$, $\epsilon = 0$, $V_{OO} = t' = 0$

a twelve-site cluster [5]. The unstable region is defined as the region where a *third* added particle has a lower energy than if added elsewhere, indicating a tendency to clustering rather than superconductivity. We expect the region of attractive interaction to grow as the cluster size increases. It is difficult to assess whether the parameter regime where pairing is found in this model is realistic for the oxide materials. As we discuss in what follows, we believe there is another primary mechanism for pairing that is operative.

IV. A New Mechanism

Let us step back for a moment and evaluate the situation. Our numerical results have shown that magnetic mechanisms don't work, and that an anion-cation charge transfer mechanism is feasible in certain parameter ranges but perhaps much too specific. Many other specific models have been proposed, involving Cu d-d excitations, out-of-plane polarization, etc. And yet the experimental information that is accumulating suggests that the mechanism is not specific but rather universal. Let us consider the following selected experimental findings (most of which were highlighted at this meeting):

1. Spectroscopic evidence strongly suggests that holes are predominantly on O sites.
2. NMR results have beautifully shown that BCS-like superconductivity is due to O holes and that Cu sites are essentially *decoupled* as far as the superconductivity is concerned.
3. Hall coefficient measurements indicate that these materials are superconductors when the conductivity is hole-like, and non-superconductors when it is electron like.
4. Transient high temperature superconductivity has been observed in the past in CuCl and CdS.

These findings suggest that the essential physics of high T_c is contained in the simple fact that *conduction in these materials occurs through holes in anions with filled shells. Anions that normally do not form conductors but highly insulating solids.*

When theorists write down model Hamiltonians such as (1) they do not usually differentiate between electrons and holes. And yet nature makes an enormous difference between them. Recall that elements with one *electron* added to a closed shell ion are simple metals, while elements with one *hole* added to a closed shell ion are halogens that solidify onto a molecular, highly insulating solid. What is it that breaks particle-hole symmetry in nature and causes these two kinds of elements (say Na and F) to behave so differently?

A simple explanation is that an electron added to a closed-shell ion changes very little its "background," the states of the other electrons in the ion. A hole added to a closed-shell ion affects substantially its "background," by modifying the states of all the remaining electrons of the outer shell. This rearrangement of electrons in the outer shell is what causes F to solidify as a molecular insulating solid rather than as a simple metal like Na. It is also, I believe, the key to high temperature superconductivity.

Imagine one could create a monatomic solid with closed-shell ions (O^{--}, Cl^-, S^{--}, etc.) in a regular lattice structure (one atom per unit cell) and dope the system with a few holes. I claim the resulting system will obviously be a high temperature superconductor. The pairing interaction will arise from polarization of the outer shell by the holes, just as in the usual electron-phonon interaction case but inverted: the conducting particles have positive charge, and the background that provides the pairing (the outer filled shell) has negative charge. The dominant interaction is the *local atomic polarization* of the outer shell by the hole that goes into that same shell. Both interaction strength and energy scale are two orders of magnitude larger than for the electron-phonon case. As we discuss elsewhere [7], I believe there is substantial evidence that suggests that this basic mechanism is not restricted to high T_c oxides but plays also an essential role in "conventional" superconductors [22].

A Hamiltonian that contains the essential physics of high T_c has to describe this interaction of the hole with the outer filled O^{--} shell. The basic components will be the kinetic energy of the holes, a Coulomb interaction between holes, and the interaction between the hole and the outer filled shell. To a first approximation, the hole-ion interaction can be neglected. The Hamiltonian is then:

$$\begin{aligned} H &= \sum_{n,\sigma} \epsilon_k c^+_{k\sigma} c_{k\sigma} + \sum_{kk'q} V(q) c^+_{k+q\uparrow} c^+_{k'-q\downarrow} c_{k'\downarrow} c_{k\uparrow} \\ &+ \sum_{kk'\lambda} I^\lambda_{k-k'} c^+_{k'\sigma} c_{k\sigma} (b_{k'-k,\lambda} + b^+_{k-k',\lambda}) + \sum_{q\lambda} \omega_k b^+_{q\lambda} b_{q\lambda} \end{aligned} \tag{5}$$

where $c^+_{k\sigma}$ creates a hole in the outer O^{--} shell, and $V(q) = 4\pi e^2/q^2$ is the Coulomb repulsion between holes. $b^+_{i\lambda}$ creates an excitation in the outer filled shell, describing a transition of an electron to an orbital in the next shell, and λ labels the different excitations of the outer shell. Off-site interactions will be much smaller than on-site interactions so that the q dependence of I^λ_q is small. The Hamiltonian (5) is formally identical to an electron-phonon Hamiltonian but the parameters are electronic energies, of order several eV. Within the conventional theory of superconductivity we can derive an effective interaction

$$V_{eff}(q,\omega) = \frac{V(q)}{\epsilon(q)} + \sum_\lambda \frac{|I^\lambda_q|^2 \omega_\lambda}{\omega^2 - \omega^2_\lambda} \tag{6}$$

with $\epsilon(q)$ the dielectric constant of the hole gas, and a critical temperature using Eliashberg theory. Because the energy scale is large, however, it is not obvious that vertex corrections will not be important. We have found, however, in recent simulation studies [23] that Eliashberg theory gives reasonable answers even for phonon frequencies of the same order as electronic energies.

An even simpler model that contains the essential physics is obtained by modelling the states of the outer filled shell of the anion by a two-level system. The holes, when they are on a given anion, induce transitions between these polarization states of the cloud. The Hamiltonian is:

$$H = \sum t_{ij}(c^+_{i\sigma} c_{j\sigma} + h.c.) + \alpha \sum_{i\sigma} \sigma^i_x n_{i\sigma} + \omega \sum_i (\cos\theta \sigma^i_x + \sin\theta \sigma^i_z) + U_0 \sum_i n_{i\uparrow} n_{i\downarrow}. \tag{7}$$

Estimates for ω, α and θ can be obtained from atomic physics calculations. ω and α are of order several eV. θ is an important parameter because it determines the band-narrowing due to the hole-cloud interaction, similarly to the situation for small polarons: for $\omega \ll \alpha$, the bare hopping of holes through the O^{--} network t_{ij} is renormalized to $\tilde{t}_{ij} = t_{ij} \cos^2 \theta/2$. The direct O-O hopping has been estimated by McMAHAN et al. [20] to be 0.65 eV, and there will be an additional contribution from hopping through the cations. For definiteness we will take $\theta = \pi/2$ in what follows. U_0 is the bare Coulomb repulsion between two holes on the same anion.

The effective interaction for two holes on the same anion is:

$$U_{site} = E_0^1(2) + E_0^1(0) - 2E_0^1(1) + U_0 \tag{8}$$

with

$$E_0^1(n) = -\sqrt{\omega^2 + \alpha^2 n^2 + 2\omega\alpha n \cos\theta} \tag{9}$$

and we know from atomic physics that $U_{site} > 0$ for any anion, in particular, as mentioned, the bare $U_{site} \sim 10$eV for O. However, when we allow the holes to hop between different anions the effective interaction between holes can be negative for parameters where $U_{site} > 0$. We have diagonalized the Hamiltonian (7) on lattices of size 2, 4 and 8 sites [24]. Figure 3 shows one example of the effective interaction between 2 holes

$$U_{eff} = E_0(2) + E_0(0) - 2E_0(1) \tag{10}$$

for various cases. We find that there is a wide range of parameters where the effective interaction is attractive, even for large on-site repulsion (up to $U_0 \sim 3\alpha$ on the 8-site system for some ω's). An extra added particle is found not to bind so that the system is stable.

The fact that the effective interaction becomes more attractive as the cluster size increases suggests that this mechanism is most effective for small doping: in our 8, 4 and 2-site clusters two particles correspond to band filling $\rho = 0.25$, 0.5 and 1 respectively. We also find that as more particles are added to the 8-site cluster the range of parameters where the effective interaction is attractive decreases [24]. This suggests that superconductivity will be lost for too high doping.

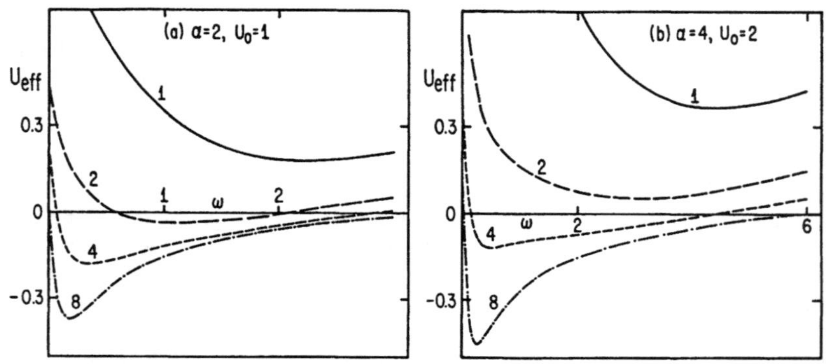

Fig. 3. Effective interaction for 2 particles for the Hamiltonian (7) on clusters of various sizes (indicated by the number next to the curve). $t = 1$, $\theta = \pi/2$

For small levels of doping and large energy scale ω we can model the system by an even simpler Hamiltonian, the Hubbard model with an *attractive* instantaneous interaction $U < 0$. Monte Carlo simulations of the two-dimensional attractive Hubbard model show large enhancement of the s-wave pair susceptibility

$$P = \int_0^\beta d\tau \langle \Delta(\tau)\Delta^+(0) \rangle \qquad (11)$$

$$\Delta = \frac{1}{N}\sum_k c_{k\uparrow}c_{-k\downarrow} \qquad (12)$$

as U becomes more negative, as expected, although not as large as predicted by BCS theory. An example is shown in Fig. 4(a), for band filling $\rho = 0.5$ and interactions $U = 0$, -2 and -4 (in units where the hopping is 1). Figure 4(b) shows the behavior of P for different values of the chemical potential and hence band filling for the case $U = -4$. To extract the two-dimensional critical temperature is difficult and requires simulations on larger lattices. However, we can obtain the transition temperature for three-dimensional superconductivity in the presence of a weak hopping between planes $t_\perp$ within a random phase approximation from the condition [25] $1 = (2t_\perp^2/|U_{eff}|)P$, where P is the in-plane pair susceptibility, (11). The inset in Fig. 4(b) shows the critical temperature versus doping obtained in this fashion for one case. Although the parameters are not realistic it illustrates the general behavior: T_c increases as $n^{1/2}$ with doping. As mentioned above, however, for too high doping the effective interaction itself will cease to be attractive and T_c will drop. Another consequence of high doping is of course that when two holes are on the same O^{--} they yield neutral O that tends to leave the sample, rendering the system unstable.

We expect also the normal state properties to be significantly altered by the hole-cloud interaction; in particular, for $\theta \to \pi$ in (7) significant band narrowing will occur, leading to a small "electronic polaron." SCALAPINO et al. [26] have discussed in detail how a polaron model can explain several of the normal state properties of the high T_c oxides.

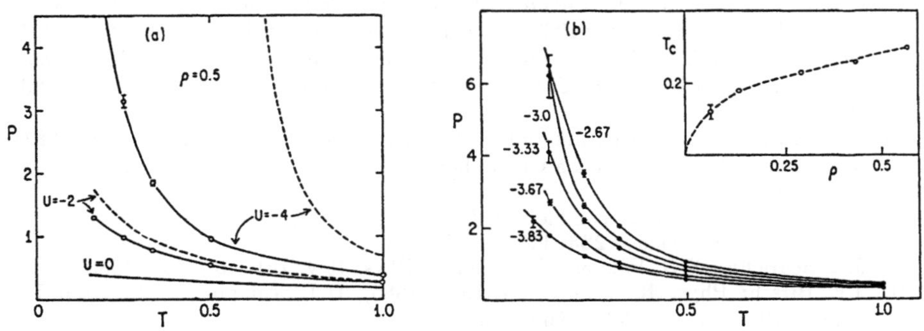

Fig. 4(a). S-wave pair susceptibility P versus temperature for an attractive Hubbard model on a 6×6 lattice, 1/4 filled band. The dashed lines are results of BCS theory. (b) P versus temperature for various band fillings for $U = -4$. The number next to each curve indicates the chemical potential. The inset shows T_c versus band filling (doping) obtained from these data as described in the text for the case $t_\perp/t = 0.9$

To summarize, we believe that the experimental evidence points towards a model of the high T_c materials where conduction occurs via holes in the O network and the only role of the cations is to make this situation possible by stabilizing an otherwise unstable structure. We have argued that conduction by holes in closed-shell anions will necessarily lead to high temperature superconductivity, and discussed some results of model calculations. We also argued that the model can explain the anomalous normal state properties of the high T_c oxides. It will be difficult to prove this model to be correct over a variety of other models involving charge fluctuations, but I believe just its simplicity and universality make it compelling. Perhaps the most convincing proof will come only when material scientists find high T_c superconductivity in a wide variety of materials whose only common characteristic will be that conduction occurs through holes in closed shell anions of elements in the right portion of the periodic table.

Acknowledgements: This work was supported by the National Science Foundation under Grant No. DMR84-51899 as well as contributions from AT&T Bell Laboratories. I am grateful to the organizers and participants of this symposium for a highly stimulating meeting; the ideas discussed in the latter part of this paper were not discussed in my presentation and were cemented to a large extent by the experimental results highlighted at this meeting. Computations were performed at the San Diego Supercomputer Center.

References

1. L.F. Mattheiss, E.M. Gyorgy and D.W. Johnson, Jr.: Phys. Rev. B **37**, 3745 (1988); R.J. Cava et al.: Nature (London) **332**, 814 (1988); B. Batlogg et al.: Phys. Rev. Lett. **61**, 1670 (1988)
2. J.E. Hirsch and H.Q. Lin: Phys. Rev. B **37**, 5070 (1988); H.Q. Lin, J.E. Hirsch and D.J. Scalapino: Phys. Rev. B **37**, 7359 (1988); J.E. Hirsch, unpublished
3. J.E. Hirsch: In *Theories of High Temperature Superconductivity*, ed. by J. Woods Halley (Addison-Wesley Pub. Co., Redwood City, CA 1988) p.241
4. V.J. Emery, Phys. Rev. Lett. **58**, 2794 (1987)
5. J.E. Hirsch, E. Loh, D.J. Scalapino and S. Tang: "Pairing Interaction in CuO Clusters," UCSD preprint, May 1988; Phys. Rev. Lett. **60**, 1668 (1988); In *Proc. of the Intl. Conf. on High-Temperature Superconductors and Materials and Mechanisms of Superconductivity*, ed. by J. Müller and J.L. Olsen (North-Holland Physics Pub. 1988) p.549
6. C.M. Varma, S. Schmitt-Rink and E. Abrahams: Sol. St. Comm. **62**, 681 (1987), had suggested early on that a Cu-O charge-transfer excitation mechanism was operative in the oxides
7. J.E. Hirsch, "Hole Superconductivity," UCSD preprint, October 1988
8. P.W. Anderson: Science **235**, 1196 (1987)
9. J.R. Schrieffer, X.G. Wen and S.C. Zhang: Phys. Rev. Lett. **60**, 944 (1988)
10. J.E. Hirsch: Phys. Rev. Lett. **54**, 1317 (1985); D.J. Scalapino, E. Loh and J.E. Hirsch: Phys. Rev. B **34**, 8190 (1986)
11. J.E. Hirsch and S. Tang: "Antiferromagnetism in the Two-Dimensional Hubbard Model," UCSD preprint, September 1988
12. D.P. Arovas and A. Auerbach: Phys. Rev. B **38**, 316 (1988); A. Auerbach and D.P. Arovas: Phys. Rev. Lett. **61**, 617 (1988)
13. J.E. Hirsch and S. Tang: "Spin-wave theory of the quantum antiferromagnet with unbroken sublattice symmetry," UCSD preprint, September 1988

14. S. Chakravarty, B.I. Halperin and D. Nelson: Phys. Rev. Lett. **60**, 1057 (1988) and Harvard preprint, 1988
15. However, M. Imada, J. Phys. Soc. Jpn. **57**, 3128 (1988) draws different conclusions from simulations of the same model.
16. F.A. Cotton and G. Wilkinson: *Advanced Inorganic Chemistry* (John Wiley & Sons, New York 1988)
17. C.E. Moore: *Atomic Energy Levels* (U.S. Gov. Printing Office, DC 1952)
18. M. Imada, these proceedings and references therein; H. Shiba, these proceedings and references therein
19. C. Balseiro, A.G. Rojo, E.R. Gagliano and B. Alascio: preprint
20. A.K. McMahan, R.M. Martin and S. Satpathy: Phys. Rev. B (to be published). See also references therein
21. The effective interaction is similar to the one that arises in models for excitonic superconductivity extensively studied in connection with quasi-one-dimensional conductors: see, for example, W.A. Little: Int. J. Quant. Chem. **15**, 545 (1981); J.E. Hirsch and D.J. Scalapino: Phys. Rev. B **32**, 117 (1985)
22. Such as Hall coefficient, chemical trends, correlation between T_c and melting point. See also B.T. Matthias, H. Suhl and C.S. Ting, Phys. Rev. Lett. **27**, 245 (1971)
23. J.E. Hirsch and F. Marsiglio, unpublished
24. J.E. Hirsch and S. Tang, unpublished
25. D.J. Scalapino, Y. Imry and P. Pincus: Phys. Rev. B **11**, 2042 (1975); R.A. Klemm and H. Gutfreund: Phys. Rev. B **14**, 1086 (1976)
26. D.J. Scalapino, R.T. Scalettar and N.E. Bickers, *Proc. of the Intl. Conf. on Novel Mechanisms of Superconductivity*, eds. S.E. Wolf and V.Z. Kresin (Plenum, New York 1987)

What Can We Learn from Small-Cluster Studies on CuO_2 and Related Models?

H. Shiba and M. Ogata

Institute for Solid State Physics, University of Tokyo,
Roppongi, Tokyo 106, Japan

It is currently an important issue on which orbital doped holes are located in high-T_c Cu-oxides. Here comparative small-cluster studies are presented on two cases: in-plane $p\sigma$ and $p\pi$ orbitals. They suggest that the $p\sigma$ orbital is more favorable for pairing of two doped holes. In this connection the interaction between doped holes and surrounding Cu spins is examined in detail. It is pointed out thereby that the spin structure factor observable by neutron scattering experiments contains information to identify the location of doped holes.

1. Introduction

As widely recognized, the high-T_c Cu-oxide superconductors discovered so far have notable common features: (1) The superconducting phase appears with doping of holes into an insulating antiferromagnet[1-4]. (2) They all contain CuO_2 layers. (3) Doped holes responsible for superconductivity mainly go onto oxygens[5-7].

It is natural then to think that these features are essential for constructing a successful model of the high-T_c superconductivity. Led by this type of reasoning, a search for mechanism of the high-T_c superconductivity is currently being made on various strongly correlated electron models[8]. The point at issue is first the relation between the magnetism and the superconductivity and secondly the orbital of oxygen, on which the holes are located, and its precise role. With this point in mind we present our small-cluster studies[9,10] on two-dimensional (2D) strongly correlated CuO_2 systems, which have been proposed in connection with the high-T_c superconductivity.

The small-cluster study is, simply speaking, an application of the exact diagonalization method to finite-size clusters, on which an appropriate boundary condition is imposed. This approach is just intermediate between local and extended ones and has some merits as well as limitations: It is expected to be useful especially for strongly correlated electrons and/or systems with short coherence length. In addition we are able to obtain much information from the ground-state wave function determined for finite-size clusters. On the other hand a weak point lies in the system size, which is limited to a fairly modest one. It is our belief that notwithstanding this limitation we can learn much on various problems concerning doped holes in high-T_c Cu-oxides.

It is a matter of controversy on which orbital extra holes are doped: in-plane $p\sigma$[7,11,12], in-plane $p\pi$[13,14] and out-of-plane $p\sigma_z$[15,16] have been proposed. Small-cluster studies are hoped to resolve, at least partly, this problem by providing with the

Springer Series in Materials Science, Vol. 11
Mechanisms of High Temperature Superconductivity
Editors: H. Kamimura and A. Oshiyama

ground-state wave function, which tells us the nature of the interaction between the doped hole and background Cu spins.

This paper is arranged as follows. The case of the in-plane $p\sigma$ orbital is examined in §2, which is followed by an analysis of the in-plane $p\pi$ orbital in §3. Concluding remarks on the results and related problems are given in §4.

2. Holes on In-Plane $p\sigma$ Orbitals

For this case we start from the following 2D CuO_2 model[7,17]:

$$\begin{aligned} H_A = &- \sum_{(ij)\sigma} t_{ij}(a^+_{di\sigma}a_{pj\sigma} + h.c.) \\ &- \sum_{(jj')\sigma} t'_{jj'}(a^+_{pj\sigma}a_{pj'\sigma} + h.c.) \\ &+ \epsilon_d \sum_{i\sigma} n_{di\sigma} + \epsilon_p \sum_{j\sigma} n_{pj\sigma} \\ &+ U_d \sum_i n_{di\uparrow}n_{di\downarrow} + U_p \sum_j n_{pj\uparrow}n_{pj\downarrow} \\ &+ V \sum_{(ij)\sigma\sigma'} n_{di\sigma}n_{pj\sigma'}, \end{aligned} \tag{1}$$

where $a^+_{di\sigma}$ and $a^+_{pj\sigma}$ are the creation operators of holes with spin σ on the i-th Cu site and j-th O site, respectively; $n_{di\sigma}=a^+_{di\sigma}a_{di\sigma}$ and $n_{pj\sigma}=a^+_{pj\sigma}a_{pj\sigma}$ are hole-number operators. t_{ij} represents a hopping integral between the Cu $d_{x^2-y^2}$ and the O in-plane $p\sigma$ orbital. The summation of (ij) and (jj') is taken over the nearest neighbor pairs. The 2D CuO_2 lattice is shown in Fig.1 of ref.9.

The model (1) itself is general enough to allow a wide variety of situations[18]; however various theoretical and experimental studies have already narrowed down our choice for high-T_c Cu-oxides:
(1) The undoped system corresponding to La_2CuO_4, $YBa_2Cu_3O_6$, ... is a charge-transfer-type insulator having a large antiferromagnetic superexchange $J_{Cu} \sim 10^3$K so that $\Delta=\epsilon_p$-ϵ_d must be small but certainly larger than a critical value Δ_c of the metal-insulator transition.
(2) Many researchers[19-21] conclude that U_d=5-10eV, $U_p = 3-6$eV, $t = 1-1.5$eV, $t' \sim$0.5eV and $V \sim$1.5eV (by assuming $\epsilon \sim$5).
Therefore we ignore t', V and U_p for simplicity. The magnitude of U_p looks large, but its effect is presumably reduced in the low-hole-density region. We note in passing that a pairing mechanism based essentially on a large V[22] is clearly unrealistic in view of the above estimation.

In such a situation it is useful to study an effective Hamiltonian, which can be derived for a small t by assuming each Cu $d_{x^2-y^2}$ orbital is singly occupied by a hole:

$$H_B = H^{(2)}_{eff} + H^{(4)}_{ex} \tag{2}$$

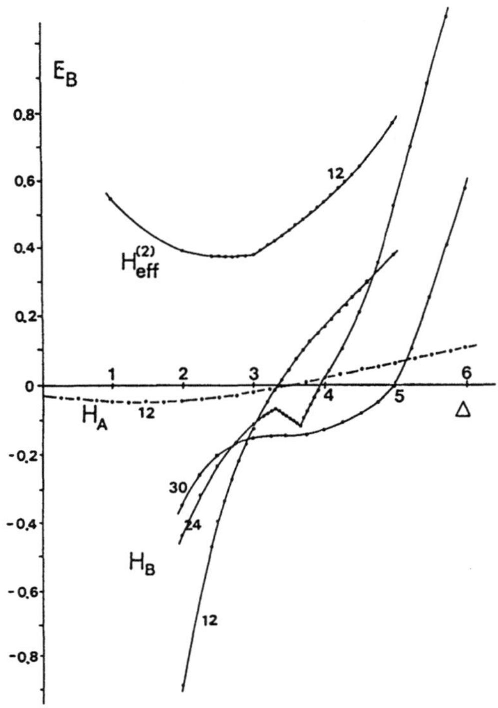

Fig. 1
E_B as a function of Δ calculated for H_B. U_d/t is taken as 8. The numbers attached on the lines denote the size of the cluster. The results for $H^{(2)}_{eff}$ and H_A are included for a comparison.
The kinks on the lines are due to switching of the ground state in one extra hole system (see also Fig.8 of ref. 9 on this point).

where

$$
\begin{aligned}
H^{(2)}_{eff} &= (\epsilon_d - \frac{4t^2}{\Delta})\sum_{i\sigma} n_{di\sigma} + (\epsilon_p + \frac{2t^2}{\Delta})\sum_{j\sigma} n_{pj\sigma} \\
&+ \sum_{(ij)\sigma}(\frac{t^2}{\Delta} + \frac{t^2}{U_d - \Delta})(a^+_{di\sigma}a_{di-\sigma}a^+_{pj-\sigma}a_{pj\sigma} - n_{di-\sigma}n_{pj\sigma}) \\
&+ \sum_{(ij)(ik)\sigma, j\neq k}[(\frac{t^2}{\Delta} + \frac{t^2}{U_d - \Delta})a^+_{di\sigma}a_{di-\sigma}a^+_{pk-\sigma}a_{pj\sigma} \\
&\qquad + (\frac{t^2}{\Delta}n_{di\sigma} - \frac{t^2}{U_d - \Delta}n_{di-\sigma})a^+_{pk\sigma}a_{pj\sigma}]
\end{aligned}
\tag{3}
$$

and

$$
H^{(4)}_{ex} = 2J_{Cu}\sum_{(ii')}(\vec{S}_i \cdot \vec{S}_{i'} - \frac{1}{4})
\tag{4}
$$

with Cu spin $\vec{S}_i$ and J_{Cu} $=2(t^2/\Delta)^2(1/\Delta+1/U_d)$. The third term in (3) represents the antiferromagnetic exchange between copper and oxygen holes and the fourth term describes the hopping of oxygen holes via copper sites.

We have determined the ground state of clusters containing Cu and O atoms up to 30 in total; they are shown in Fig.1 of ref.9. The periodic boundary condition is imposed on the boundaries. As a suitable set for high-T_c Cu- oxides, we take $\Delta/t = 3$ and $U_d/t = 8$, which give $J_{Cu} = 0.1t$ consistent with experiments[1]. t is taken as the unit of energy.

Figure 1 shows E_B as a function of Δ with U_d/t fixed at 8. E_B is defined as $E_B = E_2 + E_0 - 2E_1$ where E_n is the ground state energy for n extra holes. This quantity represents "interaction energy" of 2 doped holes. E_B becomes negative with decreasing Δ, suggesting a pairing of 2 doped holes ; it is due to J_{Cu}, since as evident in Fig.1 $H^{(2)}_{eff}$ alone does not produce negative E_B. The pairing of two doped holes can be confirmed also by looking into the density-density correlation of doped holes. The latter quantity, which has been evaluated for $U_d/t = 8$ with the ground-state wave function, is shown in Fig.2. The density-density correlation as a function of Δ is consistent with our interpretation of E_B: an attractive interaction between 2 doped holes is present for small Δ.

As discussed in detail in refs.9 and 10, each doped hole on oxygen causes a magnetic distortion of Cu spins around it through the strong antiferromagnetic coupling between neighboring Cu and O spins. Two holes attract each other by a constructive interference of the magnetic distortions. This picture renders a partial support to Aharony *et al.*'s classical description[12]. However, the quantum mechanical nature is important in this problem, since the ferromagnetic coupling between Cu and O spins is not favorable

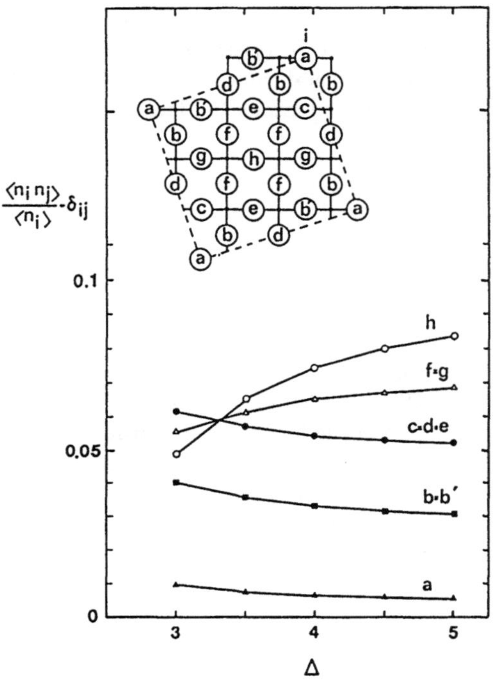

Fig. 2
The density -density correlation $< n_i n_j > / < n_i > -\delta_{ij}$ between oxygen sites as a function of Δ. n_i represents the density of hole on oxygen. The inset shows the position of j–site; the i–site is on the site a.

as shown in the next section. This sharp difference between ferromagnetic and antiferromagnetic couplings cannot be understood, if we rely on Aharony *et al.*'s classical picture.

One important aspect of the $p\sigma$ orbital is the strong antiferromagnetic coupling between the doped hole and Cu spins. We wish to point out that it shows up in a characteristic way, which can be utilized experimentally to determine the location of doped holes. Since spins are present on O sites as well as on Cu sites, the total spin structure factor $S^{zz}(q)$, which is the Fourier transform of the z-component spin-spin correlation function, consists of Cu-Cu, Cu-O and O-O spin correlations:

$$S^{zz}(q) = S^{zz}_{Cu-Cu}(q) + S^{zz}_{Cu-O}(q) + S^{zz}_{O-O}(q). \tag{5}$$

The second and third terms are nonvanishing only in doped systems.

The q dependence of $S^{zz}_{Cu-O}(q)$ has a unique feature because of the structure of the CuO_2 layer, in which O atoms are located on the bridge sites between two Cu atoms. In fact, if we assume an antiferromagnetic spin correlation between the nearest neighbor Cu and O spins, $S^{zz}_{Cu-O}(q)$ is expected to contain a contribution proportional to $-(\cos\frac{q_x}{2}+\cos\frac{q_y}{2})$. Here the minus sign is due to the antiparallel spin correlation. This simple consideration shows that the periodicity of the total spin structure factor $S^{zz}(q)$ is doubled in q space; consequently this makes $S^{zz}(q)$ asymmetric with respect to the M point (i.e. antiferromagnetic Bragg point) in the scan from Γ to Γ through M. Figure 3 demonstrates an example of $S^{zz}(q)$, which has been obtained from our small-cluster study. Evidently $S^{zz}(q)$ has the feature mentioned above. The symmetry of $S^{zz}(q)$ in q space is essentially due to the structure of the CuO_2 layer and the nature of the Cu-O spin correlation. Therefore, although the magnitude of $S^{zz}(q)$ should depend on the values of parameters, we expect the symmetry to be general. Since $S^{zz}(q)$ is observable by neutron scattering experiments, it can be used to determine the location of doped holes.

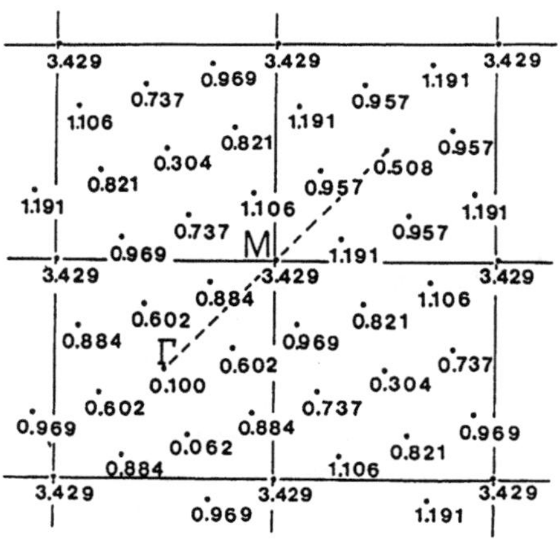

Fig. 3
$4S^{zz}(q)$ calculated for one extra hole in the 30-site cluster, which is equivalent to the hole density of 10%. The broken line corresponds to the h−scan in refs.25 and 26.

3. Holes on In-Plane $p\pi$ Orbitals

Let us now turn to the case of in-plane $p\pi$ orbitals, which was first proposed by Guo *et al.*[13] and Birgeneau *et al.*[14] This problem can be studied with the following model:

$$H = - \sum_{(jj')\sigma} t_{jj'}(a^{+}_{pj\sigma}a_{pj'\sigma} + h.c.) + 2J\sum_{(ij)} \vec{S}_i \cdot a^{+}_{pj\sigma}\vec{S}_{\sigma\sigma'}a_{pj\sigma'} + 2J_{Cu}\sum_{(ii')}(\vec{S}_i \cdot \vec{S}_{i'} - \frac{1}{4}), \quad (6)$$

where $a^{+}_{pj\sigma}$ is the creation operator of a hole on the in-plane $p\pi$ orbital of the j-th oxygen. It is assumed here again that each Cu atom is singly occupied by a hole on the $d_{x^2-y^2}$ orbital, which is represented by $\vec{S}_i$. Because of the orthogonality between $d_{x^2-y^2}$ and $p\pi$, we expect J to be ferromagnetic ($J < 0$). J_{Cu} is the same Cu-Cu superexchange as before.

We have made small-cluster studies on (6) to compare $p\sigma$ and $p\pi$ orbitals with each other[10,23]. Figure 4 shows constant-E_B lines in the $t - J$ plane by fixing J_{Cu} at 0.3. (This value of J_{Cu} is just the unit of energy in this calculation.) It demonstrates that 2 doped holes do not bind with each other for t larger than $\sim 0.2|J|$. We believe from physical grounds that the binding in small t is not relevant to superconductivity.

The presence or absence of binding can be checked by looking into the density-density correlation of two doped holes for the largest cluster. The result shown in Fig.5 is in fact consistent with the conclusions drawn from E_B. According to a recent estimation, t is $\sim$ 0.5eV so that it is much larger than J_{Cu}; therefore we think the

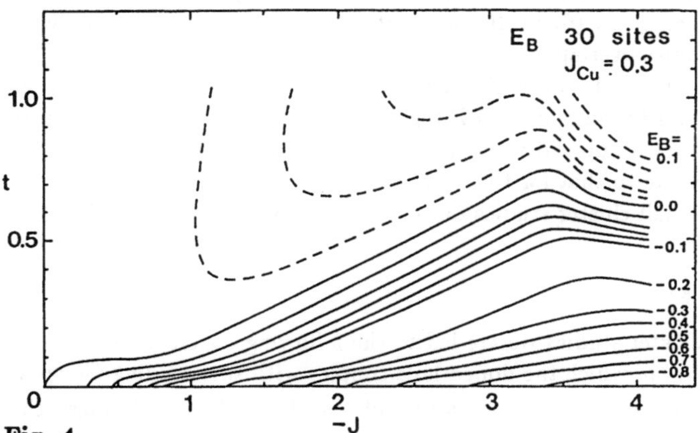

Fig. 4
The constant -E_B lines in the $t - J$ plane. The broken lines show the region with $E_B > 0$, while the region with $E_B < 0$ is shown by solid lines. The unit of energy is such that $J_{Cu} = 0.3$.

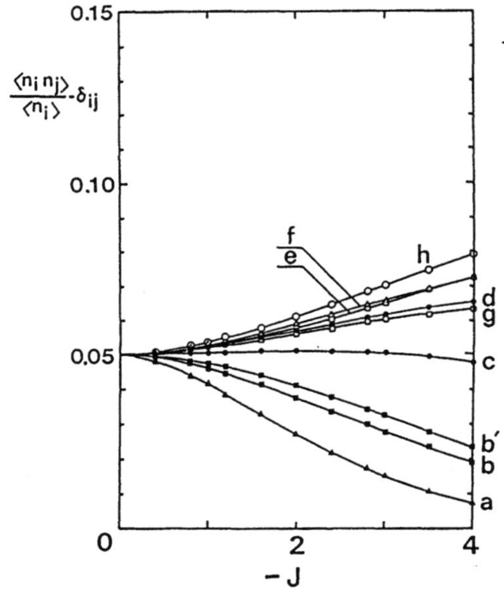

Fig. 5
The density-density correlation $< n_i n_j > / < n_i >$ $-\delta_{ij}$ calculated for two extra holes in the 30-site cluster. $t = 1$ and $J_{Cu} = 0.3$ are chosen. See the inset of Fig.2 as for the definition of sites.

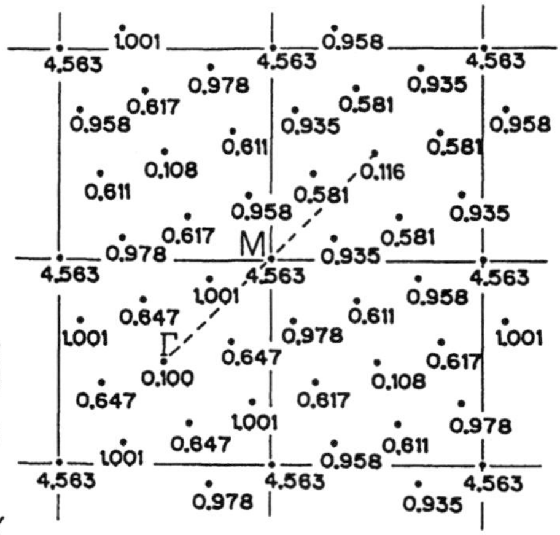

Fig. 6
$4S^{zz}(q)$ calculated for one extra hole in the 30-site cluster.
$J = -1$, $J_{Cu} = 0.3$ and $t = 1$ are chosen here.

binding of two doped holes is not likely to occur for the in-plane $p\pi$ orbital. A similar study has been made recently on (6) by Hatsugai *et al.*[24]. Their results agree with ours as far as the two studies overlap with each other.

The ferromagnetic coupling between Cu and O spins clearly appears in $S^{zz}(q)$, as shown in Fig.6. This should be compared with Fig.3, a corresponding result for the $p\sigma$ orbital. Here the asymmetry of intensity with respect to M point in the scan from Γ to Γ through M is just opposite to that for the $p\sigma$ case.

More details of our small-cluster studies on (6) will be reported separately[23].

4. Concluding Remarks

From the present small-cluster studies we are led to the following conclusions.
(1) The in-plane $p\sigma$ orbital, in which the doped hole couples with surrounding Cu spins antiferromagnetically, is more favorable for pairing of two doped holes, although the in-plane $p\pi$ orbital is not excluded completely.
(2) The magnetic correlation between Cu and O spins should show up in the spin structure factor $S^{zz}(q)$ (i.e. Fourier transform of the equal-time total spin correlation). In particular we predict an asymmetry of intensity in the scan from Γ point through

the antiferromagnetic Bragg point; the asymmetry reflects the nature of the magnetic correlation between Cu and O spins and is expected to increase with the concentration of holes. As a matter of fact, results of two recent independent experiments[25,26] on $(La,Sr)_2CuO_4$ seem to show the asymmetry of intensity, which is consistent with the antiferromagnetic Cu-O spin correlation, suggesting the doped holes mainly go onto the in-plane $p\sigma$ orbital. However, more experiments are needed admittedly to make a definite conclusion on this point.
(3) The in-plane $p\sigma$ orbital mixes with the out-of-plane $p\sigma_z$ orbital in the pyramidal structure, as recently pointed out by Fujimori[16]. Experimentally there are some indications suggesting the importance of the out-of-plane oxygen[27]. We believe they are compatible with the in-plane $p\sigma$ orbital because of the mixing mentioned above. The implications of this mixing have to be explored further.

The authors thank Y. Endoh, H. Yoshizawa and Y. Yamada for helpful discussions. This work is partly supported by Grant-in-Aid for Scientific Research on Priority Areas "Mechanism of Superconductivity" (63631007) and "New Functionality Materials-Design, Preparation and Control" (63604014) from the Ministry of Education, Science and Culture.

References

1) For $(La,Sr)_2CuO_4$, see for instance, G. Shirane *et al.*: Phys. Rev. Lett. **59**, 1613 (1987); R.J. Birgeneau *et al.*: Phys. Rev. B, in press.
2) For $YBa_2Cu_3O_{7-y}$, see J.M. Tranquada *et al.*: Phys. Rev. Lett. **60**, 156 (1988); J. Rossat-Mignod *et al.*: Physica C **152**, 19 (1988).
3) For $Bi_2Sr_2(Y,Ca)Cu_2O_y$, see Y. Nishida *et al.*: Physica C, in press; T. Fujita: this proceedings.
4) P.W. Anderson: *Proceedings of the International School of Physics "Enrico Fermi", July 1987* (North Holland, Amsterdam), in press.
5) A. Fujimori, E. Takayama-Muromachi, Y. Uchida and B. Okai: Phys. Rev. B **35**, 8814 (1987)
6) N. Nücker, J. Fink, J.C. Fuggle, P.J. Durham and W.M. Temmerman: Phys. Rev. B **37**, 5158 (1988).
7) V. Emery: Phys. Rev. Lett. **58**, 2794 (1987).
8) See for instance *Proceedings of the International Conference on High Temperature Superconductors and Materials and Mechanisms of Superconductivity* (Interlaken, Feb. 1988).
9) M. Ogata and H. Shiba: J. Phys. Soc. Jpn. **57**, 3074 (1988).
10) H. Shiba and M. Ogata: *Proceedings of 6th International Conference on Crystal-Field Effects and Heavy-Fermion Physics* (Frankfurt, July 1988), in press.
11) F.C. Zhang and T.M. Rice: Phys. Rev. B **37**, 3759 (1988).
12) A. Aharony, R.J. Birgeneau, A. Coniglio, M.A. Kastner and H.E. Stanley: Phys. Rev. Lett. **60**, 1330 (1988).
13) Y. Guo, J.M. Langlois and W.A. Goddard III: Science **239**, 896 (1988).

14) R.J. Birgeneau, M.A. Kastner and A. Aharony: Z. Phys. B **71**, 57 (1988).
15) A. Bianconi, M. de Santis, A. di Cicco, A.M. Flank, A. Fontaine, P. Lagarde, H. Katayama-Yoshida, A. Kotani and A. Marcelli: Phys. Rev. B, in press.
16) A. Fujimori: preprint.
17) C.M. Varma, S. Schmitt-Rink and E. Abrahams: Solid State Comm. **62**, 681 (1987).
18) J. Zaanen, G.A. Sawatzky and J.W. Allen: Phys. Rev. Lett. **55**, 418 (1985).
19) J. Zaanen, O. Jepsen, O. Gunnarsson, A.T. Paxton, O.K. Andersen and A. Svane: Physica C **153-155**, 1636 (1988).
20) K.T. Park, K. Terakura, T. Oguchi, A. Yanase and M. Ikeda: J. Phys. Soc. Jpn., in press.
21) T.M. Rice: private communication; F. Mila: preprint.
22) J.E. Hirsch, S. Tang, E. Loh and D.J. Scalapino: Phys. Rev. Lett. **60**, 1668 (1988).
23) M. Ogata and H. Shiba: in preparation.
24) Y. Hatsugai, M. Imada and N. Nagaosa: preprint.
25) R.J. Birgeneau *et al.*: preprint.
26) H. Yoshizawa *et al.*: preprint.
27) See for instance H. Yasuoka: this proceedings.

Recent Numerical Studies on Models for High-T_c Superconductors

M. Imada

Department of Physics, College of Liberal Arts, University of Saitama, Shimo-Ookubo, Urawa 338, Japan

Quantum simulation and exact diagonalization studies on Hubbard- like models and the coupled spin-fermion model are summarized. The pairing mechanism is discussed in the light of numerical results. The origin of the singlet cloud around a hole in two-band models and its consequences are investigated.

§1. Introduction

Because of the difficulty in theoretical treatments of strong correlation in electron systems, mechanism of high-T_c superconductivity is still far from the complete understanding. The formation of Cooper pairing may be ascribed to the attractive interaction of two mobile holes in the CuO_2 network, because it should lead to the instability at the fermi level as in the original work by Cooper[1]. However, the attractive interaction is a consequence of the subtraction of large Coulomb repulsion from the attractive part mediated by either spin or charge degrees of freedom. Since the origins of attractive and repulsive part are the same electron-electron Coulomb interaction, we have to treat them on the equal footing. The mean field theory which extracts only the attractive part as a mean field with an approximate treatment of the repulsive part should not be employed in the argument for the Cooper pair. The existence of the bound state of two holes may be a consequence of quantitative subtraction and may not be derived from a universality class.

In this circumstance, one way to speculate the existence of the Cooper pair is numerical approach. Among several basic numerical techniques, quantum simulation and exact diagonalization of finite systems provide useful information. So far, a shortcoming of the quantum simulation has been the limitation of the temperature range. However, it provides data free from biased approximations and is useful to judge the plausibility of various approximations. Because the applicable temperature range is limited only from a technical reason at this moment, we may expect a substantial progress in the near future[2].

The exact diagonalization technique is now widely used to get an insight from small cluster study. Tractable system size is limited rather severely by allowed memory size in the computer. However, it provides accurate data in the ground state. By combining appropriate extrapolation procedure, it has proven to be useful to get a physical picture. In this paper, I summarize our recent results of quantum simulation and the exact diagonalization study done for models of the CuO_2 network. In §2, several models for the high-T_c oxides are presented. Their interrelation and experimental relevance are discussed. In §3, several additional comments are given with the summary of the singlet cloud mechanism obtained from the numerical studies.

Springer Series in Materials Science, Vol. 11
Mechanisms of High Temperature Superconductivity
Editors: H. Kamimura and A. Oshiyama

§2. Models

A general hamiltonian for the CuO_2 plane has the form:

$$H = \sum_{k\sigma fg} \epsilon_{fg}(k)(c^{\dagger}_{fk\sigma} c_{gk\sigma} + h.c.) + \sum_{<fg>} U_{fg} \sum_{<ij>} n_{fi} n_{gj}, \qquad (1)$$

where the fermion creation (annihilation) operators $c^{\dagger}_{fk\sigma}(c_{fk\sigma})$ represent the holes in one of Cu-$3d$, O-p_σ or O-p_π orbitals with the wave number k and the spin σ. The intra and inter-band transfer term $\epsilon_{fg}(k)$ depends on the symmetry of the location of holes.

It is now generally believed that the holes occupy Cu-$3d$ orbitals as $3d^9$ states in the half-filled band, which leads to the antiferromagnetic long range order through the superexchange interaction. There is an accumulation of experimental evidences which support that the doped holes beyond the half-filling go into oxygen p-orbitals[3-5]. The symmetry of its orbital has not been established yet. If we assume that the holes are located in the p_σ orbital in the CuO_2 plane, the hamiltonian is reduced to a two-band $d - p_\sigma$ model[6-9]:

$$\begin{aligned} H = & -t_1 \sum_{<ij>,\sigma} (d^{\dagger}_{i\sigma} p_{j\sigma} + d^{\dagger}_{i\sigma} q_{j\sigma} + h.c.) \\ & - t_2 \sum_{<ij>\sigma} (p^{\dagger}_{i\sigma} q_{j\sigma} + h.c.) \\ & + U_d \sum_i n_{di\uparrow} n_{di\downarrow} + U_p \sum_j (n_{pj\uparrow} n_{pj\downarrow} + n_{qj\uparrow} n_{qj\downarrow}) \\ & + V \sum_{<ij>} n_{di}(n_{pj} + n_{qj}) \\ & + \varepsilon_d \sum_i n_{di} + \varepsilon_p \sum_j (n_{pj} + n_{qj}). \end{aligned} \qquad (2)$$

When the doped holes are assumed to be located in p_π orbitals at the oxygen site in the CuO_2 plane, the hamiltonian for the $d - p_\pi$ model may be obtained from eq.(2) by putting $t_1 = 0$. The fermion operators d, p and q represent holes in Cu-$3d_{x^2-y^2}$, p_x and p_y orbitals, respectively.

It has been clarified from the photoemission experiments[3] that the high-T_c oxides have the character of the charge transfer type. This means $U_d > \delta \equiv \varepsilon_p - \varepsilon_d$. Although it has not been fully clarified yet, some experimental indications [3-5] seem to support that the local magnetic moment on the Cu-site is preserved even in the doped material, which is realized when $\delta > t_1, t_2$. This region is characterized by the Kondo limit, where the Cu-$3d$ band is reduced to spin-1/2 localized spins. The charge fluctuation at the Cu-sites is suppressed in this case. Then the effective hamiltonian may be derived from eq.(2) by using the perturbational expansion with respect to t/U_d and t/δ. The relevant effective hamiltonian is given as $H_0 + H_1 + H_2$ from

$$H_0 = -t \sum_{<lm>} (c^{\dagger}_l c_m + h.c.)$$

and

$$\begin{aligned} H_1 &= -2J^{(1)}_K \sum_i \vec{S}_i \cdot \vec{\sigma}^{(1)}_i, \\ H_2 &= -2J^{(2)}_K \sum_i \vec{S}_i \cdot \vec{\sigma}^{(2)}_i \end{aligned} \qquad (3)$$

up to the second order, where $\vec{S}_i$ is the spin -1/2 operator at the i-th Cu-site. The fermion creation operator $c_l^\dagger$ is defined at the oxygen sites. The summation with respect to $< lm >$ is over the nearest neighbor pairs. The parameters are given by $t = t_2$ and $J_K^{(1)} = t_1^2(1/(U_d - \delta - V) + 1/(\delta + U_p - V))$ and $J_K^{(2)} = t_1^2(1/(U_d - \delta - V) + 1/(\delta - V))$ for p_σ holes. The operators are defined by

$$\vec{\sigma}_i^{(1)} = \frac{1}{2}\sum_{\vec{\delta}} c^\dagger_{\vec{i}+\vec{\delta},\sigma}(\vec{\sigma})_{\sigma\sigma'}c_{\vec{i}+\vec{\delta},\sigma'} \tag{4}$$

and

$$\vec{\sigma}_i^{(2)} = \frac{1}{2}(\sum_{\vec{\delta}} c_{\vec{i}+\vec{\delta},\sigma})(\vec{\sigma})_{\sigma\sigma'}(\sum_{\vec{\delta}'} c_{\vec{i}+\vec{\delta}',\sigma'}) - \vec{\sigma}_i^{(1)}, \tag{5}$$

where the summation over $\vec{\delta}$ and $\vec{\delta}'$ denotes that over the nearest neighbor oxygens of the $i-th$ copper site. For p_π orbitals, $J_K^{(2)} = 0$ and $J_K^{(1)}$ is ferromagnetic. When the hole concentration is small, the fourth order term has the form

$$H_3 = -2J_S\sum_{<ij>}\vec{S}_i \cdot \vec{S}_j$$

up to the lowest order of the hole concentration. The superexchange coupling constant J_S is given in the case of the p_σ orbital as

$$J_S = \frac{2t^4}{(\delta + V)^2}\left[\frac{1}{U_d} + \frac{2}{U_p + 2\delta}\right]. \tag{6}$$

The total effective hamiltonian $H = H_0 + H_1 + H_2 + H_3$ provides a starting point in the Kondo limit.

The key idea is that a doped hole has a strong tendency to form a singlet with one of the localized spins, when J_K is antiferromagnetic. It was pointed out by the author[7,8] in the argument of the two-band $d - p_\sigma$ model. Later on, the symmetrized singlet state around the Cu-site was stressed to form a local singlet with the Cu spin[10]. Although it is frequently discussed to justify the effective Hamiltonian of the single band Hubbard model from the two-band point of view, two important points are neglected in this argument.

The first point concerns the site-off-diagonal term included in $H_1 + H_2$. If we neglect U_p for simplicity, the Kondo-like term $H_1 + H_2$ with the definition of $\vec{\sigma}_i^{(1)}$ and $\vec{\sigma}_i^{(2)}$ in eqs.(4) and (5) may be written as

$$H_1 + H_2 = -2J_K\sum_{\vec{i}}\vec{S}_{\vec{i}}(\vec{\sigma})_{\sigma\sigma'}a^\dagger_{\vec{i}\sigma}a_{\vec{i}\sigma'} \tag{7}$$

and

$$\begin{aligned} a^\dagger_{\vec{i}\sigma} = & f^\dagger_{\vec{i}\sigma} + \frac{1}{2}(f^\dagger_{\vec{i}-2\vec{\delta}_1,\sigma} + g_{\vec{i}-2\vec{\delta}_1,\sigma} \\ & + f^\dagger_{\vec{i}-2\vec{\delta}_2,\sigma} - g^\dagger_{\vec{i}-2\vec{\delta}_2,\sigma}) \end{aligned} \tag{8}$$

with $J_K = J_K^{(1)}$, where f and g are symmetrized and antisymmetrized hole operator in the p_σ orbitals defined as

$$f^\dagger_{\vec{i}\sigma} = \frac{1}{\sqrt{2}}(c^\dagger_{\vec{i}+\vec{\delta}_1,\sigma} + c^\dagger_{\vec{i}+\vec{\delta}_2,\sigma}) \tag{9}$$

and

$$g_{i\sigma}^{\dagger} = \frac{1}{\sqrt{2}}(c_{\vec{i}+\vec{\delta}_1}^{\dagger} - c_{\vec{i}+\vec{\delta}_2}^{\dagger}), \tag{10}$$

respectively. The vectors are defined as $\vec{\delta}_1 = (1/2, 0)$ and $\vec{\delta}_2 = (0, 1/2)$ in the unit of Cu-Cu distance.

If we assume that t is larger than J_K, only the symmetrized orbitals may contribute to the problems at low concentration of holes. Then eq.(8) may be reduced to

$$a_{\vec{i}\sigma}^{\dagger} = f_{\vec{i}\sigma}^{\dagger} + \frac{1}{2}(f_{\vec{i}-2\vec{\delta}_1,\sigma}^{\dagger} + f_{\vec{i}-2\vec{\delta}_2,\sigma}^{\dagger}). \tag{11}$$

The operators $\{a_{\vec{i}\sigma}^{\dagger}\}$ are not orthogonal between the nearest neighbors. The properly orthogonalized Wannier set is given by[10)]

$$\tilde{a}_{\vec{i},\delta}^{\dagger} = \sum_{\vec{j}} Q(\vec{i} - \vec{j}) a_{\vec{j}\sigma}^{\dagger}, \tag{12}$$

where

$$Q(\vec{r}) = \int e^{-i\vec{k}\cdot\vec{r}} \frac{d\vec{k}}{\sqrt{1 - (cosk_x + cosk_y)/2}}. \tag{13}$$

Then the Kondo-like term may be written as

$$H_1 + H_2 = -2J_K \sum_{\vec{i}} \vec{S}_{\vec{i}}(\vec{\sigma})_{\sigma\sigma'} \sum_{lm} \tilde{Q}(\vec{i} - \vec{l})\tilde{Q}(\vec{i} - \vec{m})\tilde{a}_{\vec{l}\sigma}^{\dagger}\tilde{a}_{\vec{m}\sigma'}, \tag{14}$$

where

$$\tilde{Q}(\vec{r}) = \int e^{i\vec{k}\cdot\vec{r}} \sqrt{1 - (cosk_x + cosk_y)/2}\, d\vec{k}. \tag{15}$$

This suggests the extended character of the coupling J_K with a long tail proportional to $1/r$. It may be extended to a more general case to allow the treatment in the region $|J_K| > t$ by taking account of antisymmetric operator g. It introduces another set of Wannier state with another long-ranged coupling to the localized spins. This consideration suggests that the singlet pair formed by the hole and the substrate spin necessarily has extended character and may not be described by a local singlet. If the superexchange interaction J_S is introduced, the extension of the Kondo coupling is effectively cut off and an extend singlet cloud may appear in the sea of antiferromagnetic correlation of the substrate.

The second point neglected in ref.10 is related to the competition between t and J_K. In the relevant experimental situation, it is likely that t and J_K have comparable strengths. The hole's spin cannot form a complete on-site singlet because of the itinerancy of the hole. The extended hole's wave function forms an extended singlet with droplet of localized spins.

Both of two aspects mentioned above seem to bring similar effect, i.e., formation of extended singlet cloud. The importance of this extended character has been pointed out in the numerical analysis[11-13)].

To discuss the essence of superconductivity in the simplest model, we have simplified eq.(5) by neglecting the site-off-diagonal hopping term, which corresponds to neglect the term proportional to $J_K^{(2)}$[11-13)]. The extended character of the singlet cloud is retained after this simplification for $t \sim |J_K|$ and the qualitative feature is expected to be the same. Quantitative comparison of results between in the presence and in the absence of $J_K^{(2)}$ for the p_σ orbital is currently investigated. The hamiltonian we finally obtain is the coupled spin-fermion hamiltonian given by

$$H = -t \sum_{<i,j>\sigma} (c_{i\sigma}^{\dagger} c_{j\sigma} + c_{j\sigma}^{\dagger} c_{i\sigma}) + U_h \sum_i n_{i\uparrow} n_{i\downarrow}$$

$$- 2J_K \sum_{<i,l>} \vec{S}_l \cdot \vec{\sigma}_i - 2J_S \sum_{<l,m>} \vec{S}_l \cdot \vec{S}_m \qquad (J_S < 0)$$

$$\vec{\sigma}_i = \frac{1}{2} c_{i\sigma}^{\dagger} (\vec{\sigma})_{\sigma\sigma'} c_{i\sigma'}$$
$$n_{i\sigma} = c_{i\sigma}^{\dagger} c_{i\sigma} \tag{16}$$

To investigate the role of specific lattice structure of the oxygen conduction band, a more simplified lattice structure may also be investigated to extract the essence. A possible simplification is to take the fermion lattice of the same structure as the square lattice of the localized spins. In this case, the summation over $\vec{\delta}$ in eq.(4) is dropped. The fermions and the spins are coupled at the same site through J_K. This case is refered to as the $z_K = 1$ case in comparison with eq.(4) which is refered to as $z_K = 2$ case. The $z_K = 1$ class is also related to a possibility of holes doped in $d_{3z^2-r^2}$ orbital hybridized with the p_z orbital out of the CuO_2 plane[14], when we take ferromagnetic J_K. Another connection of the $z_K = 1$ class is seen in the $t - J$ model[15]:

$$H = -t \sum_{<ij>} (1 - n_{i,-\sigma}) c_{i\sigma}^{\dagger} c_{j\sigma} (1 - n_{j,-\sigma}) + h.c.$$
$$- 2J_S \sum_{<ij>} \vec{S}_i \cdot \vec{S}_j, \tag{17}$$

originally derived as an effective hamiltonian of the single band Hubbard model. We have shown the equivalence of this $t-J$ model to the $z_K = 1$ class in the limit $J_K \to -\infty$[11,12]. The $t - J$ model is the extreme limit, where the on-site local singlet state is completely formed. Although $J_K \to -\infty$ does not seem to be a realistic situation for the exchange coupling of the spin and hole in the CuO_2 network, it would provide an interesting problem. The on-site singlet is surrounded by a domain of enhanced quantum mechanical fluctuation of the antiferromagnetic correlation. Therefore, it is conceivable that the formation of the quantum cloud could be a common feature even in this case. This comes from the significant retardation effect in the localized spin system due to the situation $|J_S| \ll t$.

§3. Pairing Mechanism

We first briefly summarize the results obtained in ref.7,8 and 11-13 for the model of holes in the p_σ orbitals. In the quantum simulation[7,8] performed for the two-band $d - p_\sigma$ model in the case of $t_2 = U_p = V = 0$, the superconducting pairing susceptibilities for the pairing of two oxygen holes in a certain range of distance showed small but finite enhancement as compared to the noninteracting system with each of the oxygen and the copper hole concentrations kept at the same values as the interacting system. The interacting and noninteracting cases show almost the same amplitude of pairing susceptibilities in the temperature range of the transfer t_1 in eq.(2) and the enhancement gradually increases with the decrease of temperature. Although the results are not conclusive as for the possibility of the superconductivity because of the limited temperature range, the results are in contrast with the case of the single band Hubbard model. The enhancement is seen in an extended zone of the oxygen-oxygen pairing. It suggests that the pairing has an extended character. The nearest neighbor pairing seems to have relatively small amplitude. Another aspect of the pairing susceptibilities is that they have a peak around 0.2 of doping away from the half-filling as a function of the hole concentration at low temperatures[16].

The parameter region mainly investigated in ref.8 is in the mixed valence region, where the level difference δ is comparable to t_1. Recent experimental indications seem to suggest that the high-T_c oxides are located in the Kondo regime as mentioned in §1. We expect that the coupled spin-fermion model provides a good description in this case. However, it should also be noted that it is not clear enough whether the real parameter values assure the Kondo regime. Partly inspired by the experimental indications, the coupled spin-fermion model was investigated by the exact diagonalization technique in the ground state[11-13]. Here I summarize the results and comment on several additional aspects of the singlet cloud mechanism.

Spin and charge correlation functions have been calculated in the coupled spin-fermion systems with one hole. The spin correlation shows that the cloud of extended magnetic distortion developes in the region $t \sim |J_K| \gg |J_S|$. In the cloud, the nearest neighbor spin-spin correlation in the substrate spin system is reduced as compared to the pure Heisenberg system. The spin-spin correlations between the hole's spin and localized spins in the cloud are always antiferromagnetic. This shows that the hole's spin and the extended spin polarization in the cloud form the totally singlet state. It should be contrasted with the picture of static magnetic polaron or soliton[17,18]. Dynamical exchange of spins in the time scale of J_K^{-1} leads to the picture that the itinerant hole is represented by the singlet-type linear combination of up hole's spin with down polarization and down hole's spin with up polarization. This singlet cloud mechanism is in sharp contrast with the ferromagnetic polaron and charge polaron where the polarization in the cloud is static. The dynamical exchange of spins in the singlet cloud makes it free from the self-trapping due to the mass enhancement. The singlet cloud is also free from the magnetic instability even in the extremely strong coupling case. They are important differences from the charge polaron case.

When we take $J_K^{(2)} = 0$, the ground state of one hole system for $|J_K| \gg |J_S|$ and $t = 0$ may be obtained from the diagonalization of 3-site Cu-O-Cu problem. The ground state shows strong ferromagnetic correlation between neighboring two Cu spins[19]. It has been argued that it may introduce the frustration in the system[20]. The existence of the ferromagnetic correlation is crucially important in the frustration mechanism of the pairing. It, however, turned out[12,13] that this ferromagnetic correlation decreases with the increase of the transfer term. This is due to the extended character of the hole wave function. The Cu-Cu spin correlation even changes its sign to antiferromagnetic when the singlet cloud is developed. In the singlet cloud region of the parameter space, the Cu-Cu spin correlation around the hole is small or antiferromagnetic. It suggests that the frustration mechanism becomes ineffective with the developement of the singlet cloud.

Two singlet clouds have an attractive interaction in the sea of stronger antiferromagnetic correlation. The attractive interaction exists even in the one-dimensional system. In fact, for the singlet cloud mechanism, the dimensionality is not crucially important.

It is likely that the spin symmetry of the pairing is singlet. The order parameter has the form of

$$\Delta = \sum_k f(k) < c_{k\sigma}^\dagger c_{-k\sigma'}^\dagger > . \tag{18}$$

If Δ were of the triplet symmetry, we needed the preexistence of the polarization of opposite spin to create a pair of singlet cloud, because the pair of singlet cloud should not have any spin polarization. This preexistence of spin polarization is unlikely to be the case. In the case of singlet pairing we have no such a difficulty.

The bound state of two holes is seen not only in the region $t \sim |J_K| \gg |J_S|$ but also in a wide range of parameter space. This suggests a continuous relationship to other regions and may provide a possibility for a unified understanding of the superconductivity induced by the magnetic mechanism. Weak coupling region is characterized by $|J_K| \ll |J_S|$ and t, where the spin wave analysis may provide a good description and the coexistence of antiferromagnetism and superconductivity would be characteristic. In the small transfer region, the existence of the bound state is rather obvious, because the doped hole forms a local defect with the nearest neighbor Cu spins. The attractive interaction is derived from the count of wrong antiferromagnetic bonds. Roughly speaking, this mechanism is valid for $t < |J_K|, |J_S|$. The binding energy may well be reproduced from the spin wave analysis at $t = 0$[21]. However, the relevance of this local defect mechanism to the superconductivity is questionable, because the condensation in real space (segregation) occurs for sufficiently small t in the coupled spin-fermion model.

The order parameter given by eq.(18) is expected to be valid in all the regions of the parameter space. An interesting relation is that the superconducting order parameter

$$\Delta = \sum_k f(k) < d_{k\sigma} d_{-k,\sigma'} > \tag{19}$$

employed in the RVB theory[15,22] is also expressed in the form (18) in the limit $J_K \to -\infty$ of $z_K = 1$ class. This is easily shown by the charge conjugation. The creation of two fermions in the coupled spin-fermion model is equivalent to the annihilation of two holes in the Cu-d-band of the $t - J$ model. However, in the region of singlet cloud mechanism, the form factor may not be the type of nearest neighbor pairing in constrast to the case frequently employed[22] in the $t - J$ model. It is conceivable that the pairing symmetry changes between the "*t − J region*" and the "*singlet cloud region*". This possibility should be examined in the future.

Acknowledgements

The author would like to thank Institute for Scientific Interchange in Torino, where this manuscript was prepared, for the hospitality extended to him.

References

1) L.N. Cooper: Phys. Rev. 104, 1189 (1956).
2) S. Sorella, E. Tosatti, S. Baroni, R. Car and M. Parrinello: Proceedings of the Adriatico Research Conference "*Towards the Theoretical Understanding of the High − T_c Superconductors*" ed. by S. Lundquist et al. (World Scientific, Singapore, 1988) to appear in Modern Physics B.
3) A. Fujimori, E. Takayama-Muromachi, Y. Uchida and B. Okai: Phys. Rev. B 35, 8814 (1987).
T. Takahashi, H. Matsuyama, H. Katayama-Yoshida, Y. Okabe, S. Hosoya, K. Seki, H. Fujimoto, M. Sato and H. Inokuchi: Nature 334, 691 (1988).
4) G. Shirane et al.: Phys. Rev. Lett. 59, 1613 (1987).
Y. Endoh et al.: Phys. Rev. B 37, 7443 (1988).
R.J. Birgenau et al.: preprint.
J.M. Tranquada et al.: Phys. Rev. Lett. 60, 156 (1988).
M. Sato et al.: preprint.
5) Y. Kitaoka et al.: preprint.
6) V.J. Emery: Phys. Rev. Lett. 58, 2794 (1987).
7) M. Imada: J. Phys. Soc. Jpn. 56, 3793 (1987).
8) M. Imada: J. Phys. Soc. Jpn. 57, 3128 (1988).
9) M. Ogata and H. Shiba: J. Phys. Soc. Jpn. 57, 3074 (1988).
10) F.C. Zhang and T.M. Rice: Phys. Rev. B 37, 3759 (1988).
11) M. Imada, N. Nagaosa and Y. Hatsugai: J. Phys. Soc. Jpn. 57, 2901 (1988).
12) M. Imada, Y. Hatsugai and N. Nagaosa: Proceedings of the Adriatico Research Conference "*Towards the Theoretical Understanding of High − T_c Superconductors*" ed. by S. Lundquist et al. (World Scientific, Singapore, 1988) to appear in Modern Physics B.
13) Y. Hatsugai, M. Imada and N. Nagaosa: submitted to J. Phys. Soc. Jpn.

14) E. Takayama-Muromachi et al.: J. Appl. Phys. 27 , L223 (1988).
H. Kamimura et al.: Solid State Comm. 67, 363 (1988).
A. Bianconi et al.: preprint.
15) P.W. Anderson: Science 235, 1196 (1987).
G. Baskaran, Z. Zou and P.W. Anderson: Solid State Commn. 63, 973 (1987).
16) to be published.
17) P.B. Wiegmann: Phys. Rev. Lett. 60, 821 (1988).
18) B.I. Shrainman and E.D. Siggia: Phys. Rev. Lett. 61, 467(1988).
19) V.J. Emery and G. Reiter: preprint.
20) A. Aharony et al.: Phys. Rev. Lett. 60, 1330 (1988).
21) N. Nagaosa, Y. Hatsugai and M. Imada: submitted to J. Phys. Soc. Jpn.
22) H. Yokoyama and H. Shiba: J. Phys. Soc. Jpn. 57, 2482 (1988).

Fermi Liquid and Non Fermi Liquid Phases of the Extended Hubbard Model

G. Kotliar

Massachusetts Institute of Technology, Cambridge, MA 02139, USA

We outline the auxiliary boson approach to the one band and two band Hubbard model. In this framework we find Fermi Liquid and Non Fermi Liquid behavior in different regions of the phase diagram. Differences and similarities with the heavy Fermion problem are emphasized.

The discovery of high temperature superconductivity[1] has led to an intensive experimental investigation of the physical properties of the rare earth based copper oxides. While there are many theoretical proposals as to the mechanism of the superconductivity, there is not yet consensus in how to model their normal and superconducting phase. In this talk I will summarize some aspects of the auxiliary boson[2] approach to this problem. The rational for this approach is the belief that the copper oxides are strongly correlated systems and that the proximity to metal insulator transition is an essential feature that the ultimate theory of the copper oxides should contain. The auxiliary boson technique provides a coherent framework for addressing the strong correlation problem and has been very useful in understanding many aspects of the heavy fermion systems.[3] Modelling the heavy fermions and the high temperature superconductors with the same Hamiltonian and the same technique gives us some clues of what are the essential differences, and similarities between these two systems.

The starting point of the investigation is Anderson's[4] observation that the magnetic state of the insulating parent of the high temperature superconductor is a spin liquid with a wave function:

$$|\Phi\rangle = P_G \prod_k (u_k + v_k\, d^\dagger_{k\uparrow} d^\dagger_{-k\downarrow})\, |0\rangle \qquad (1)$$

which resembles the projection of a BCS state with Bogolubov coefficients $u_k, v_k = 1/2(1 \pm e_k/E_k)$; $E_k = \sqrt{e_k^2 + \Delta_k^2}$, and gap parameter Δ_k. $d^\dagger_{k\sigma}$ are copper creation operators. At half filling this wave function describes a Fermi Liquid or a superconductor whose charge degrees of freedom have been totally eliminated. The excitation spin spectrum relative to that state is that of particle hole pairs with quasiparticle dispersion E_k. Upon doping a small amount of charge fluctuations of the order of the doping δ is allowed and the state smoothly evolves into a superconducting state.[4] Variational wave functions describing liquids with strongly suppressed charge fluctuations are well known in the context of the heavy fermion problem. In fact most of our current understanding of this problem is based on approximate ground states of the form[5]

Springer Series in Materials Science, Vol. 11
Mechanisms of High Temperature Superconductivity
Editors: H. Kamimura and A. Oshiyama

$$|\psi\rangle = P_G \Pi_{k\sigma} [u_k + v_k d^\dagger_{k\sigma} p_{k\sigma}] \prod_{\substack{k<k_F \\ \sigma}} p^\dagger_{k\sigma} |0\rangle \tag{2}$$

where $p^\dagger_{k\sigma}$, $d^\dagger_{k\sigma}$ are creation operators of conduction electrons and localized electrons respectively. The Gutzwiller projection acts to eliminate the doubly occupancy of localized electrons.

Very shortly after Anderson's proposal it was established that at $\delta = 0$ the ground state of the Heisenberg model[6] and of the insulating parents of the high temperature superconductors[7] has antiferromagnetic long range order. I take the view that in the regime of doping where the copper oxides do not exhibit magnetic long range order they can be thought of as superconductors or fermi liquids whose charge fluctuations are strongly suppressed. From this point of view the origin of high temperature superconductivity is the proximity to the metal insulator transition which renormalizes the kinetic energy to a value small or comparable with the pairing energies. In this regime of strong correlation high temperature superconductivity is possible. At the same time the presence of strong correlations imply the existence of other competing ground states with different forms of long range order, like antiferromagnetism and dimerization, and a small but finite doping, is necessary to eliminate these instabilities.

The study of Gutzwiller projected wave functions is closely related to the auxiliary boson description of strongly correlated systems. The mean field approach to the large U Hubbard model starts with the Hamiltonian:

$$H = -t \sum_{\langle ij\rangle,\sigma} (d^\dagger_{i,\sigma} b_i b^\dagger_j d_{j,\sigma} + d^\dagger_{j,\sigma} b_j b^\dagger_i d_{i,\sigma}) - \mu_0 \sum_{i,\sigma} d^\dagger_{i,\sigma} d_{i,\sigma}$$

$$+ J \sum_{\langle ij\rangle} (\sigma_i \cdot \sigma_j - (1 - b^\dagger_i b_i)(1 - b^\dagger_j b_j))$$

$$+ \sum_i \lambda_i (\sum_\sigma d^\dagger_{i,\sigma} d_{i,\sigma} + b^\dagger_i b_i - 1) \tag{3}$$

with b_i a slave boson which ensures the single occupancy constraint on each site and $\sigma_i{}^a = d^\dagger_{\sigma i} \sigma^a{}_{\sigma\sigma'} d_{i\sigma'}$ the copper spin operator. Mean field theory is done by performing a consistent Hartree Fock Bogolubov factorization of this Hamiltonian.[8] This factorization respects the global SU(2) symmetry of the half filled system. The order parameters are

$$b_i = \langle b_i \rangle, \quad \Delta_{ij} = \langle d^\dagger_{i\sigma} d^\dagger_{j-\sigma} \rangle, \quad \chi_{ij} = \langle d^\dagger_{i\sigma} d_{i\sigma} \rangle . \tag{4}$$

The first solutions of the mean field equations are due to Bascaran Zou and Anderson[9] who found, at half filling, an excitation spectrum of the form

$$E_k = |\cos k_x + \cos k_y| \tag{5}$$

which vanishes along lines in the two dimensional Brillouin zone called the pseudofermi surface. Later Kotliar[10] and Affleck and Marston[10] found a lower energy phase with an excitation spectrum of particle hole pairs with a quasiparticle dispersion relation

$$E_k = \sqrt{\cos^2 k_x + \cos^2 k_y} \tag{6}$$

vanishing on 4 Fermi points in the Brillouin zone. This s + id, or flux phase has an intrinsically generated flux of per plaquette. Finally there is a spin Peirls phase in which the ground state is a valence bond state and the excitations are localized and fully gapped. The SU(2) symmetry[10] results in an infinite number of degenerate ground states which can be classified by their flux. Kotliar and Liu[8] addressed the question of which spin liquid phases can be continued upon doping into superconductors. In the framework of mean field theory they found the uniform state evolves into a fermi liquid upon doping while the states with flux per plaquette like the s + id phase evolve upon doping into a strongly coupled d wave superconductor with a gap function proportional to $\Delta_k = \cos k_x - \cos k_y$. As the doping increases the ground state and the mean field equations from which it is derived evolve into a conventional weakly coupled d wave BCS superconductor.

In the slave Boson technique the physical copper electron creation operator is represented by the product

$$d^\dagger_{phys\ i} = d^\dagger_i b_i \tag{7}$$

therefore the physical pairing order parameter decomposes into a product

$$\langle d^\dagger_i d^\dagger_j \rangle \langle b_i b_j \rangle = \Delta_{phys\ ij} \ . \tag{8}$$

Hence, in this approach there are two characteristic temperatures, the critical temperature of bose condensation T_{cb}, below which $\langle b_i b_j \rangle \neq 0$ which can be thought of as a Kondo temperature signaling the onset of coherence, and T_{cRVB}, the pairing temperature below which $\langle d^\dagger_{i\sigma} d^\dagger_{j-\sigma} \rangle \neq 0$. In Fig.(1) we show a calculation of the critical temperatures using mean field theory and assuming a weak interlayer coupling to stabilize the Bose condensation. The approach to calculate $T_c(b)$ is very primitive and the consistent incorporation of finite temperature effects in the strong correlation problem remains one of the important unsolved problems in this field. In particular, there are reasons to believe that the mean field theory overestimates the slope of the curve $T_c(b)$ vs. δ. Very close to half filling the mean field solutions found are unstable against antiferromagnetism and dimerization but this is not indicated in the phase diagram.

b plays the role of quasiparticle residue and it is non zero in the normal metallic and superconducting phases. When the quasiparticle residue b is non vanishing, Δ measures the quasiparticle pairing and is proportional to the usual superconducting order parameter. As the doping increases the renormalized fermi energy increases relative to the pairing energies and the

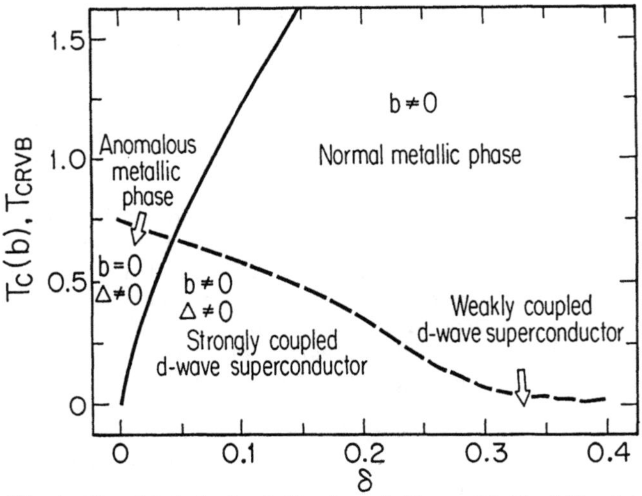

Fig 1. T_b, (dot-dashed line) and T_{rvb} (dotted line) in units of J vs. filling factor. From the mean field solutions of Hamiltonian[3] for t = 10J.

superconducting state evolves continuously into a conventional weakly coupled d wave superocndluctor. The phase with $b = 0$ $\Delta \neq 0$ is an anomalous non Fermi Liquid phase where there are no quasiparticles or metallic coherence since $b = 0$ $\Delta \neq 0$ indicating that pairs are present but for the system to be superconducting coherence between the pairs needs to be established. This occurs below $T_c(b)$.

The same techniques can be used to study the two band model that treats explicitly copper and oxygen holes with creation operators $d^\dagger_{\sigma i}$, $P^\dagger_{\sigma i}$ respectively. Here we generalize the spin index to run from 1 to N, N = 2 being the spin 1/2 realistic case to allow for a systematic 1/N expansion. The Hamiltonian is given by[13,14]

$$H = \varepsilon^0_d \sum_i d^\dagger_{i\sigma} d_{i\sigma} + \frac{J}{N} \sum_{(ij)\sigma\sigma'} d^\dagger_{i\sigma} d_{i\sigma'} d^\dagger_{j\sigma'} d_{j\sigma}$$

$$+ [\lambda_i [b^\dagger_i b_i + \sum_\sigma d^\dagger_{i\sigma} d_{i\sigma} - 1] + \varepsilon_p \sum_{i\sigma} p^\dagger_{i\sigma} p_{i\sigma}$$

$$- \frac{2t}{\sqrt{N}} \sum_\sigma \gamma_k (d^\dagger_{k+q\sigma} b^\dagger_q p_{k\sigma} + p^\dagger_{k\sigma} d_{k+q\sigma} b^\dagger_q) \tag{9}$$

ε^0_d and ep are the bare d and p levels $t\gamma_k$ is a hybridization matrix element between the copper and the oxygen while J is a direct copper-copper super-exchange which is generated by integrating our high energy degrees of freedom to generate and effective low energy Hamiltonian. b is again a slave boson introduced to take care of the single occupancy constraint on the copper site. In the large N approach we introduce order parameters $b = \langle b_i \rangle$ and $\chi_{ij} = \langle d^{\dagger}_{i\sigma} d_{j\sigma} \rangle$. The boson b plays again the role of coherence or "fermi liquid order parameter". It measures the quasiparticle residue and its vanishing signals the disappearance of the Fermi liquid phase. The phase with b=0 is a non fermi liquid phase in which, in the presence of an intrinsic oxygen bandwidth, the oxygen holes form a fermi liquid decoupled to first order from the copper spins.

The phase diagram of this model in the large N limit is very rich and can be summarized in Fig.(2). We will assume that $J \ll t$. Along the $\delta = 0$ axis there is a Brinkman Rice transition when the charge transfer energy $U \equiv \varepsilon_p - \varepsilon_d^0$ becomes comparable to the hopping t. The order parameter b jumps discontinuously to zero and at the same time the order parameter χ_{ij} jumps from an uniform state for small U to a dimerized state for large U.

Below the Brinkman Rice point the chemical potential μ is continuous while the effective mass m^* is finite at $\delta = 0$. Above the Brinkman Rice point the chemical potential jumps discontinuosly at $\delta = 0$.

There is a first order line separating a Fermi liquid regime where χ_{ij} is uniform and $b \neq 0$ to a non fermi liquid phase where the χ_{ij} are dimerized.

The transition occurs when the renormalized fermi energy is of the order of the exchange energy so the location of the first order line separating the Fermi Liquid phase ($\chi_{ij} = \chi$ uniform independent of ij) from the dimerized phase is given approximately by $J = \delta\, t^2/(\varepsilon_p - \varepsilon_d^0)$.

The dimerized phase behaves very differently upon doping depending on the ratio of the spin exchange J to the Kondo exchange $t^2/(\varepsilon_p - \varepsilon_d^0)$. When $J > t^2/(\varepsilon_p - \varepsilon_d^0)$, $b = 0$ even at $\delta \neq 0$ and the holes go on the oxygen sites, while the magnetic background remains inert. When $J < t^2/(\varepsilon_p - \varepsilon_d^0)$ but δ is small the

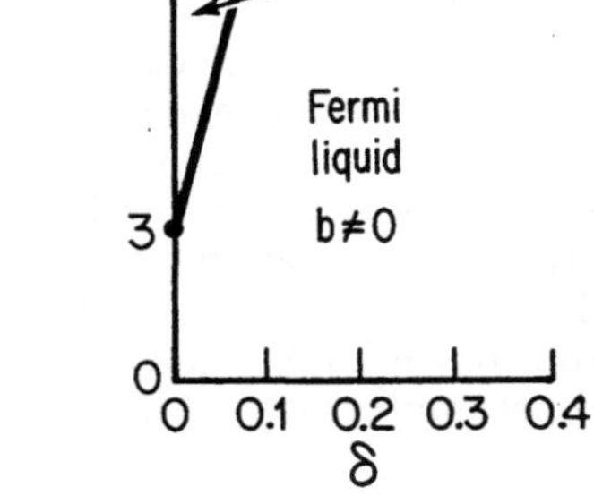

Fig. 2 Schematic phase diagram of Hamiltonian (9) in the $N = \infty$ limit, and for small but finite J. $U \equiv \varepsilon_p - \varepsilon_d^0$

dimerized copper band mix strongly with the oxygen and the carriers are mixtures of the upper dimer band and the oxygen band. These two regimes (dimer I and dimer II) in the phase diagram of Fig.(2) are separated by a line of first order transitions.

Approaching the uniform to dimer transition from the metallic side one has a strongly correlated fermi liquid with antiferromagnetic spin correlations. Close to this transition the linear coefficient of specific heat γ and the susceptibility χ behave as

$$\gamma \sim \chi = \frac{1}{J + \dfrac{\delta t^2}{\varepsilon_p - \varepsilon_d{}^0}} . \tag{10}$$

In the large N approach superconductivity arises from fluctuations around mean field theory. The pairing originates from a screened superexchange interaction between the quasiparticles. The maximum superconducting temperatures indeed occur on the Fermi Liquid side but close to the metal insulator when $\delta\, t^2/(\varepsilon_p - \varepsilon^0{}_d) \sim J$. The superconductivity is in the d wave $\cos k_x - \cos k_y$ channel in agreement with the previous mean field one band results.[8]

The main difference between the copper oxide problem and the heavy fermion problem lies in the different parameter range of the Kondo lattice Hamiltonians. In the heavy fermion problem the oxygen bandwidth is the largest scale in the problem, and it is well in the Kondo regime. The copper oxide systems the oxygen bandwidth is very small compared with the Kondo exchange and the fact that the direct copper copper superexchange is large indicates that one is closer to the mixed valence regime. The treatment of Refs.(13-14) correspond to the strong coupling limit of the Kondo lattice problem in which the resonant impurity level band is pulled out of the continuum of the conduction band as the conduction electron bandwidth is reduced. In the heavy fermion system $T_c(b)$, the Kondo temperature or the fermi liquid coherence temperature, is much larger than the pairing energies which drive it superconducting at low temperatures. This inequality is probably reversed in the high temperature superconductors. The superconductivity would then occur at $T_c(b)$ without ever passing thru a fermi liquid regime. This conjecture is appealing in the light of recent NMR experiments which show that above T_c the temperature dependence of the nuclear spin relaxation rate on the copper and oxygen sites is completely different suggesting that above T_c one is in a phase where $b = 0$ and copper and oxygen are to leading order decoupled while below T_c the spin relaxation rate drops on both sites indicating that the superconductivity affects both oxygen holes and copper spins. Therefore below T_c coherence between copper spins and oxygen holes is established. The existence of very large mass renormalizations in the fermi liquid phase of the heavy fermions and its absence in the copper oxides can be accounted by the fact that the copper resonant level is pulled from the continuum and the fact that close to half filling the renormalized fermi energy is of the order of the exchange energy. As shown by Eq.(10), in this limit the fermi liquid parameters do not depend very strongly on filling factor.

In Refs.(12-13) a zero oxygen bandwidth was assumed. When the oxygen bandwidth becomes comparable with the hybridization energy $t^2/(\varepsilon_p - \varepsilon_d^0)$ a crossover between the behavior found in Refs.(12-13) and heavy fermion behavior is expected to occur as the doping is increased away from half filling.

References

1. J.G. Bednorz and K.A. Muller, Z. Phys. B64, 188 (1986).
2. P. Coleman, Phys. Rev. B29, 3035 (1984); S. Barnes, J. Phys. F6, 1375 (1976).
3. A. Millis and P.A. Lee, Phys. Rev. B35, 3394 (1987); A. Auerbach and K. Levin, Phys. Rev. Lett. 57, 877 (1986).
4. P.W. Anderson, Science 235, 1196 (1987).
5. C. Varma and Y. Yafet, Phys. Rev. B13, 2950 (1976); T.M. Rice and K. Ueda, Phys. Rev. Lett. 55, 995 (1985).
6. J. Reger and P. Young, Phys. Rev. B18, 5978 (1988).
7. D. Vaknin, S.K. Sinha, D. Moncton, D. Johnston, J. Newsam, C. Safinya, and J. King, Jr., Phys. Rev. Lett. 58, 2802 (1987).
8. G. Kotliar and J. Liu, Phys. Rev. B38, 5142 (1988).
9. G. Baskaran, Z. Zou, and P.W. Anderson, Sol. Stat. Commun. 63, 973 (1987).
10. G. Kotliar, Phys. Rev. B37, 3664 (1988).
11. I. Affleck and Marston, Phys. Rev. B37, 3774 (1988).
12. G. Baskaran and P.W. Anderson, Phys. Rev. B37, 580 (1988); I. Affleck, Z. Zou, T. Hsu, and P.W. Anderson, Phys. Rev. B38, 745 (1988).
13. G. Kotliar, P. Lee, and N. Read, Physics BC, 153-542, 538 (1988).
14. G. Kotliar, Proceedings of the Adriatico Research Conference "Towards the Theoretical Understanding of High T_c Superconductors", to appear in Int. J. Mod. Phys.; G. Kotliar and C. Castellani, to be published in Phys. Rev. B, Rapid Communications.

Motion of Holes in Magnetic Insulators

S. Maekawa, J. Inoue, and M. Miyazaki

Department of Applied Physics, Nagoya University, Nagoya 464-01, Japan

Starting with the Hubbard model with strong electron correlation, we examine motion of holes in the square lattice in the moment expansion method. Special emphasis is put on effects of the exchange interaction between spins on the motion of holes.

1. Introduction

The purpose of this paper is to examine motion of holes in two-dimensional magnetic insulators. The motivation of this work has come from the following facts: The parent compounds of the high T_c superconducting oxides such as $La_{2-x}Sr_xCuO_4$ and $YBa_2Cu_3O_{7-x}$ are antiferromagnetic insulators. When hole carriers are introduced in the insulators, the Neel temperatures are depressed quickly and the high T_c superconducting states appear. Thus, it seems natural to consider that the superconductivity is based on the insulators.

Antiferromagnetic insulators are described by the half-filled Hubbard model with strong electron correlation. In the square lattice, which is a unit element of the high T_c oxides, the magnetic ground state is the Neel state with large quantum spin fluctuation or the resonating valence bond (RVB) state [1] because of the low dimensionality and the low spin value ($S=1/2$).

When a hole is introduced in the insulators, the hole may carry the electric current. However, a question is how the hole propagates in the lattice. We consider that answers to the question will provide a key information to the superconductivity. In the following sections, we review our recent work of the motion of holes in the magnetic insulators studied by applying the moment expansion method to the hole Green's function [2,3].

2. Hole Green's function

The Hubbard Hamiltonian with strong electron correlation may be rewritten near half-filling as

$$H = -t \sum_{<ij>} \sum_{\sigma} (1-n_{i-\sigma}) c^{\dagger}_{i\sigma} c_{j\sigma} (1-n_{j-\sigma}) + J \sum_{<ij>} (\vec{S}_i \cdot \vec{S}_j - \frac{1}{4} n_i n_j), \qquad (1)$$

Springer Series in Materials Science, Vol. 11
Mechanisms of High Temperature Superconductivity
Editors: H. Kamimura and A. Oshiyama
Springer-Verlag Berlin Heidelberg © 1989

where $c_{i\sigma}$ is the annihilation operator of an electron with spin σ at site i, $n_{i\sigma}=c^{+}_{i\sigma}c_{i\sigma}$, $n_i=\Sigma_\sigma n_{i\sigma}$, and $\vec{S}_i$ is the spin operator with S=1/2 at site i. The first term on the right hand side in eq. (1) shows that electrons propagate to the nearest neighbor sites only when they are empty, whereas the second term shows the antiferromagnetic exchange interaction between the nearest neighbor spins (J>0). We note that t is much larger than J.

Let us first examine a hole introduced in the classical Neel state. When the hole propagates to gain the kinetic energy, it leaves behind the string of overturned spins as shown in Fig.1. Therefore, the further the hole goes, the more the exchange energy increases. As a result, the hole will be localized as a self-trapped magnetic polaron. Its spatial size (ξ) and the numbers of the spins in a polaron (N) are estimated to be $\xi \simeq \pi a\sqrt{t/J}$ and $N \simeq (\xi/a)^2$, respectively, where a is a lattice constant [4]. We note that the classical spins act on the holes as potentials. However, in the quantum spin systems, the polaron may not be localized, since the potentials fluctuate quantum-mechanically.

To take into account such quantum effects on the motion of holes, we introduce the hole Green's function given by

$$G_{ii}(\omega) = \sum_\sigma \langle S|c^{\dagger}_{i\sigma}(\omega - H)^{-1}c_{i\sigma}|S\rangle , \tag{2}$$

where $|S\rangle$ denotes a magnetic state at half-filling. It is convenient to introduce the continued fraction representation of the Green's function [5],

$$G_{ii}(\omega) = \cfrac{1}{\omega - a_0 - \cfrac{b_1{}^2}{\omega - a_1 - \cfrac{b_2{}^2}{\omega - a_2 \ddots}}} , \tag{3}$$

In the representation, the coefficients, a_n and b_n^2, have the following meanings: We introduce a hole at site i in the magnetic state $|S\rangle$ and write the state as ψ_0 . Then, operating Hamiltonian H on ψ_0 and create a new state as $\psi_1=H\psi_0-a_0\psi_0$. Here, a constant a_0 is subtracted to make ψ_1 orthogonal to ψ_0. Operating H on ψ_1, we create another new state as $\psi_2=H\psi_1-a_1\psi_1-b_1^2\psi_0$. In the same way, we can create a new state successively as $\psi_n=H\psi_{n-1}-a_{n-1}\psi_{n-1}-b^2_{n-2}\psi_{n-2}$. The coefficients, a_n and b_n^2, are expressed by using these states as

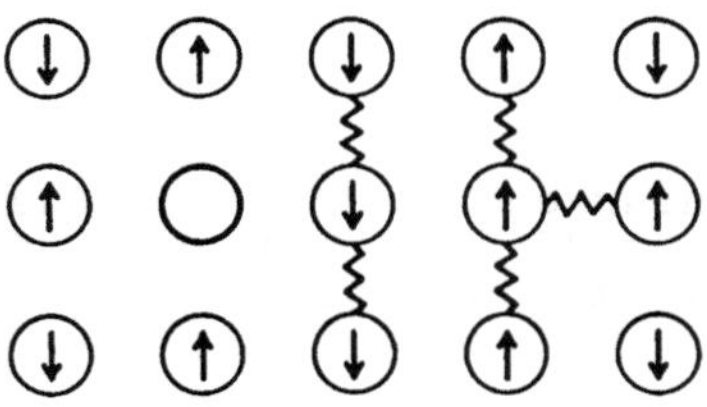

Fig.1. When the hole propagates in the classical Neel state, the magnetic energy increases.

$$a_n = \langle\psi_n|H|\psi_n\rangle/\langle\psi_n|\psi_n\rangle \, , \qquad (4)$$
$$b_n{}^2 = \langle\psi_n|\psi_n\rangle/\langle\psi_{n-1}|\psi_{n-1}\rangle \, .$$

Thus, a_n and b_n^2 can be understood as the energy of the n-th state and the expansion of the n-th state from the (n-1)-th state. We note that the coefficients are also expressed by using the moments of the density of states of holes. Therefore, the method to utilize eqs. (3) and (4) is called the moment expansion method.

We have calculated a_n and b_n^2 with $n \leq 2$ in the two-dimensional square lattices with the classical Neel (Neel), quantum antiferromagnetic (QAF), and RVB states by taking into account the first order effects of the exchange interaction (J) on each a_n and b_n^2. RVB states are given by linear combinations of the spin singlet pairs (SP). Here, we consider two RVB states given by

$$\Phi_{RVB1} = (1/3)^{1/2}(\Phi_a - \Phi_b) \, ,$$
$$\Phi_{RVB2} = (2/11)^{1/2}(\Phi_c - \Phi_d - \Phi_e) \, , \qquad (5)$$

where Φ_a and so on denote the SP states defined in Fig.2 and expressed by the fermion operators. In the Figure, each SP is shown by a circle. The coefficient a_0 denotes the formation energy of a hole in the insulators. Since we are interested in the propagation of a hole, a_0 is taken to be an origin of the energy in each magnetic state.

In Table I, the values of a_n and b_n^2 in the magnetic states are listed. The numerical results of the lowest pole, ω_0, of the Green's function calculated as a function of the exchange

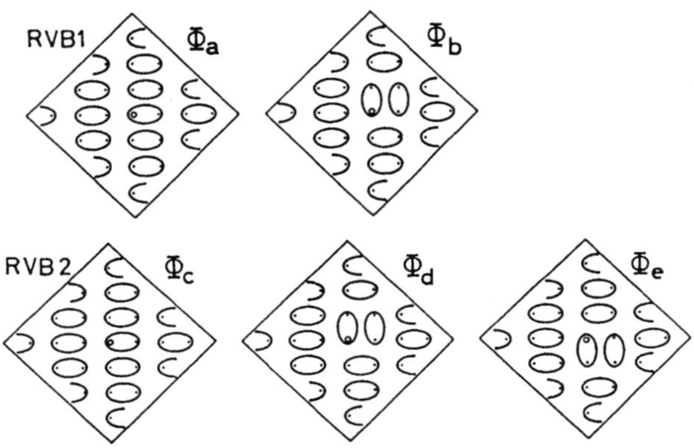

Fig.2. Two RVB states are given by linear combinations of the nearest-neighbor spin singlet pairs. A small circle in one of the pair states denotes the origin of the hole Green's function.

Table.1. the values of a_n and b_n^2.

	a_0/J	a_1/J	a_2/J	b_1^2/t	b_2^2/t
Neel	4.00	7.00	9.00	4.00	3.00
QAF	4.63	7.74	9.18	4.00	2.58
RVB1	4.00	5.50	6.18	4.00	2.75
RVB2	4.32	6.06	6.55	4.00	2.75

parameter J are plotted in Fig.3. The values of ω_0 may be a measure of the stability of a mobile hole in a magnetic insulator. As seen in Fig.3, the hole is stabilized in the RVB states more than in the Neel or QAF states, when the exchange interaction is finite. This is because the magnetic coherence is weaker in the RVB states with nearest neighbor singlet pairs [6-8]. Since holes disturb the magnetic coherence when they propagate, the RVB states are favorable for holes. Several authors [9-11] have studied the stability of a hole in the magnetic states in the limit of $J \to 0$ and concluded that the hole was the most stable in the Neel state if the ferromagnetic state is neglected [12]. Our conclusion with $J \neq 0$ is in contrast with the case with $J=0$.

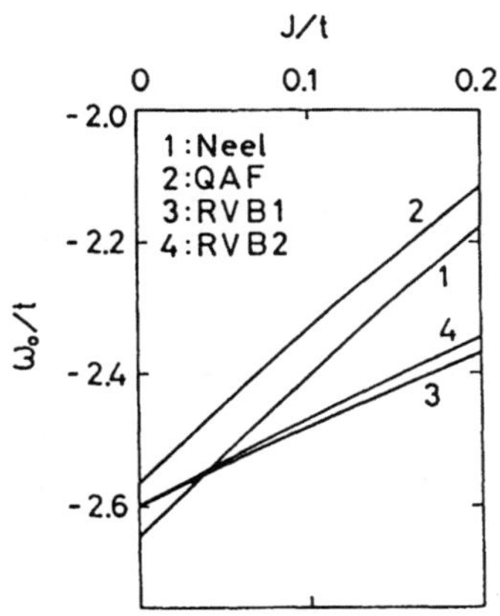

Fig.3. The lowest pole ω_0 as a function of J.

3. Spectral Function of Holes

In the preceding section, the Green's function of holes was studied in square lattices. However, the calculation was limited in the finite terms of the continued fraction representation and of the exchange interaction. In order to examine effects of the higher order terms, we introduce the Bethe lattice, which corresponds to the retraceable path approximation neglecting paths with closed loops. One of the advantages of this lattice is that all coefficients, a_n and b_n^2, can be given exactly in the classical Neel states as

$$a_n - a_0 = (2n+1)J ,$$
$$b_n^2 = \begin{cases} zt^2 , & \text{for } n=1 \\ (z-1)t^2 , & \text{for } n>1 , \end{cases} \tag{6}$$

with z=4. Here $a_n - a_0$ is the energy stored in the path when a hole propagates. Inserting eq.(6) into eq.(3), we can solve the density of states of holes numerically. Thus,the Bethe lattice provides a reference system to study the motion of holes in the quantum spin systems.

In Fig.4, the numerical results of the density of states of holes are plotted as a function of energy with J=0.0 and 0.1 in the classical Neel state. The results with J=0 have been studied by BRINKMAN and RICE [4] and RICE and ZHANG [13]. In this case, a hole propagates diffusively in the lattice and the motion is incoherent to spins. Once J is finite, it makes a self-trapped magnetic polaron state and is localized in space. Thus, the density of states is given by the delta functions in all energy regions.

We next consider effects of quantum spin fluctuation on the hole. In the quantum antiferromagnetic state, spins fluctuate quantum-mechanically around the Neel state. Since two spins can change the directions simultaneously by the term $(S_i^+ S_j^- + h.c.)$ in the Hamiltonian, the string of overturned spins created in the path of a hole may partly relax [14-17]. In our continued fraction representation, we start with the classical Neel state and operate the Hamiltonian on the states successively. Then, we find that when n becomes of the order

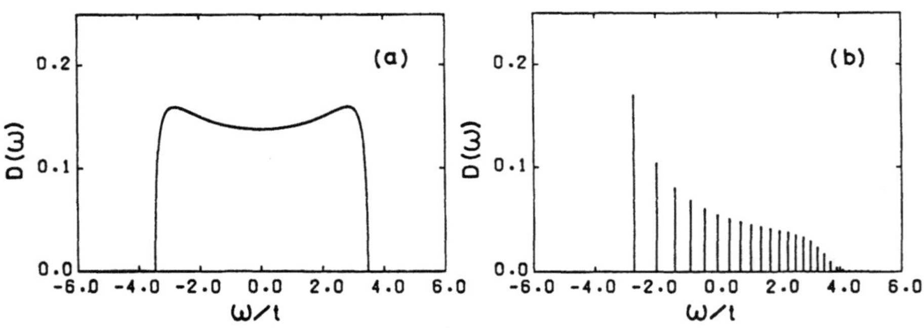

Fig.4. Density states of holes in the Bethe lattice with the Neel state. (a) J/t=0.0 (b) J/t=0.1.

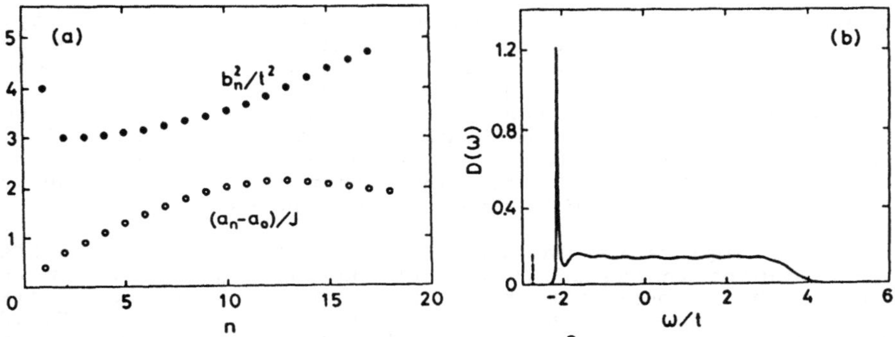

Fig.5. (a) The coefficients a_n and b_n^2 in a model system are plotted with J/t=0.1. (b) Density states of holes with J/t=0.1 in this model system given.

of t/J, the term in the Hamiltonian works strongly and the string is divided into two pieces. We have neglected the piece which is not connected to the hole, assuming the fluctuation causes its relaxation. In this case, a_n and b_n^2 with J/t=0.1 are given in Fig.5(a).

In Fig.5(b), the numerical result of the density of states for this case is plotted. As seen in the Figure, there exists a sharp peak of the density of states at the low energy region. The width of this peak is found to be of the order of J. We consider that the peak corresponds to the mobile magnetic polaron states which are released by quantum spin fluctuation from localization. Several authors have discussed that in the quantum systems a hole propagates with the effective mass renormalized by spin fluctuation [14-17]. The low energy peak in Fig.5(b) is consistent with these arguments. The broad band at high energy region is similar to that with J=0 in Fig.4(a), and the band represents the diffusive motion of holes incoherent to spins.

In Fig.5(b), there also exists a peak isolated from the bands shown by a broken line. We consider that this peak has probably occurred because of the incomplete treatment of the spin fluctuation started with the classical Neel state. The more detailed study is in progress.

4. Discussions

We have found that the antiferromagnetic exchange interaction tends to make a self-trapped magnetic polaron, and that the quantum spin fluctuation releases holes from the trapped states and creates mobile magnetic polarons. When the exchange interaction is not negligibly small, RVB states seem to stabilize holes more than the Neel and QAF states. Some implications to the experimental data may result from these theoretical conclusions.

THOMAS et al. [18] have observed the optical conductivity, $\sigma(\omega)$, of $YBa_2Cu_3O_{7-x}$ and found a characteristic structure in $\sigma(\omega)$ at around 0.2 eV which is of the order of the exchange energy, J. According to RICE and ZHANG [13], the optical conductivity is given by the density of states of holes divided by the frequency, where the density of states is measured from the band edge. When we see Fig.5(b) it is easy to identify the structure in $\sigma(\omega)$ as the crossover from the mobile magnetic polaron states to the diffusive states of holes.

Another implication may be the stability of the Neel state by the doping in $La_{2-x}Sr_xCuO_4$ and $YBa_2Cu_3O_{7-x}$. It is known that in these compounds 1% of holes per magnetic ion seems sufficient to suppress the magnetic ordering. As seen in Sec.2 each hole makes magnetic polaron and disturb $N \simeq \pi^2 t/J$ spins. In the oxides, N can be of the order of a few tens. Therefore, almost all spins are disturbed by 1% of holes and the magnetic ordering is suppressed.

In this paper, we have studied the motion of holes in the square lattice starting with the half-filled Hubbard model. In the oxide superconductors, the more detailed model may be used by taking into account O and Cu ions precisely. Although it is interesting to extend the study to the model, we consider that the essential results obtained in this paper do not change [19-22].

5. Acknowledgments

This work has been supported by Grant-in-Aid for Scientific Research on Priority Areas "Mechanism of Superconductivity".

References

1. P.W. Anderson, Science 235, 1196 (1987).
2. J. Inoue, M. Miyazaki and S. Maekawa, to be published in Physica C.
3. J. Inoue, M. Miyazaki and S. Maekawa, in preparation.
4. W.F. Brinkman and T.M. Rice, Phys. Rev. B2, 1324 (1970).
5. R. Haydock, V. Heine and M.J. Kelly, J.Phys. C: Sol. St. Phys. 8 2591 (1975).
6. B. Sutherland, Phys. Rev. B37, 3786 (1988).
7. M. Kohmoto, Phys. Rev. B37, 3812 (1988).
8. K. Takano and K. Sano, to be published.
9. R. Joynt, Phys. Rev. B37, 7979 (1988).
10. P. Lederer and Y. Takahashi, preprint.
11. Y. Takahashi, Z. Phys. B67, 503 (1987).
12. Y. Nagaoka, Phys. Rev. 147, 392 (1966).
13. T.M. Rice and F.C. Zhang, preprint.
14. S.A. Trugman, Phys. Rev. B37, 1598 (1988).
15. B.I. Shraiman and E.D. Siggia, Phys. Rev. Lett. 60. 740 (1988).
16. S. Schmitt-Rink, C.M. Varma and A.E. Ruckenstein, Phys. Rev. Lett. 60, 2793 (1988).
17. C.L. Kane, P.A. Lee and N. Read, to be published.

18. G.A. Thomas, J. Orenstein, D.H. Rapkine, M. Capizzi, A.J. Millis, R.N. Bhatt, L.F. Schneemeyer and J.V. Waszczack, Phys. Rev. Lett. 61, 1313 (1988).
19. F.C. Zhang and T.M. Rice, Phys. Rev. B37, 3759 (1988).
20. S. Maekawa, T. Matsuura, Y. Isawa, and H. Ebisawa, Physica C 152, 133 (1988).
21. N. Andrei and P. Coleman, to be published.
22. F. Mila, to be published.

Fractional Quantization in High-Temperature Superconductivity

R.B. Laughlin

Department of Physics, Stanford University, Stanford, CA 94305, USA, and
Lawrence Livermore National Laboratory, PO Box 808, Livermore, CA 94550, USA

I would like to thank Prof. Maekawa for providing such an apt introduction to my talk. In my remarks today, I shall make the case that the mechanism of high-temperature superconductivity is pairing by a "gauge" force, first identified by Halperin [1] to occur in the fractional quantum hall problem, and referred to in 2-dimensional systems as fractional statistics.

Shortly after high-temperature superconductivity was reported by Bednorz and Müller [2], it became clear to many of us that such high transition temperatures probably could not be understood within the context of the traditional fermi liquid theory of superconductivity, and that the materials in question were Mott insulators. Partly as a consequence of the complexity Solid State Physics, these "clear" features of the problem are controversial, although they are becoming less so with time. For the purposes of this lecture, I shall assume they are true. For those unwilling to agree with this assumption, I strongly recommend reviewing the Cohen and Anderson [3] paper on limits to transition temperatures, recent work by Weber [4] on this subject, the work Brinkman and Rice [5] on the Mott problem and references therein, and the article I recently wrote on this subject for Science magazine [6].

Mott insulators, systems which insulate solely as a result of coulomb interactions, have never been well understood. Having been interested in Mott insulators for much of his professional career, Anderson understood this at once and argued [7], persuasively, that the occurrence of high-temperature superconductivity in this class of materials could not be an accident. Something fundamental, a generic feature of the Mott insulating state we have missed all these years, must cause the effect.

Most of us are familiar with the paradoxes associated with motion of charge carriers in a Mott insulator. This problem is usually considered within the context of the Hubbard model

$$\mathcal{H} = T \sum_{\langle j,k \rangle} \sum_{\sigma} c_{j\sigma}^{\dagger} c_{k\sigma} + U \sum_{j} c_{j\uparrow}^{\dagger} c_{j\downarrow}^{\dagger} c_{j\downarrow} c_{j\uparrow} \quad . \tag{1}$$

although there is nothing special about this model. It is just the simplest system one can think of that exhibits the Mott phenomenon. In the limit that U is large and positive, the ground state at half filling has one electron per site and, as illustrated in Fig. 1, is antiferromagnetically ordered. The system insulates because there is an energy gap of order U, associated with double occupancy of one of the sites. If we ask how a carrier forced into the system moves, we are forced to conclude that it should leave a wake of unpaired spins behind. The motion of charges, in other words, cannot be understood without first solving the problem of the dynamics of the spins because the spins get in

Springer Series in Materials Science, Vol. 11
Mechanisms of High Temperature Superconductivity
Editors: H. Kamimura and A. Oshiyama
Springer-Verlag Berlin Heidelberg © 1989

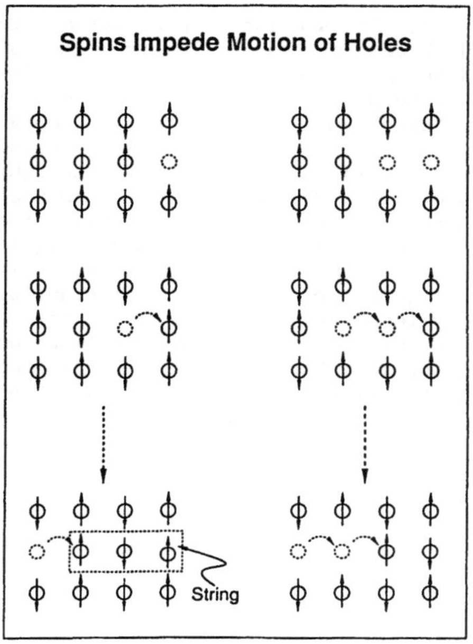

Fig. 1: Illustration of the motion of charge carriers in a Mott insulator. Isolated holes cannot move without leaving a wake, but pairs of holes can.

the way. It was recently pointed out by Hirsch [8] that a *pair* of charges can move freely through the Neel ordered state without a wake, and that this effect might be a good candidate for the force that pairs charges in the superconducting state. This idea, taken literally, cannot be right because the antiferromagnetic order on which it relies to be defined does not exist in high-temperature superconductors. However, it is appropriate to keep it in mind because it is a simple way to understand pairing by fractional statistics.

One of Anderson's early insights in this problem was to realize that the vacuum in which the carriers in a doped Mott insulator move was probably not the Neel state at all, but rather a state in which the spins disorder quantum mechanically. This is illustrated in Fig. 2. A system of spin-1/2 electrons on a lattice may be thought of as a lattice gas of bosons. One thinks of of a site with a "down" electron as being empty, a site with an "up" electron as being occupied, and thus of the various spin configurations of the system as the various ways bosons can be deployed on a lattice. The ground state is simply a rule assigning a complex number to each of these configurations. A possible ground state of this system is the antiferromagentically ordered one, which corresponds to a crystal of bosons. However, it is conceivable that under certain circumstances the crystal should melt quantum mechanically, as occurs, for example, in liquid helium. The spin ground state to which this corresponds is called the "spin liquid" or Anderson "resonating valence-bond" state. The latter term is somewhat unfortunate, as the state has nothing to do with the resonating valence bond in benzene.

My interest in this subject actually begins with the spin liquid state. Once it became established that particles carrying fractional charge were legitimate excitations of the fractional quantum hall vacuum, it seemed

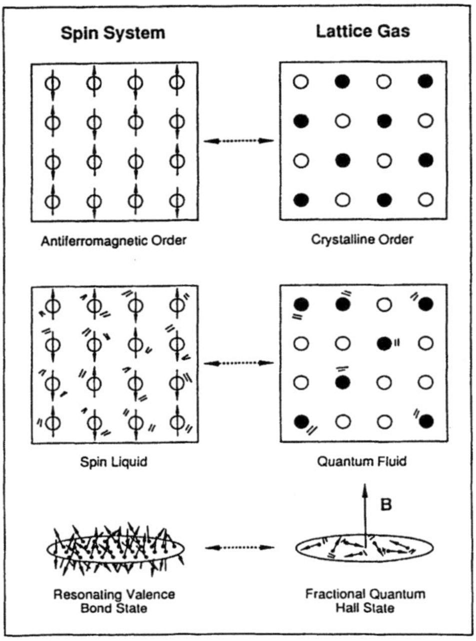

Fig. 2: A Heisenberg magnet may be though of as a lattice gas of bosons, with an "up" site interpreted as occupied and a "down" site interpreted as empty. The antiferromagnetically ordered state of the spins may be thought of as a the crystalline state of these bosons. Given that the bosons also have a quantum liquid state, the spin analog is the "spin liquid" or Anderson "resonating valence bond" state. The theory discussed in this paper is based on the idea that this state exists and is physically similar to the fractional quantum hall state.

likely to me that behavior of this kind occurred elsewhere in nature and was worth finding. I was directed by D. H. Lee and J. D. Joannopoulos to Anderson's [9] early work on the spin liquid state and became convinced by it that his state constituted another example of the effect. The reason, illustrated in Fig. 3, is quite simple. Suppose there exists a rotationally invariant spin Hamiltonian for which the ground state is an antiferromagnetic liquid. Let us get rid of boundary effects by placing the spins on a torus. If the number of spins is *even*, the total spin of the ground state must be zero, since a total spin other than zero means that the state is ferromagnetically polarized. Let us now add a row of spins so as to make the total number of sites *odd*. The total spin cannot now be less than 1/2. However, since the ground state has no long-range order, even and odd must be equivalent in the thermodynamic limit. It must therefore be the case that the state has neutral spin-1/2 excitations, commonly termed "spinons". This reasoning, which for all its simplicity is actually correct, leads us to the following problem. In the lattice gas language, an "up" spinon is a region of the sample containing, on average, *half* a boson excess. If we think of bosons as carrying electric charge e, then a spinon is a particle with a net "charge" of $e/2$. In other words, we conclude that whenever the spin vacuum lacks long-range order, its elementary excitations necessarily carry a fractional quantum number. The fractional quantum hall effect is the only precedent for behavior of this kind in nature. I find it inconceivable that the two states could be unrelated.

The mathematical case that the resonating valence bond state and the fractional quantum hall state are one in the same is becoming stronger with time. Kalmeyer and I [10] established that the quantum hall state, taken

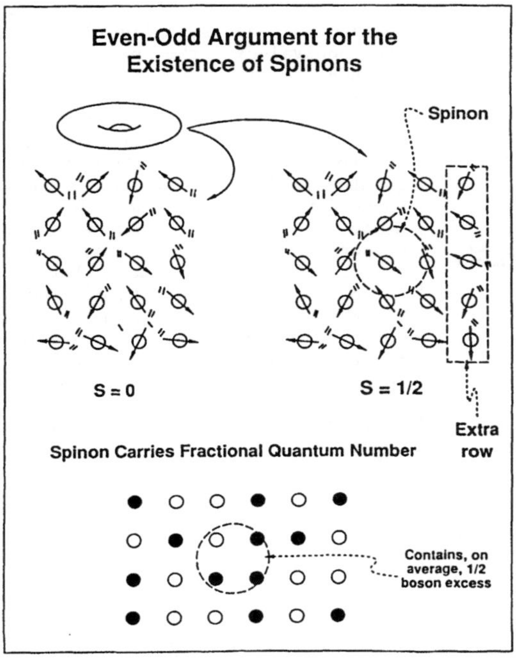

Fig. 3: If the spin liquid state exists, it must possess elementary excitations that have spin 1/2 and electric charge 0. This is because a system with an even number of spins must have a spin singlet ground state while a system with an odd number of spins can only have a spin 1/2 ground state. However, because the state is a liquid, even and odd must be equivalent in the thermodynamic limit. This excitation, the "spinon" carries fractional "charge" in the sense of being a region of the sample containing, on average, 1/2 boson excess.

as a variational ansatz for the spin problem, gave a competetive variational energy for the frustrated Heisenberg antiferromagnet and was an exact spin singlet. Zou, Doucot, and Shastry [11] recently showed that this state is nearly identical to a Gutzwiller projection of the Affleck-Marston [12] flux phase, one of the candidates for the spin liquid state. I [13] have recently established a proof of principle by showing that the quantum hall wavefunction is the exact ground state of the Hamiltonian

$$\mathcal{H} = \sum_{\alpha} D_{\alpha}^{\dagger} D_{\alpha} \quad , \tag{2}$$

where α denotes a lattice site and where

$$D_{\alpha} = \sum_{\beta \neq \alpha} \frac{\sigma_{\alpha} \cdot \sigma_{\beta}}{z_{\alpha} - z_{\beta}} \quad , \tag{3}$$

with z denoting the location of a lattice site expressed as a complex number and σ denoting a Pauli matrix. Work on this subject is still in progress.

The analogy between the fractional quantum hall and spin liquid states enables us to make several strong statements about the latter which are insensitive to details. The first of these, illustrated in Fig. 4, is that the spectrum of spin waves, the excitation probed by magnetic neutron scattering, *must* have an energy gap, as this gap is the measure of how liquid the spin are. The analogous excitation in the fractional quantum

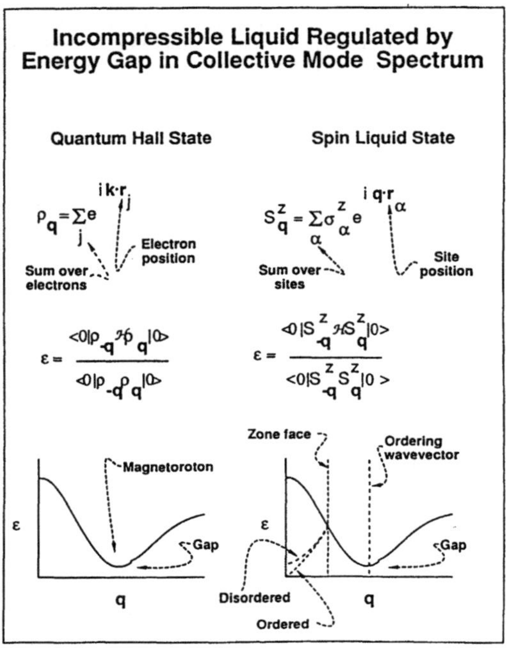

Fig. 4: Given that the fractional quantum hall and spin liquid states are equivalent, the presence or absence of antiferromagnetic order in the latter must be regulated by an energy gap in the spin wave spectrum. This is because the gap in the analogous excitation in the quantum hall state, a longitudinal sound wave, measures how liquid the state is. The deep minimum in the dispersion curve of this mode occurs at the reciprocal lattice vector of the ordered state. Thus, if the curve is redrawn in the reduced zone scheme of the ordered state, this minimum may be identified with the acoustic phonon or magnon, this linear dispersion of which at small momentum reflects order.

hall state is a compressional sound wave. It was established by Girvin, Platzman, MacDonald [14] that the dispersion curve of this sound wave has a deep minimum, referred to as the "magnetoroton", at a wavevector corresponding to the interparticle spacing. As the Hamiltonian is varied so as decrease this energy gap, the quantum hall state become increasingly difficult to distinguish from a Wigner crystal. If this curve is redrawn in the reduced zone scheme of the Wigner crystal, it becomes apparent that the presence or absence of long-range positional order is regulated by the presence or absence of this gap. In the limit that the gap vanishes, the magnetoroton become the acoustic phonon of the crystal. By analogy, then, the presence or absence of antiferromagnetic order should be regulated by the presence or absence of a gap in the acoustic *magnon* of the spins. Given that the gap is essential, the spin liquid state, and by inference high-temperature superconductors, have *no* low-lying magnetic excitations other than those induced by disorder.

By far the most important feature of this type of vacuum, however, is the long-range force it propogates between the particles. As illustrated in Fig. 5, this force in 2 dimensional systems takes the form of fractional statistics. Particles obeying ν fractional statistics act as though they were bosons carrying a magnetic solenoid containing a fraction ν of a flux quantum. The other particles interact in the usual way with the vector potential generated by this solenoid, the net result being to cause the many-body wavefunction to acquire a phase $\exp(i\nu\pi)$ when two particles are adiabatically interchanged. Fractional statistics does not correspond to an attractive potential between the particles, but rather to a fundamental alteration in the way they avoid one another quantum mechanically. This is

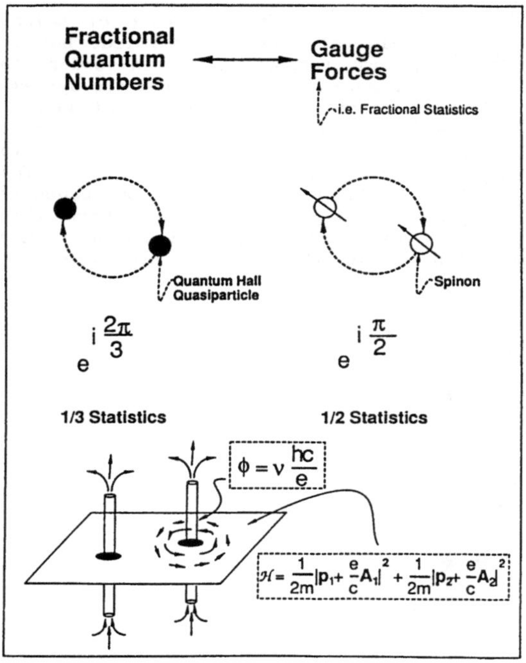

Fig. 5: The most important consequence of the analogy with the quantum hall problem is the occurrence of a long-range force between the particles carrying fractional quantum numbers. In two spatial dimensions, this takes the form of fractional statistics, which means that the many-body wavefunction acquires a phase $\exp(-i\pi\nu)$ when two particles are adiabatically interchanged. This may also be thought of as the possession by each particle of a magnetic solenoid containing a fraction ν of a flux quantum.

why numerical searches for a magnetically mediated attractive force in the Hubbard model have produced such ambiguous results. Fractional statistics only makes sense in 2 dimensions. It is not completely clear to me what the analog of this interaction is in the context of a 3-dimensional spin liquid, but I believe that each particle should act as though it were a Dirac monopole. In this problem the fraction $\nu=1/2$ is very special because the "charge" of the spinons, *i.e.* the z-component of the total spin, is half the indivisible "charge" of the bosons of the lattice gas. That this must must be so is imposed on the problem by rotational invariance. The only irreducible representations of the full rotation group are states with integral or half-integral spin. Thus, if ν is anything other than 1/2, the system must have a preferred direction in spin space, which is to say it must be ferromagnetically ordered.

This brings me to the subject of holons. It was first suggested by Kivelson, Rokhsar, and Sethna [15] that the charged excitations of the Mott insulator might be spinless. While this has not been demonstrated clearly one way or the other, it is almost certainly true. These particles cannot be bosons, however, as these authors originally suggested, because the electrons in question can lose their spin only by donating a spinon to the vacuum. The reasoning leading to the existence of holons is illustrated in Fig. 6. In order to remove an electron from a site, it is first necessary to determine whether the electron is up or down. This is necessary because a given electron is fluctuating between the up and down states. The most economical way to do this is to create a spinon at the site. With the spin on the site thus defined, the electron can be removed in an unambiguous way. The particle thus created obeys 1/2 fractional statistics because it is a bound state of an electron and a spinon.

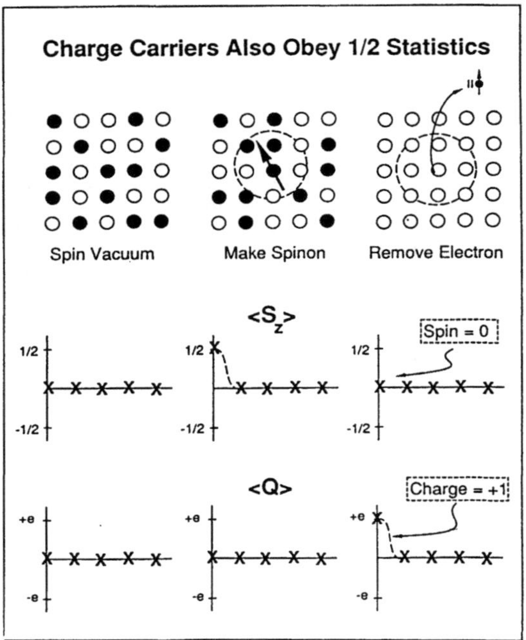

Fig. 6: The existence of holons has not been demonstrated, but is very reasonable. In order to remove an electron from the system, the spin state of the electron must first be established. The most economical way to do this is to create a spinon at a site. The particle made by subsequently removing the electron from this site has charge but no spin, and obeys 1/2 fractional statistics.

Let us now explore the possibility that the 1/2 fractional statistics obeyed by the charge carriers is the cause of high-temperature superconductivity. For the purposes of this I shall introduce [16] the model Hamiltonian

$$\mathcal{H} = \sum_j \frac{1}{2m} |\mathbf{p}_j + \mathbf{A}_j|^2 \quad , \tag{4}$$

where

$$\mathbf{A}_j = (1-\nu) \sum_{k \neq j} \frac{\hat{z} \times (\mathbf{r}_j - \mathbf{r}_k)}{|\mathbf{r}_j - \mathbf{r}_k|^2} \quad , \tag{5}$$

and where the particles are assumed to be spinless *fermions*. This constitutes the fermi representation for the fractional statistics problem. If ν were 1, we would have noninteracting fermions, the ground state of which we know to be a fermi sea with degeneracy pressure. If ν were 0, we would have the fermi representation for a noninteracting bosons, which has no degeneracy pressure. Since the case of $\nu=1/2$ is half way between these extremes, it must have roughly half the degeneracy pressure of the equivalent fermi gas. Thus, if we mistakenly assumed the particles to be fermions, we would conclude that there was an enormous attractive force in the problem, comparable in scale to the fermi energy. On the other hand, if we mistakenly assumed the particles to be bosons, we would conclude that there was an enormous repulsive force that somehow prevented them from bose condensing. We now observe that even though particles obeying 1/2 statistics are not bosons, and thus cannot bose condense, *pairs* of them, like pairs of electrons, are bosons and can bose condense. This, of course, is exactly what happens in an ordinary superconductor. We next observe [16], as is illustrated in Fig. 7, that a solution of this problem in the Hartree-Fock approximation leads to a Lagrange multiplier spectrum for the holons that contains a gap. At the mean-field level, a gas of particles, each of which carries a fraction $(1-\nu)$ of a flux quantum, produces a magnetic field

$$B_{eff} = (1-\nu)\,\frac{hc}{e}\,\rho \quad , \tag{6}$$

where ρ is the density of the particles. This produces Landau levels, which, in turn, must be filled to the level

$$n = \frac{1}{1-\nu} \quad , \tag{7}$$

to achieve self-consistency. Thus, the fractions ν = 1/2, 2/3, 3/4, ... are special cases for which low-lying "fermionic" excitations do not exist. For the case of 1/2 fractional statistics, self-consistency requires 2 Landau levels to be filled. A full solution of the Hartree-Fock problem reveals that the energy gap between the second and third Landau levels is logarithmically large, so that the energy to make an isolated "electron" or "hole" is formally infinite. The reason for this, illustrated in Fig. 7, is that these excitations are actually charged vortices, the circulation of which is appropriate for a bose condensate of *pairs*.

The Hartree-Fock ground state I reported [16] obviously lacks the broken symmetry characteristic of the superfluid state. This, however, is well-known [17] to follow from perturbation theory, once the longitudinal

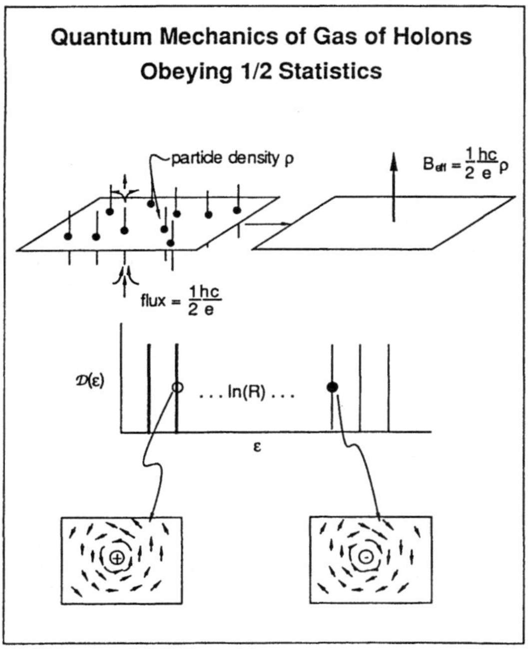

Fig. 7: A gas of particles obeying 1/2 fractional statistics is a charge-2 superconductor. This is indicated by a Hartree-Fock solution of the problem in the fermi representation. In the mean-field sense, each particle sees a uniform magnetic field tied to the density. The Hartree-Fock orbitals are thus Landau levels. Two of these are filled to achieve self-consistency. The fraction 1/2 is one of the special values for which an energy gap opens up in the "fermionic" spectrum. This gap turns out to grow logarithmically with sample size, as a consequence of the fact that the "particle" and "hole" are really charged vortices containing 1/2 of a flux quantum, as appropriate for a charge-2 superfluid.

sound wave of a quantum fluid is established to disperse linearly. That this is implicit in the Hartree-Fock calculation follows from the fact that the total variational energy per particle is proportional to ρ. I stated incorrectly in my paper [16] that the bulk modulus this led to was zero. It is in fact finite. I also stated incorrectly that a vertex correction to the longitudinal response function, *i.e.* taking into account the attraction of the particle and hole, accounted properly for this mode. As illustrated in Fig. 8, however, this causes the excitation to have a finite energy but does not destroy the gap. The persistence of the gap is a consequence of the presence in this problem of long-range potentials, which need to be accounted for properly in the perturbation theory. Fetter, Hanna and I [17] have recently found that a renormalization of this reponse function using the random phase approximation cures this problem and produces a linearly dispersing collective mode compatible with the bulk modulus. In addition, since the bare response function does not dissipate at low frequencies, as a consequence of the gap, the renormalized collective mode has an infinite lifetime. Since superfluid flow is the same thing as a long-wavelength sound wave that never decays, we became suspicious that the ground state implicit in the random phase approximation actually was superconducting, and in particular exhibited broken symmetry. We have since verified this [17] by computing the Meissner kernel in the random phase approximation and showing it to have no paramagnetic part in the $\mathbf{q} \longrightarrow 0$ limit. The coherence length in this calculation came out to be roughly the interparticle spacing, which is not surprising since this is the only length in the problem. Let me remind you that one of the key experimental puzzles in high-temperature superconductivity is why the coherence length is so short.

Let me now discuss some experimental consequences of these ideas. I mentioned that the Hamiltonian of Eqn.(4) gives rise to a fermionic

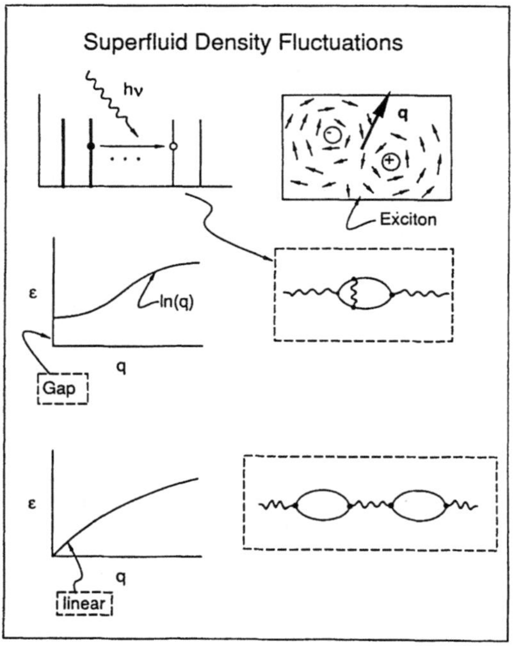

Fig. 8: The energy to make a particle-hole pair, or density fluctuation is finite at the mean-field level because the "particle" and "hole" attract each other with a logarithmic potential. Correcting the vertex in this manner, however, does not cause the collective mode to disperse to zero, as it does in a conventional superconductor, because of the presence of long-range potentials in the problem. A random phase approximation treatment of this problem produces a linearly dispersing Goldstone mode, a full Meissner effect, and a coherence length of order the interparticle spacing.

spectrum with an infinite energy gap. This implies that tunneling of holons into the superconducting holon fluid is impossible. Since tunneling of real electrons into the superconductor has been observed, the theory cannot be right unless spinons are central to the spectroscopy of these materials. The large energy cost to inject a holon into the fluid occurs because the holon effectively carries with it a magnetic solenoid which produces a vector potential, and thus current in the fluid, at large distances. That is, it is caused by the fractional statistics. Since spinons also obey fractional statistics, they also should polarize the fluid at large distances and carry a large energy cost. Given that this is the case, the particles carrying fractional quantum numbers, the holons and spinons, cannot be observed as spectroscopic particles, much the way quarks cannot be observed as free particles, but can only be observed in *pairs*. As illustrated in Fig. 9, there are three possible pairings:

1. **Spinon-Spinon**: When coupled to a spin-1 state, this excitation is a **spin wave**, analogous to the antiferromagnetic magnon of the magnetically ordered state. Excitations of this type have been seen in magnetic raman [18] and neutron [19] scattering. When coupled to a spin-0 state, this is the **resonon** of Kivelson and Rokhsar [20].

2. **Holon-Holon**: This is a density fluctuation or a longitudinal sound wave. My current understanding is that this should hybridize with the Goldstone mode of the superfluid state and thus be equivalent to superfluid flow, which has been observed.

3. **Spinon-Holon**: For want of anything else, let us call this excitation an "electron". An object with these quantum numbers has been observed in tunneling.

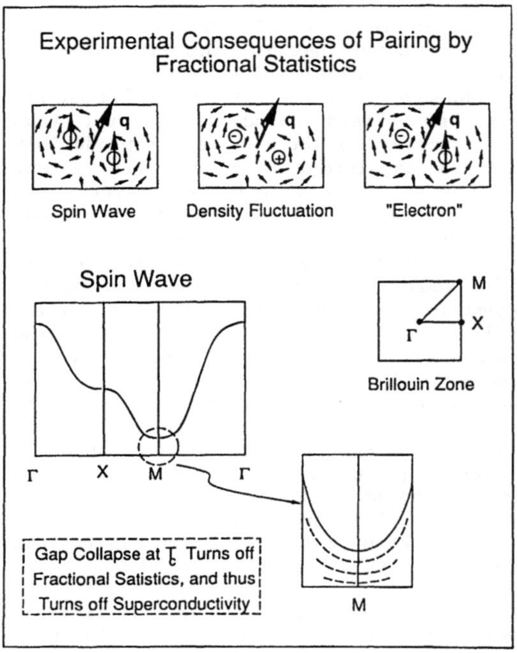

Fig. 9: Because the fractional statistics obeyed by particles with fractional quantum numbers polarizes the holon fluid into vortices, such particles cannot be observed spectroscopically. The observable particles consist of *pairs*, of which there are three principle kinds. The spin wave, which is a pair of spinons, is of particular interest because it regulates the superconductivity. It should have a minimum somewhere near the M-point of the Brillouin zone, and collapse to zero at the superconducting transition temperature.

Thus, all but one of the spectroscopic particle implicit in the theory have already been observed experimentally.

Let us now concentrate on the spin wave. As illustrated in Fig. 9, this is expected to have its minimum neat the M-point in the Brillouin zone, as this is the reciprocal lattice vector of the magnetically ordered state. This minimum is analogous to the magnetoroton minimum [14] of the fractional quantum hall state and is therefore the regulator of the problem. As long as this gap is nonzero, the elementary excitations of the spin liquid state must carry a fractional quantum number and must obey 1/2 statistics exactly. This, in turn, implies that the ground state is a superconductor, unless thermal fluctuations destroy the long-range order, as occurs when liquid helium is heated. Given that the latter does not occur, as is indicated by experiment, the only other possiblilty is that thermal fluctuations destroy the fractional statistics by collapsing the spin-wave gap. That there is a tendency for this is clear from studies [21] of the effects of disorder on the magnetoroton gap. Furthermore, one can say without knowing any details that the temperature at which the gap should collapse is roughly equal the gap itself. Thus, I predict that there should be a gap in the spin-wave spectrum, that this gap should be comparable to the superconducting transition temperature, and that the gap should collapse at the transition.

Let me now mention some speculations based on calculations that are either under way or being contemplated. An appropriate place to begin is the electron. Not having performed any calculations for this excitation, I imagine using the variational procedure, much as was done for the magnetoroton, to compute its dispersion curve. As it did for the magnetoroton, this procedure would yield a curve with a minimum somewhere in the Brillouin zone. Let us guess this to be at the X-point, as this would lead to a large magnetic susceptibility at M, consistent with the previous discussion. As illustrated in Fig. 10, this would imply the existence of a fermi *point*, as opposed to a fermi surface, for this excitation, and a tunneling density of states that is constant above the gap. However, it is well known [22] that quantum fluctuations, in this case those associated with the superconductivity, tend to linearize quadratic dispersion obtained from mean-field calculations, and thus produce a linear density of states, similar to the one very reproducibly observed in experiment. This idea is consistent with the observation, reported by the Stanford group (M. Lee *et. al.*) at this conference, that the slope of this density of states *decreases* as the temperature is raised. Thus, I predict that both the linearity of this density of states and its temperature dependence are consequences of the presence of a fermi point. Other consequences in *clean* samples would be that transport meaurements would be inherently temperature dependent, and that spin-lattice relaxation times would decay exponentially below the transition and violate the Korringa law above the transition. Also, the material would have an abnormally large dielectric susceptibility at a momentum transfer corresponding to the M-point. One expects, mostly from common sense, that the tunneling gap should be half the spin wave gap, which in turn should equal the optical gap, and that all of these should collapse simultaneously at the transition. In addition, these gaps should be extremely sensitive to carrier density, and in particular should go to zero in the limit of extremely low or high doping. It is clear that the experimental transition temperature has doping dependence of this kind, but whether the gap follows these trends remains to be established.

It was pointed out recently by Kivelson [23] that fractional statistics, and thus the theory of superconductivity I have outlined in this lecture,

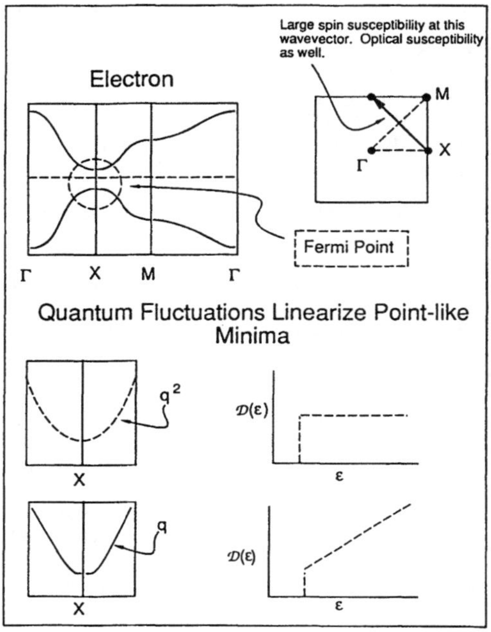

Fig. 10: It is hypothesized that the "electron", consisting of a holon and spinon, has no internal quantum number other than spin, and disperses roughly as shown. This is the outcome of a "gedanken" variational calculation. The minimum is placed at the X-point because this is consistent with a large susceptibility at M. A calculation of this kind gives a fermi *point*, as opposed to a fermi surface. Quantum fluctuations tend to linearize the quadratic mean-field dispersion about the minimum, giving a tunneling density of states above threshold similar to that commonly observed.

makes no sense in a system with time-reversal invariance. It has subsequently been pointed out by Wen, Wilczek, and Zee [24] that tacit in the theory is *spontaneous* breaking of time-reversal and parity invariance. In other words, there are two spin liquid ground states, related to one another by complex conjugation, that the system must decide between when it condenses into the superconducting state. Given that this is the case, which I now believe, the superconducting state is predicted to be parity violating. Since the BCS state conserves parity, the observation of a transport coefficient that requires intrinsic handedness to be nonzero would establish clearly whether or not the theory is correct. I do not presently know what kind of measurement to suggest, nor am I in a position to predict the magnitude of such an effect. However, Wilczek's remark is correct in principle and constitutes a qualitative test of the theory.

This work was supported primarily by the National Science Foundation under Grant No. DMR-88-16217 and by the NSF-MRL program through the Center for Materials Research at Stanford University. Additional support was provided by the U.S. Department of Energy through the Lawrence Livermore National Laboratory under Contract No. W-7405-Eng-48.

References

1. B. I. Halperin, Phys. Rev. Lett. 52, 1583 (1984).
2. G. Bednorz and K.A.Müller, Z. Phys. B66, 189 (1986).
3. M.L. Cohen and P.W. Anderson, in **Superconductivity in d- and f-Band Metals**, D. Douglas, ed. (AIP, New York, 1972).
4. W. Weber, Phys. Rev. Lett. 58, 1371 (1987).

5. W.F. Brinkman and T.M. Rice, Phys. Rev. **B2**, 1324 (1970); Phys. REv. **B2**, 4302 (1970).
6. R.B. Laughlin, Science, **242**, 525 (1988).
7. P.W. Anderson, Science **235**, 1196 (1987).
8. J.E. Hirsch, Phys. Rev. Lett. **59**, 228 (1987).
9. P.W. Anderson, Mater. Res. Bull. **8**, 153 (1973).
10. V. Kalmeyer and R.B. Laughlin, Phys. Rev. Lett. **59**, 2095 (1987).
11. Z. Zou, B. Doucot, and B.S. Shastry, (*To be published in Phys. Rev. B*).
12. I. Affleck and J.B. Marston, Phys. Rev. **B37**, 3774 (1988); G. Kotliar, Phys. Rev. **B37**, 3664 (1987).
13. R.B. Laughlin, (*To be published in Ann. Phys.*).
14. S.M. Girvin, A.H. MacDonald, and P.M. Platzman, Phys. Rev. Lett. **54**, 581 (1985).
15. S.A. Kivelson, D.A. Rokhsar, and J.P. Sethna, Phys. Rev. **B35**, 8865 (1987).
16. R.B. Laughlin, Phys. Rev. Lett. **60**, 2677 (1988).
17. A. Fetter, C. Hanna, and R.B. Laughlin, (*Submitted to Phys. Rev. B.*)
18. K.B.Lyons, P.A.Fleury, L.F. Schneemeyer, and J.V. Waszczak, Phys. Rev. Lett. **60**, 732 (1988).
19. G. Shirane *et. al.*, Phys. Rev. Lett. **59**, 1613 (1987); Y. Endoh *et. al.*, Phys. Rev. **B37**, 7443 (1988).
20. S. A. Kivelson and D.A. Rokhsar (*To be published in Phys. Rev. Lett.*)
21. S.M. Girvin, A.H. MacDonald, and P.M. Platzman, Phys. Rev. **B33**, 2481 (1986).
22. C. Kittel, **Quantum Theory of Solids** (Wiley, New York, 1963), p.58.
23. S.A. Kivelson and D.S. Rohksar, Phys. Rev. Lett. **61**, 2630 (1988).
24. X.G. Wen, F. Wilczek and A. Zee (*Preprint*).

Recent Studies of the CU D-D Excitation Model

W. Weber[1], *A.L. Shelankov*[2,3], *and X. Zotos*[2]

[1]Kernforschungszentrum Karlsruhe, INFP, D-7500 Karlsruhe, Fed. Rep. of Germany
[2]Inst. f. Theorie der Kond. Materie, Universität, D-7500 Karlsruhe, Fed. Rep. of Germany
[3]A.F. Ioffe Physico-Technical Institute, SU-194021 Leningrad, USSR

In the $d-d$ excitation model, it is proposed that the Jahn-Teller levels of Cu^{2+} represent excitonic centers for the pairing of oxygen p holes. In this paper, recent work on the model is reviewed, both the treatment of an effective Hamiltonian in a weak coupling limit and finite size studies for strong coupling. Furthermore, studies of optical spectra are presented.

1. INTRODUCTION

The electronic structure of the copper oxide superconductors appears to be described best by a localized picture for the $3d$ electrons of $Cu^{2+}(d^9)$, and by an itinerant one for the oxygen $2p$ electrons (better: holes) in case of doping. The main experimental arguments for this view are: i) in the undoped limit, the copper oxides are antiferromagnetic insulators like most of the other oxides and halides of the $3d$ transition metals [1], and ii) with doping, the holes are predominantly found on the oxygen lattice [2]. A schematic picture of the electronic structure of the copper oxides is displayed in Fig. 1.

Thus, the theoretical treatment of the copper oxides employs Hubbard-type models, all of which yield anti-ferromagnetism in the undoped limit, as the two-fold spin degeneracy of an isolated Cu d hole is removed by superexchange coupling to neighbors. Further, a suitable choice of the difference between the Cu and oxygen orbital energies leads to doping into the oxygen p hole band. The two-orbital Hubbard model (or call it periodic Anderson model) is being widely investigated both concerning "magnetic" and "excitonic" coupling mechanisms for superconductivity [3-5].

Our work is concerned with a further property of Cu^{2+}, its additional orbital degeneracy in octahedral ligand symmetry which makes it a - very strong - Jahn-Teller ion. All superconducting copper oxides have perovskite-type crystal lattices, with O octahedra, or pieces thereof, around the central Cu^{2+} ions. In case of a perfect octahedron the orbital degeneracy is two-fold, as the d hole exhibits e_g symmetry, either $d(x^2-y^2)$ or $d(3z^2-r^2)$. It is removed by deformations of the O octahedra, in an analogous way as the spin degeneracy is removed by exchange coupling. We have argued elsewhere [6] that the crystal structures of the copper oxides show two signatures due to a big Jahn-Teller effect. One is either a very large elongation of the O octhahedra or even the removal of O atoms to form incomplete octahedra. The second is partial orbital ordering of neighbor Cu d holes, apparent for instance in the staggering of O pyramides or squares in $YBa_2Cu_3O_7$ or in the non-superconducting $La_4BaCu_5O_{13}$.

The basic idea of the $d-d$ excitation model is that the Jahn-Teller levels of Cu^{2+} can act as very effective excitonic centers to mediate the pairing of O p holes for

Springer Series in Materials Science, Vol. 11
Mechanisms of High Temperature Superconductivity
Editors: H. Kamimura and A. Oshiyama

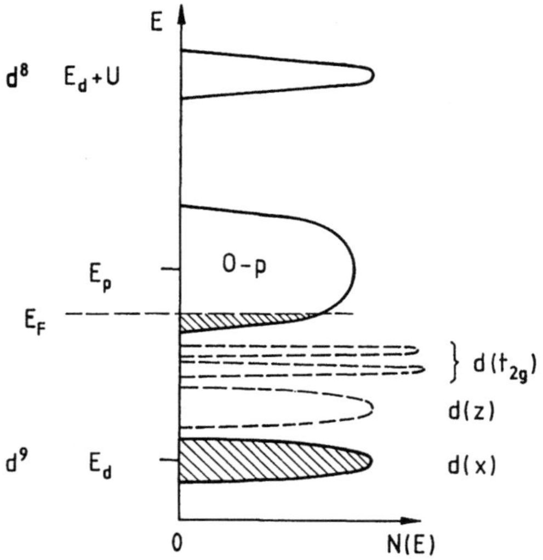

Fig. 1: Schematic picture of the electronic structure of the copper oxides (in hole representation). Solid lines represent positions of the Cu d^9 and d^8 levels and of the $O2p$ band. The d^9 ground state corresponds to the $d(x) \equiv d(x^2-y^2)$ hole orbital Hubbard band. Dashed lines indicate excited states of the d^9 configuration, in particular $d(z) \equiv d(3z^2 - y^2)$ orbital band (from Ref. 9).

superconductivity. In our view, this is one peculiar property of Cu^{2+}, which makes this ion different from other $3d$ ions, such as Ni^{2+}, whose oxides can also be made metallic by doping, but are not superconductors. As of now, no further $3d$ oxides have been found superconducting, except some Ti oxides, known since long [7]. It should be noted that, again unlike most of the other $3d$ ions, Cu^{2+} is also a spin 1/2 ion, which leads to very large spin fluctuations. This peculiarity of Cu^{2+} is stressed in Anderson's resonating valence bond model [8].

2. THE HAMILTONIAN

The $d-d$ excitation model is represented by a three-orbital Hamiltonian of the form

$$\begin{aligned} H &= \sum_{i\sigma}[\Delta_z n^z_{i\sigma} + U^+_{xz} n^x_{i\sigma} n^z_{i\sigma} + U^-_{xz} n^x_{i\sigma} n^z_{i-\sigma}] \\ &+ \sum_{i}[U_{xx} n^x_{i\sigma} n^x_{i-\sigma} + U_{zz} n^z_{i\sigma} n^z_{i-\sigma}] \\ &+ \sum_{j\sigma} E_p n_{j\sigma} + \sum_{jj'\sigma} t_{pp}(p^+_{j\sigma} p_{j'\sigma} + \text{c.c}) \\ &+ \sum_{j} U_{pp} n_{j\sigma} n_{j-\sigma} + \sum_{jj'\sigma\sigma'} V_{pp} n_{j\sigma} n_{j'\sigma'} \\ &+ \sum_{ij\sigma}[t_{px} x^+_{i\sigma} p_{j\sigma} + t_{pz} z^+_{i\sigma} p_{j\sigma} + \text{c.c}] \\ &+ \sum_{ij\sigma\sigma'}[V_{px} n^x_{i\sigma} n_{j\sigma'} + V_{pz} n^z_{i\sigma} n_{j\sigma'}] \\ &= H^{Cu} + H^{O} + H^{I} \end{aligned} \tag{1}$$

Here, $x_{i\sigma}$, $z_{i\sigma}$, and $p_{j\sigma}$ are respectively the annihilation operators for the $d(x^2-y^2)$, $d(3z^2-r^2)$, and the p hole orbitals at Cu sites i and O sites j, with spin index $\sigma = \pm 1/2$. $n^x_{i\sigma}$, $n^z_{i\sigma}$ and $n_{j\sigma}$ are the corresponding number operators. We put the orbital energy $E_x = 0$, so that $E_z = \Delta_z$. The various quantities $U_{\alpha\beta}$ and $t_{\alpha\beta}$ represent nearest neighbor inter-site repulsion and hopping terms, respectively. Estimates for the model parameters are [9] $t_{px} \approx \sqrt{3}t_{pz} \approx 1.6$, $t_{pp} \approx 0.5$, $\Delta_z \approx 0.5$, $U_d \approx U_{xz}^{+-}$, $U_{xx}, U_{zz} \approx 8$, $E_p = 4$ (all in eV). Further is $V_{px}, V_{pz} \approx 1-2$, and it is essential that $V_{px} > V_{pz}$ (plausible as both $d(x^2-y^2)$ and $p(\sigma)$ holes are lying in the CuO_2 planes). $\Delta V = V_{px} - V_{pz} \approx 0.3 - 0.5$. Because of the small p doping, the parameters U_{pp} and V_{pp} are not very important, estimates are $U_p \approx 3-5$, $V_p \approx 0-1$. Realistically $U_{xz}^{-} > U_{xx}, U_{zz} > U_{xz}^{\pm}$, with $| U_{xx} - U_{xz}^{\pm} | \approx 2$, in accordance with Hund's rules. Note that the $x \equiv d(x^2-y^2)$ and $z \equiv d(3z^2-r^2)$ orbitals have different symmetries in the basal CuO_2 planes. As a result, the $p-d$ hybridizations are different for x and z. These features have been properly incorportated in the following.

The Hamiltonian (1) can be seen as a member of a series of Hubbard-type Hamiltonians with increasing complexity. The simplest one in this series is the one-orbital Hubbard model, the next is the (two-orbital) periodic Anderson model, as employed in the Emery model [3]. Addition of pd intersite repulsion terms will lead to the model of Varma et al. [4]. The one orbital Hubbard model may also be extended to a model with orbital degeneracy, as used; e.g., to describe the anti-ferrodistortive order of Jahn-Teller systems such as K_2CuF_4 [14].

Although the form of the Hamiltonian (1) appears complicated, conceptually the origin of pairing is very simple. If the $d-d$ excitation occurred without the mediation of the oxygen p orbital, the model would be identical to the classical Little model [15], which has been extensively studied [16]. As the direct $d-d$ excitation is symmetry forbidden, the indirect $d-d$ excitation through the oxygen orbitals mixes the spins of the oxygen carriers and the excited d holes, complicating the analysis. As a consequence, the model cannot be understood without knowledge of the underlying simpler models. Our strategy of treating the model was to investigate as simple limiting cases as possible. A side effect was that we also studied problems related to the simpler models.

3. THE EFFECTIVE HAMILTONIAN

In the limit $U, E_p \to \infty$, the system consists of two noninteracting parts, the Cu^{++} ions with one d hole and a partially filled oxygen p hole band of width t_{pp}. A d-hole on site i is specified by its spin σ_i; and a pseudospin $\mathbf{I}_i$ for the $x-z$ orbital degree of freedom. For finite values of U and E_p, virtual hopping leads to an interaction between the Cu and oxygen subsystems described by an effective Hamiltonian $H_{eff} = H_o + H_{d-d} + H_{\text{spin}}$. Here,

$$H_o = \sum_{jj's} \tilde{t}_{jj'} p_{js}^{+} p_{j's} + \sum_i \mathbf{h}\mathbf{I}_i \tag{2}$$

$$H_{d-d} = \sum_{ijj's} \left(\mathbf{W}_{i,jj'}\mathbf{I}_i\right) p_{js}^{+} p_{j's} \tag{3}$$

$$H_{\text{spin}} = \sum_{ijj'\alpha\beta} J_{i,jj'}^{\alpha\beta} \sigma_i^{\alpha\beta} \mathbf{S}_{jj'} \tag{4}$$

H_o is the renormalized single particle Hamiltonian of the Cu and O subsystems. $\tilde{t}_{jj}$ represents the on-site p orbital energy renormalized by the interactions V_x and V_z, and $\tilde{t}_{jj'}$ describes the hopping t_{pp} modified by indirect hopping via Cu. The pseudospin term is a matrix in $x-z$ space with $(\mathbf{hI})_{xx} = -(\mathbf{hI})_{zz} = -\Delta_z/2$ and

$$(\mathbf{hI})_{xz} = (\mathbf{hI})_{zx} \equiv t_{xz} = -\sum_j t_{jx}t_{jz}/E_p \tag{5}$$

where t_{jx}, t_{jz} are the standard $p-x$, $p-z$ hopping terms. H_{d-d} describes the spin-independent part of the interaction between p holes and the Cu $d-d$ excitations (pseudospins). We have

$$\begin{aligned}(\mathbf{W}_{i,jj'}\mathbf{I}_i)_{xx} &= -(\mathbf{W}_{i,jj'}\mathbf{I}_i)_{zz}\\ &\equiv V_1 = \Delta V/2 + (t_{jx}t_{j'x} - t_{jz}t_{j'z})/D\\ (\mathbf{W}_{i,jj'}\mathbf{I}_i)_{xz} &= (\mathbf{W}_{i,j'j}\mathbf{I}_i)_{zx}\\ &\equiv V_2 = t_{jz}t_{j'x}/2D\end{aligned} \tag{6}$$

with $D^{-1} = (U - E_p)^{-1} + E_p^{-1}$. Finally, H_{spin} gives the exchange interaction of Cu and O spins. In standard spin notation, e.g. the z component of $\sigma_i^{\alpha\beta}$ is given by $\alpha_\uparrow^+\beta_\uparrow - \alpha_\downarrow^+\beta_\downarrow$, with α, β representing any x, z orbitals on the i-th site (a corresponding definition holds for the O spins $\mathbf{S}_{jj'}$). Then

$$J_{i,jj'}^{\alpha\beta} = t_{j\alpha}t_{j'\beta}/2D \tag{7}$$

with the orbitals α, β on site i. H_{eff} (Eqs. 2-4) neglects all processes higher than second order in p-d hopping, e.g. of superexchange type. The relative order of corrections to H_{eff} is $t_{pd}^2/(U, E_p)$ and $\Delta V/(U, E_p)$ where t_{pd} stands for t_{px} and t_{pz}. For realistic model parameters, H_{eff} (2-4) describes a strongly correlated system as $t_{pp} \approx t_{pd}^2/(E_p, U)$. For simplicity, we assume $t_{pp} \gg t_{pd}^2/(E_p, U)$. This corresponds to a weak coupling limit. In this case, and neglecting the term H_{spin} for the moment, superconductivity can be studied in analogy to electron-phonon interaction (see also [16]). Diagonalization of H_o leads to O p band states and to d levels split by $\widetilde{\Delta} = (\Delta_z^2 + 4t_{xz}^2)^{1/2}$. The ground state is $\widetilde{x} = ux + vz$ with $u^2 = \Delta_z^2/\widetilde{\Delta}^2$, $v^2 = 1 - u^2$. The interaction term H_{d-d} leads to an effective attraction

$$V_{eff} = -\Big(< V_1^2 > v^2 + < V_2^2 > u^2\Big)/\widetilde{\Delta} \tag{8}$$

where $<>$ represents Fermi surface averages of the quantities V_1^2 and V_2^2. This attraction would lead to a BCS-like superconductivity with a transition temperature

$$T_{c0} \simeq \widetilde{\Delta} \exp\Big(-1/(N(0)V_{eff})\Big) \tag{9}$$

where $N(0)$ stands for p hole density of states and the prefactor $\widetilde{\Delta}$ represents the cut-off energy for attraction. As this expression includes only electronic parameters, one may expect T_{c0} to be quite large.

Note that V_{eff} is very sensitive to the symmetry of the lattice. For a perfect $Cu-O$ square plane, $v^2 = 0$, due to different rotational symmetry of the x and z orbitals.

For the same reason $< V_2^2 > \approx 0$ when E_p is close to the bottom of the p band, so that $< V_2^2 > \propto n$, the p hole concentration. These restrictions do not hold for the $Cu-O$ chains or for the orthorhombic $Cu-O$ planes. Furthermore, any intersite $x-z$ coupling, superexchange-like, would remove these symmetry restrictions.

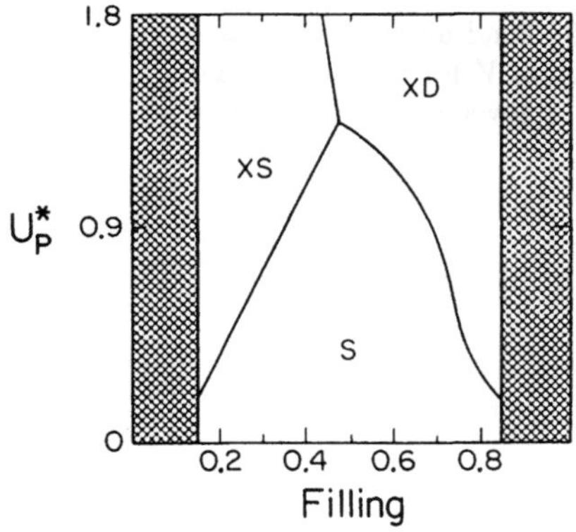

Fig. 2: Superconducting phase diagram for the d-d excitation model as a function of U_{pp} (here denoted as U_p^*) in a weak coupling limit (from Ref. 12). S, XS, and XD denote s-wave, extended s-wave, and extended d-wave regions, respectively.

A more detailed BCS analysis of the $d-d$ excitation model in the weak coupling limit has been carried out by Jarrell et al. [12]. In their study, they have also included U_{pp} and V_{pp}, treated as local and extended Coulomb pseudopotentials, respectively. Furthermore, they have studied a variety of gap harmonics of s and d wave character within spin singlet pairing. They find that for reasonable values of U_{pp} and V_{pp} and for low filling rate of the p hole band, s-wave pairing dominates. As is expected, the increase of U_{pp} will suppress on-site s wave pairing, yet extended s-wave symmetry still prevails. At higher band filling, a transition to d-wave pairing is observed (see Fig. 2).

Let us now briefly discuss the influence of the Cu spins on T_c. They can either form a coherent state of a Kondo-lattice type with a typical energy scale T_K, and/or short range spin ordering with a characteristic energy ('Néel temperature') T_N. In the simplest (and least probable) case, when $T_{c0} > T_N, T_K$, Cu spins act as magnetic impurities causing p spin-flip with rates $1/\tau_s \approx J^2/t_{pp}$. Superconductivity will survive as long as $1/\tau_s < T_{c0}$. If the superconducting transition happens in the Kondo lattice regime; i.e. $T_{c0} < T_K$, and spin correlations between Cu and O sites are the most important ones, we expect $T_c \approx T_{c0} \cdot T_K/\widetilde{\Delta}$, in accordance with arguments suggested for phonon-induced superconductivity in heavy fermion systems [17].

If however, the Cu spin correlations are mainly controlled by superexchange interaction (not included in H_{eff}), and $T_N > T_{c0}$, we do not expect any BCS-like superconductivity to sustain the spin exchange. Estimates show that the repulsive interaction between the p holes mediated by spin excitation can be as large as U. Yet superconductivity with a more complex order parameter is not excluded for this case.

4. FINITE SIZE SYSTEM STUDIES

To obtain some insight on the model in the realistic (strong coupling) region of parameters, we have investigated a molecule consisting of two Cu and three O atoms in the linear configuration $O - Cu - O - Cu - O$ by exact diagonalization of the Hamiltonian matrix. We used the Hamiltonian (1) with the coupling parameters of section 2 and included the appropriate symmetry of the $p-x$ and $p-z$ hybridization elements. Because of this symmetry, which is an important feature of the cuprate energy bands [18], the x (z) level couples more strongly with the bottom (top) of the oxygen hole 'band'.

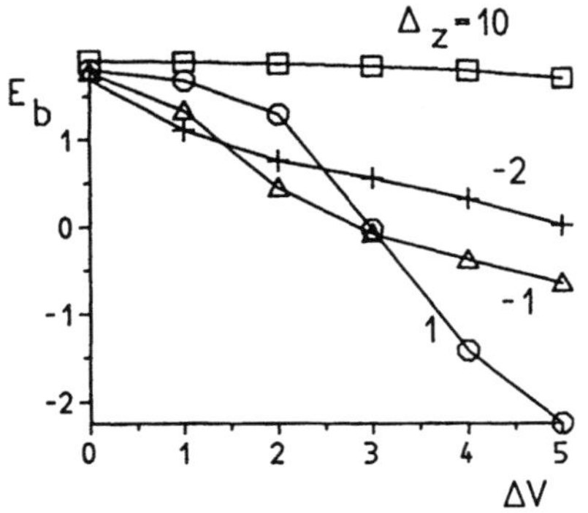

Fig. 3: Binding energy E_b as a function of V for different values of the z level energy Δ_z (from Ref. 11).

We have studied in particular the binding energy for a pair of oxygen holes as a function of ΔV and Δ_z. The binding energy is defined as $E_b = E_o(4) + E_o(2) - 2E_o(3)$ where $E_o(n)$ is the ground state energy for n holes ($Cu-d$ holes included). Attraction corresponds to negative values of E_b. It is clear that in the atomic limit (all hopping terms $t_{pp} = t_{pz} = t_{px} = 0$) there is binding for sufficiently large ΔV accompanied by a complete occupation of the z(upper) level.

As shown in Fig. 3, for finite hopping, introduction of the z level (variation of Δ_z) favors binding up to values of $\Delta_z \simeq -1$ where the $x-z$ levels invert in energy). It should be noted that for finite $p-d$ hopping terms, the $x-z$ level splitting, as determined by the excitation spectrum of the system with two Cu holes, also depends on t_{px}, t_{pz} and on E_p. We found that a $d-d$ excitation energy of order unity ($\simeq 0.5eV$) as suggested by experiment [19] is achieved by chosing a 'bare' value $\Delta_z = -1$. This value of Δ_z is specific to this particular molecule, as the $p-d$ hybridization depends on the number of oxygen neighbors per Cu site. From Fig. 3 it follows that for the 'realistic' values $\Delta_z = -1$ and $\Delta V \simeq 1$ the values of E_b are still positive, although there is considerable reduction of E_b compared to the results for large $\Delta_z \simeq U$. This result suggests that a linear model system with realistic parameters is stable against formation of small bipolarons. In two dimensions it is easier to minimize the kinetic energy and maximize attraction on a scale of one unit cell. We also found that the values of E_b can be lowered further when it is assumed that $U_{zx} \leq U_{zz} = U_{zz} = U_{xx}$, as is suggested from Hund's rules.

5. OPTICAL SPECTRA

The optical spectra of the metallic copper oxides, in particular of $YBa_2Cu_3O_7$, exhibit a rather structureless, yet relatively large optical absorption up to $\approx 4eV$, with some broad features near 0.5, 1.5 and $3eV$ [19]. These absorption features have been interpreted as $d-d$ excitations, in particular the $0.5eV$ feature, with the provision that the $d-d$ transitions - dipole forbidden in the atomic limit - are strongly enhanced because of the coupling to the p conduction holes [19]. Alternate interpretations are that the feature near $1.5eV$ represents a transition via the charge transfer gap [20], and the $0.5eV$ structure - if present at all [21] - is an excitonic absorption split off from the charge transfer gap [22].

Presently, theoretical work on optical spectra of Hubbard-type models is rather limited [23]. In our work we have focused on two approaches. One is the use of the Gutzwiller method [24] to calculate a generalized density of states $\widetilde{N}(E)$, and the other employs the evaluation of excited states in finite size calculations to estimate oscillator strengths for $d-d$ transitions as a function of doping.

5.1 Gutzwiller Method

In this method, a variational many-body wave function is used to find a minimum ground state total energy E_{tot} as a function of band filling N. In a paramagnetic one orbital model, the only variational parameter is the site double occupancy ν, so that

$$E_{tot} = q(\nu)\bar{\epsilon} + \nu U, \qquad \text{with} \qquad \partial/\partial\nu E_{tot} = 0 \tag{11}$$

Here $\bar{\epsilon}(N)$ is the average one-particle energy and q is the step at Fermi momentum. In a two-orbital model further variational parameters can be introduced to mimic the effective orbital hybridization [25]. With $E_{tot}(N)$ we can also obtain the Fermi energy

$$E(N) = \partial/\partial N E_{tot}(N)$$

and

$$\widetilde{N}(E) = [\partial E/\partial N]^{-1} \tag{12}$$

$\widetilde{N}(E)$ is the number of electrons per energy unit, which can be filled into the band at a certain value of N, or correspondingly, at a Fermi energy $E(N)$. For $U = 0$, $\widetilde{N}(E)$ represents the single particle density of states $N(E)$.

In Fig. 4, $\widetilde{N}(E)$ is shown as a function of U. For this specific calculation we have used two-dimensional one particle bands

$$\epsilon(k) = 2[cos(k_x a) + cos(k_y a)] + cos(k_x a)cos(k_y a) \tag{13}$$

and have carried out minimizations (Eq. 11) and differentations (Eq. 12) numerically. In Eq. 13 a second neighbor hopping term has been introduced to move the density-of-states singularity away from half-filling.

With increasing U, $\widetilde{N}(E)$ broadens and eventually splits in two subbands at $U_c = 8\bar{\epsilon}$, the Brinkman-Rice condition for the metal insulator transition [26]. With further increasing U, the two subbands, the 'lower' and the 'upper' Hubbard bands, shrink in width and a wide gap of order U develops. However, most remarkably the very pronounced singularity in $\widetilde{N}(E)$ for $U = 0$ vanishes rapidly, even for moderate values of $U \approx 4$, where the metal insulator transition is still far away.

We interpret the $\widetilde{N}(E)$ curves in the following way: At a certain band filling N, the occupied part represents the range of energies where electrons can be excited from, such as by photoemission experiments. The unoccupied part spans the energy range where electron can be deposited into, for instance by inverse photoemission. This interpretation can be confirmed by E_{tot} calculations using many-body wave functions with particles added or removed in certain regions of momentum space and by varying the double occupancy ν within the appropriate limits. Further, a convolution of the occupied and unoccupied parts of $\widetilde{N}(E)$ will provide the basic features of the optical spectra.

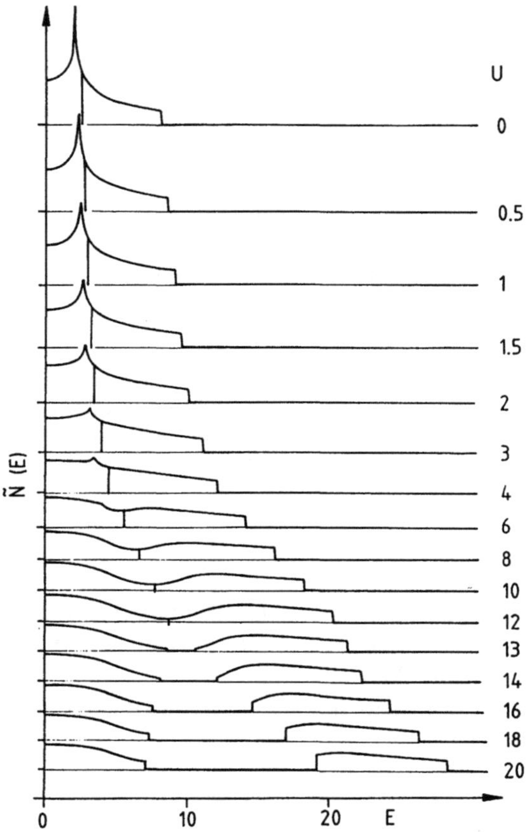

Fig. 4: Generalized density of states $\tilde{N}(E)$ obtained by the Gutzwiller method for a square planar Hubbard model as a function of U. Note that the singularity in the $\tilde{N}(E)$ curve vanishes for values of U much smaller than U_c, where the Brinkman-Rice metal-insulator transition occurs.

Certainly, matrix element effects for the transition probabilities are neglected. However, from the experience with results of one-electron band theory, we expect these effects not to cause additional structure in the spectra, but either to enhance or to reduce existing structure.

Our results suggest that the copper oxides exhibit intermediate values of U, just sufficient to enforce the metal-insulator transition for half- filling. In this range one does not expect any sharp features in the optical data, nor a pronounced 'upper' Hubbard band in inverse photoemission. Also, normal photoemission should provide $N(E)$ curves for the occupied band which are reduced in intensity and tailing off to higher Fermi energies as compared to results from one particle energy band theory.

5.2 Results from Finite Size Systems

In this study, we have calculated numerically the conductivity

$$\sigma(\omega) = |< n \mid \hat{j} \mid 0 >|^2 \, (\epsilon_n - \epsilon_o)^{-1} \delta(\omega - \epsilon_n + \epsilon_o) \tag{14}$$

for all excited states $\mid n >$ with energies ϵ_n of the $Cu-O$ molecule described in section 4. $\hat{j}$ is the current operator. As the system is finite, we obtain only discrete values of

σ at the energies $\epsilon_n - \epsilon_o$. The $\sigma(\omega)$ spectrum can be depicted as a sequence of lines having different intensities. To identify the nature of the various lines, we have also calculated different correlation functions of the excited states $\mid n >$, e.g. occupation numbers and spin-spin correlations.

For the case with two holes (undoped situation), the first perceptible lines are found for $\omega \geq E_p$, corresponding to Cu-oxygen charge transfer excitations. As expected for an insulator, the $d-d$ excitation at low energies is very weak. Upon doping; i.e. with the third hole, more lines appear at low energies, corresponding to 'intra-band' excitations of the oxygen hole. Further, the $d-d$ excitation line is strongly enhanced. In addition, the new oxygen hole 'intra-band' excitations show considerable admixture of the Cu z orbital. These results are in qualitative agreement with experiment, yet the small size of the system prohibits any quantitative comparison.

6. CONCLUDING REMARKS

The recent studies of the $d-d$ excitation model have demonstrated the possibility for 'excitonic' superconductivity in the copper oxides, with the Jahn-Teller ions Cu^{2+} acting as excitonic centers to mediate the pairing of oxygen p holes. The $d-d$ model thus emphasizes a property of Cu^{2+} which is distinct from many other $3d$ ions. The results on the effective attractive interaction V_{eff} turn out to be rather sensitive on local symmetry, at least in second order expansion of $p-d$ hopping. It is important that the $d(3z^2 - r^2)$ orbital is to some extent admixed in the ground state level - as in fact seen experimentally [27]. Furthermore, V_{eff} depends both on the difference ΔV of the intersite $p-d$ Coulomb repulsions and the difference of the $p-d$ hopping terms. The latter contribution is relevant only, when the p holes are of σ type, which presently is still a matter of debate. For $p(\pi)$ holes, ΔV leads to attraction only when the π holes also lie in the basal $Cu-O$ plane, but probably not for a π_z hole. On the other hand, pair-breaking spin flip processes should also be largely reduced for π holes.

In our opinion, the role of the Cu spins remains the biggest open question for the $d-d$ model - actually for any 'excitonic' model in the limit of large U. For obvious reasons, this is also the central question for the 'magnetic' models. Thus a more detailed understanding of the Cu spin liquid as a function of doping appears to be the most prominent problem to investigate.

REFERENCES

1. D. Vaknin et al., Phys. Rev. Lett. **58**, 2802 (1987) and references therein; J.M. Tranquada et al., Phys. Rev. Lett. **60**, 156 (1988).
2. See, e.g., N. Nücker, J. Fink, J.C. Fuggle, P.J. Durham and W.M. Temmerman, Phys. Rev. **B37**, 5158 (1988).
3. V. Emery, Phys. Rev. Lett. **58**, 2794 (1987).
4. C.M. Varma, S. Schmitt-Rink, and E. Abrahams, Solid State Commun. **62**, 681 (1987).
5. Proceedings of the Int. Conf. on High Temperatures Superconductors and Materials and Mechanisms of Superconductivity, Interlaken (1988), (Eds. J. Müller and J.L. Olsen), North-Holland, Amsterdam, Physica **C153-155**.
6. W. Weber, in Festkörperprobleme (Advances in Solid State Physics) Vol. **28**, 241 (Ed. U. Rössler), Vieweg, Braunschweig (1988).
7. D.C. Johnston et al., Mater. Res. Bull. **8**, 777 (1973).
8. P.W. Anderson, Science **235**, 1196 (1987).

9. W. Weber, Z. Phys. **B70**, 323 (1988).
10. A.L. Shelankov, X. Zotos, and W. Weber, Physica C **153-155**, 1307 (1988).
11. X. Zotos, A.L. Shelankov, and W. Weber, Physica C **153-155**, 1309 (1988).
12. M. Jarrell, H.R. Krishnamurty, and D.L. Cox, Phys. Rev. **B38**, 4584 (1988).
13. Yu. B. Gaididei and V.M. Loktev, phys. stat. sol. **146**(2), 154 (1988).
14. D.I. Kugel and K.I. Khomskii, Solid State Commun. **13**, 763 (1973).
15. W.A. Little, Phys. Rev. **134A**, 1416 (1964).
16. J.E. Hirsch and D.J. Scalapino, Phys. Rev. **B32**, 117 (1985).
17. H. Razafimandimby, P. Fulde, and J. Keller, Z. Phys. **B54**, 111 (1984).
18. L.F. Mattheiss, Phys. Rev. Lett. **58**, 1028 (1987); J. Yu, A. J. Freeman, and J.H. Xu, Phys. Rev. Lett. **58**, 1035 (1987).
19. H.P. Geserich, G. Scheiber, J. Geerk, H.C. Li, G. Linker, W. Assmus, and W. Weber, Europhys. Lett. **6**, 277 (1988).
20. M. Garriga et al., Physica **C153-155**, 643 (1988).
21. I. Bozovic et al., Phys. Rev. Lett. **59**, 2219 (1987).
22. J. Orenstein et al., Phys. Rev. **B36**, 729 (1987).
23. A.M. Oleś et al., Phys. Rev. **B32**, 2167 (1985).
24. M.C. Gutzwiller, Phys. Rev. **137**, A 1726 (1965).
25. C.M. Varma, W. Weber, and L.J. Randall, Phys. Rev. **B33**, 1015 (1986).
26. W.F. Brinkman and T.M. Rice, Phys. Rev. **B2**, 4302 (1970).
27. N. Nücker, H,. Romberg, X.X. Xi, J. Fink, B. Gegenheimer, Z.X. Zhao, Phys. Rev. B, submitted.

Electronic Structure, Fermi Liquid and Excitonic Superconductivity in the High T_c Cu-Oxides

A.J. Freeman, Jaejun Yu, and S. Massidda

Department of Physics and Astronomy, Northwestern University, Evanston, IL 60208, USA

1 Introduction

It need not be stressed to this audience that the discovery of the high T_c superconductors $La_{2-x}M_xCuO_4$ [1] and $YBa_2Cu_3O_{7-\delta}$ [2] has generated excitement among scientists and technologists on an unprecedented scale. The recent discoveries [3-5] of superconductivity above 85 K in Bi-Sr-Ca-Cu-O and above 120 K in Tl-Ba-Ca-Cu-O, which do not have a rare-earth element, have added a new dimension to the important subject of high T_c. A particularly exciting aspect of having added a third and fourth oxide superconducting material lies in the opportunity for seeking out common features in all four materials which may be relevant to determining the mechanism of their high T_c. One of the starting points is certainly a detailed picture of the electronic structure of the compound, a goal which is achievable by present day supercomputers in combination with highly precise numerical methods to solve the local density functional (LDF) Kohn-Sham equations in a self-consistent way. Even today, the origin of superconductivity in the new metallic oxides remains a challenge despite some intriguing hints obtained from experiment and electronic structure calculations. Detailed high resolution LDF band structure results have served to demonstrate what has been our major emphasis, namely the close relation of the physics (band structure) and chemistry (bonds and valences) to the structural arrangements of the constituent atoms; they may also provide insight into the basic mechanism of their superconductivity. The successes of these LDF studies include excellent agreement of their predictions of their anisotropic Fermi surface [6], and transport and thermopower properties [7-8] with experiment.

For the electronic structure calculations, we used the highly precise full-potential linearized augmented plane wave method (FLAPW) [9-10] within the local density approximation and the Hedin-Lundqvist form for the exchange correlation potential. In the FLAPW approach no shape approximations are made to either the charge density or the potential. Results obtained on the systems we have studied - La_2CuO_4, $YBa_2Cu_3O_7$, $GdBa_2Cu_3O_7$, $Bi_2Sr_2CaCu_2O_8$, $Tl_2Ba_2CaCu_2O_8$ and $Tl_2Ba_2Ca_2Cu_3O_{10}$ - indicate a number of common chemical and physical features, especially the role of intercalated layers such as the CuO chains, and Bi_2O_2 and Tl_2O_2 rock-salt type layers. In this paper, we provide a brief summary of the results on the detailed electronic structures of the $La_{2-x}M_xCuO_4$, $RBa_2Cu_3O_{7-\delta}$ (R = Y and Gd), $Bi_2Sr_2CaCu_2O_8$ and $Tl_2Ba_2CaCu_2O_8$ ($Tl_2Ba_2Ca_2Cu_3O_{10}$) systems, compare them, and point out their relations to an excitonic mechanism of high T_c superconductivity.

2 Electronic Structure of the Normal Metallic States

2.1 $La_{2-x}M_xCuO_4$

Early on, the results of our highly precise all-electron local density full potential linearized augmented plane wave [9-10] (FLAPW) calculations of the energy band structure, charge densities, Fermi surface, etc., for $La_{2-x}M_xCuO_4$ (M = Sr, Ba) [11-14] demonstrated: (i) that the material consisted of metallic Cu-O(1) planes separated by insulating (dielectric) La-O(2) planes and (ii) that this 2D character and alternating metal/insulator planes would have, as some of their most important consequences, strongly anisotropic (transport, magnetic, etc.) properties. Thus, the calculated band structure along high symmetry directions in the Brillouin zone shows only flat bands, i.e., almost no dispersion, along the c axis, demonstrating that the interactions between the Cu, O(2) and La atoms are quite weak. However, along the basal plane directions there are very strong interactions between the Cu-O(1) atoms leading to large dispersions and a very wide bandwidth (~9 eV).

The band structure near E_F has a number of interesting features [11-12]. What is especially striking is that, in contrast to the complexity of its structure, only a single free electron-like band crosses E_F and gives rise to a simple Fermi surface [13]. Since this band originates from the Cu $d_{x^2-y^2}$-O(1) $p_{x,y}$ orbitals confined within the Cu-O(1) layer, it exhibits clearly all the characteristics of a two dimensional electron system. Particularly striking is the occurrence of a van Hove saddle point singularity (SPS). Such an SPS is expected, and found to contribute strongly, via a singular feature, to the density of states (DOS). Interestingly, the variation of T_c with x, in metallic $La_{2-x}Sr_xCuO_4$ shown esperimentally by Torrance et al. [15] is very consistent with the variation of the DOS at E_F on x, N(E_F; x), and can be explained by our calculations [12]. This dominance of the DOS near E_F by the SPS contribution is responsible for many of the striking properties of this material with divalent ion (M_x) additions (including variations of T_c and other properties).

2.2 $YBa_2Cu_3O_{7-\delta}$

For the 90K superconductor $YBa_2Cu_3O_{7-\delta}$, discovered by Chu et al. [2], we presented [16-17] detailed high resolution results on the electronic band structure and density of states derived properties as obtained from the same highly precise state-of-the-art local density approach. These results demonstrated the close relation of the band structure to the structural arrangements of the constituent atoms and have helped to provide an integrated chemical and physical picture of the interactions.

The important structural features of the $YBa_2Cu_3O_{7-\delta}$ compounds arise from the fact that (2+δ) oxygen atoms are missing from the perfect triple perovskite, $YCuO_3(BaCuO_3)_2$. The O vacancies in the Cu plane (between two BaO planes) give rise to the formation of a linear chain of Cu and O ions (labelled Cu1-O1-Cu1). The total absence of O ions in the Y plane leads to the two Cu ions (called Cu2) in five-coordinated positions - as shown in Fig. 1. The double layers of Cu2-O planes in $YBa_2Cu_3O_7$ yield a 2D structure, corresponding to the single CuO_2 plane in $La_{2-x}M_xCuO_4$.

The calculated band structure of stoichiometric $YBa_2Cu_3O_7$ along high symmetry directions in the bottom (k_z = 0) plane of the orthorhombic BZ is

Figure 1

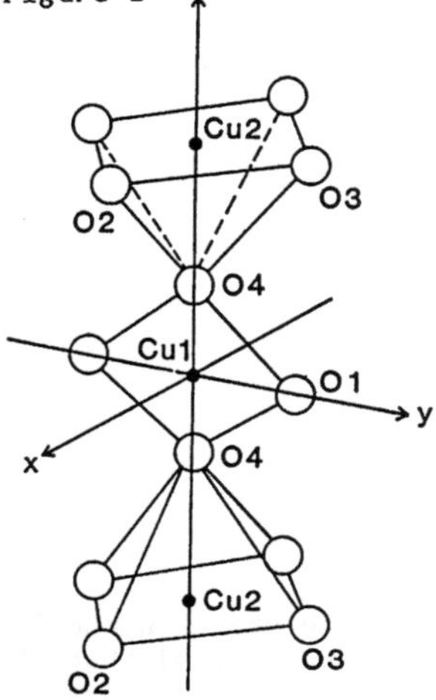

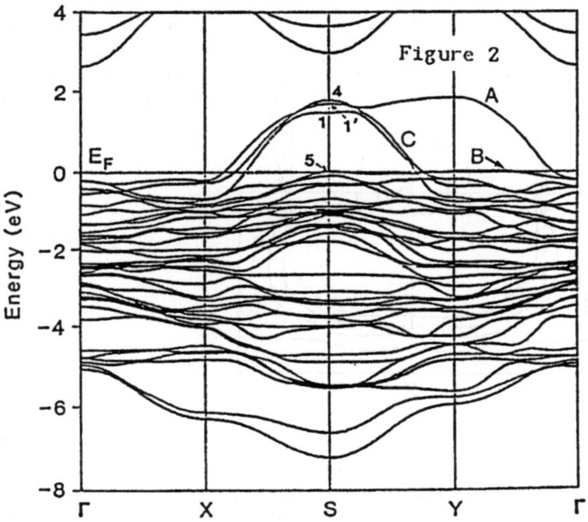

Figure 1 A local environment for the Cu1 and Cu2 atoms in $YBa_2Cu_3O_7$, following the Y-Cu2-Ba-Cu1-Ba-Cu2-Y ordering along z.

Figure 2 Band structure of $YBa_2Cu_3O_7$ along symmetry directions in the k_z = 0 plane of the orthorhombic Brillouin zone.

shown in Fig. 2. The very close similarity in the band structure for the k_z = 0 and $k_z = \pi/c$ planes [16] indicates the highly 2D nature of the band structure. It is seen from Fig. 2, that as in the case of La_2CuO_4, a remarkably simple band structure near E_F emerges from this complex set of 36 bands (originating from three Cu (3d) and seven O (2p) atoms). Four bands - two consisting of Cu2(3d)-O2(p)-O3(p) orbitals and two consisting of Cu1(d)-O1(p)-O4(p) orbitals - cross E_F. Two strongly dispersed bands C ($S_1{}'$ and S_4 in Fig. 2; the labelling is given by their character at S) consist of Cu2($d_{x^2-y^2}$)-O2(p_x)-O3(p_y) combinations and have the 2D character which proved so important for the properties of $La_{2-x}M_xCuO_4$. Significantly, the Cu1($d_{z^2-y^2}$)-O1(p_y)-O4(p_z) anti-bonding band A (S_1 in Fig. 2) shows the (large) 1D dispersion expected from the Cu1-O1-Cu1 linear chains but is almost entirely unoccupied. This band is in sharp contrast to the π-bonding band B (formed from the Cu1 (d_{zy})-O1(p_z)-O4(p_y) orbitals) which is almost entirely occupied in the stoichiometric (δ = 0) compound.

We have predicted the Fermi surfaces (FS) of $YBa_2Cu_3O_7$ determined from our band structure (c.f., Fig. 3). Two 2D Cu-O dpσ bands yield two rounded square FS's (C in Fig. 3) centered around S. These 2D FS show strong nesting features along (100) and (010) directions. In addition, the 1D electronic structure also gives a 1D FS (A in Fig. 3) with possible nesting features along the (010) direction. There are two additional hole pockets (A in Fig. 3) around Y(T) and S(R) which come from the flat dpπ bands at E_F discussed before. Our predictions of the FS for $YBa_2Cu_3O_7$ have been confirmed recently by positron annihilation experiments [6-7]. The dot-dashed lines in Fig. 3 correspond to the experimentally observed FS by Smedskjaer et al. It is important to note that confirmation of the FS results has significant impact on several theories (e.g., the so-called resonant valence band or RVB theory)

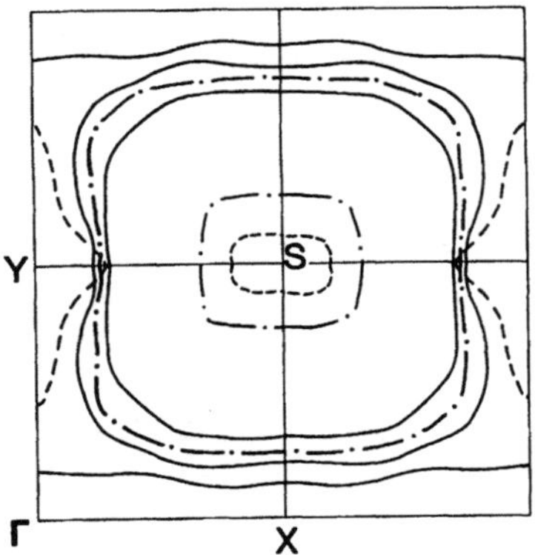

Figure 3 Calculated Fermi Surface for $YBa_2Cu_3O_7$. The dot-dashed lines are experimentally measured Fermi surfaces (Smedskaer et al.)

[18] which deny the Fermi liquid nature of the normal ground state in the Cu-oxide superconductors. Significantly, P.W. Anderson has stated [19] that the proven existence of a Fermi surface would necessitate "withdrawal" of his RVB theory.

Another important confirmation of the Fermi liquid nature of the normal ground state was given recently by Arko et al.'s [20] low temperature photoemission (UPS) experiments. These experiments on single crystals of $EuBa_2Cu_3O_{6.7}$ showed that only when cleaved and measured at 20 K will a stable Fermi edge (larger than that in Cu metal) appear - thus demonstrating metallic behavior in agreement with our band calculations [21]. In addition, these experiments provide evidence for both Cu-d and O-2p occupation at E_F as predicted by our calculations. This result is at variance with other theoretical models which assume either the dominance of Cu-3d or O-2p at E_F. Moreover, the predictions of transport properties (Hall coefficient and thermopower) of $YBa_2Cu_3O_7$ by Allen et al. [7,8] with the use of the LDF energy band results show good agreement with experiment [22]. These successful verifications of the predictions made by the LDF band theory confirms the Fermi liquid nature of the normal metallic ground state. These facts reinforce the use of the LDF ground state as a reasonable starting point for the investigation of the origin of the superconductivity in these materials.

Here too, charge density calculations [16-17] reflect the structural properties of the material. Charge density plots for the individual states near E_F demonstrate the 2D nature of Cu2-O2-O3 dpσ bands and the 1D nature of the Cu1-O1-O4 dpσ bands. The ionic Y (or R = rare earth) atoms act as electron donors and do not otherwise participate. Also, the partial DOS at E_F for Y give extremely low values for the conduction electrons (the same is true for Gd). These results give an immediate explanation for the observed [23] coexistence of the high T_c superconductivity and magnetic ordering in the $RBa_2Cu_3O_{7-\delta}$ structures. The lack of conduction electron density around the R-site [24] means that the unpaired rare-earth f-electrons are decoupled from the Cooper pairs (i.e., magnetic isolation) and so cannot pair-break.

However, there are several experimental observations [25-26] of the antiferromagnetic insulators of $YBa_2Cu_3O_6$ as well as La_2CuO_4. These lead to the question whether the antiferromagnetic insulating ground state can be described by a (charge-) spin-density wave state within a band picture. Such a failure of LDF is well known for the case of CoO [27], for example. Later on, there were several reports of a stable magnetic ground state being found (in a band picture), but most of them are not convincing. Although there are still unresolved problems as to the relation of the O vacancies to the antiferromagnetic ordering in $La_2CuO_{4-\delta}$ and $YBa_2Cu_3O_{6+x}$, there is now a large effort to overcome this short-coming of the LSDA (local spin-density approximation). We believe that this part of the phase diagram has no relation to the superconductivity observed for the metallic phases.

2.3 $Bi_2Sr_2CaCu_2O_8$

For the new high T_c superconductor Bi-Sr-Ca-Cu-O, we presented [28] results of a highly precise local density determination of the electronic structure (energy bands, densities of states, Fermi surface, and charge densities). As in the case of the other high T_c Cu-O superconductors, we found a relatively simple band structure near E_F and strongly anisotropic highly 2D properties. One of the interesting points in the Bi-Sr-Ca-Cu-O system is that the Bi-O planes contribute substantially to $N(E_F)$ and to the transport properties.

A proposed structure of $Bi_2Sr_2CaCu_2O_8$ by Sunshine et al. [29] shows the presence of two CuO_2 layers (separated by a Ca layer) and of rock-salt type Bi_2O_2 layers; the $(CuO_2)\cdot Ca\cdot (CuO_2)$ layers are separated by single SrO layer from the Bi_2O_2 layers. It is striking that this new system has no rare-earth elements; instead, it has Bi atoms replacing those strongly electro-positive trivalent ions.

The calculated band structure of $Bi_2Sr_2CaCu_2O_8$ has many points in common with those of the other high T_c Cu-oxide compounds [11, 16-17]: above a set of fully occupied bands (in this case 48) with predominant Cu d-O p character, we find a relatively simple band structure at E_F, which in this compound consists of only three bands crossing E_F. Two (almost degenerate) bands with strong Cu-O dpσ character cross E_F and have two dimensional character. They do not cross E_F at the midpoint of the Γ-Z direction because of the existence of the Bi-O band which also crosses E_F. Their quasi-degeneracy proves the weakness of the interplane interactions for these states.

At energies (mostly) above E_F we find a set of six bands corresponding to the antibonding hybrids of the p orbitals of the two Bi atoms in the unit cell with the O2 and O3 p states. These bands form electron pockets near the L point and at the midpoint between Γ and Z (which will be referred to as $\bar{M}$). Their dispersion across the BZ is quite different from that of the Cu-O dpσ bands, as a consequence of the different bonding character (ppσ versus dpσ) and local coordination (rock-salt versus perovskite-like). The doubly periodic dispersion of Bi-O ppσ bands can be understood on the basis of simple tight-binding arguments.

The total density of states at the Fermi level, $N(E_F)$, is 3.03 states/(eV-cell). Large contributions to $N(E_F)$ come from both the Cu-O1 and the Bi-O2 layers. The Bi-O2 contributions are from the pσ bands which create small electron pockets around L (and $\bar{M}$). Therefore (and significantly), both Cu-O1 and Bi-O2 layers provide conduction electrons in this material. This result contrasts with the case of $La_{2-x}M_xCuO_4$ and of $YBa_2Cu_3O_{7-\delta}$, where the cations do not contribute to $N(E_F)$ but give rise to conduction bands which lie 2-3 eV above E_F.

2.4 $Tl_2Ba_2Ca_{n-1}Cu_nO_x$

New high T_c superconductors of the Tl-Ba-Ca-Cu-O system have been discovered [5] and found to have two different but related superconducting phases [30-31], with compositions $Tl_2Ba_2CaCu_2O_8$ (which we refer to as "Tl/2212") and $Tl_2Ba_2Ca_2Cu_3O_{10}$ ("Tl/2223"), with T_c ~112 K and ~125 K, respectively. For both Tl/2212 and Tl/2223, we presented [32] results of highly precise local density calculations of the electronic structure. A relatively simple band structure is found near E_F and strong 2D properties are predicted - again as in the case of the other high T_c materials.

The crystal structures of Tl/2212 and Tl/2223 determined by Subramanian et al. [31] show essentially the same features as that of $Bi_2Sr_2CaCu_2O_8$. The structure of Tl/2212 consists of two CuO_2 layers (separated by a Ca layer) and of rock-salt type Tl_2O_2 layers, where the (CuO_2)-Ca-(CuO_2) layers are separated by single BaO layers from the Tl_2O_2 layers. Similarly, the Tl/2223 structure is related to the Tl/2212 structure by an addition of extra Ca and CuO_2 layers, where the (CuO_2)-Ca-(CuO_2)-Ca-(CuO_2) layers are separated by single BaO layers from the Tl_2O_2 layers.

The calculated energy bands of Tl/2212 (in an extended zone scheme) are shown in Fig. 4. These bands present, as one would expect, strong similarities with those of all the other high T_c Cu-oxide superconductors [11, 13-14, 16-17, 27]. As in $Bi_2Sr_2CaCu_2O_8$, we have in Tl/2212 two Cu-O dpσ bands (one per Cu-O sheet) crossing E_F, while three Cu-O dpσ bands crossing E_F are present in the Tl/2223 compound.

Despite these common features, the Tl systems present some interesting new points. In both the Tl/2212 and Tl/2223 compounds, there exists the presence of electron pockets around the Γ and Z points. A careful analysis of the

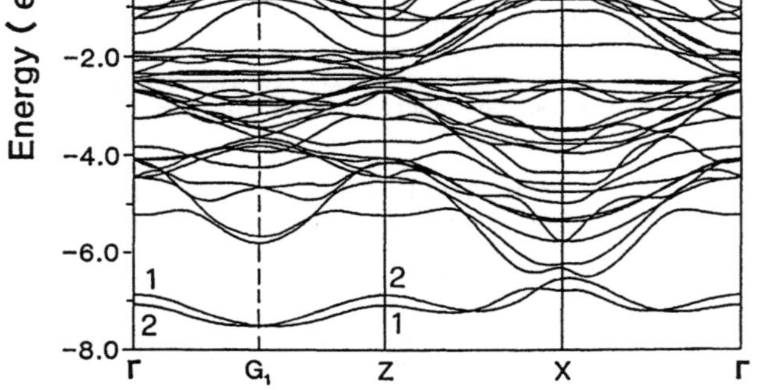

Figure 4 Energy bands of $Tl_2Ba_2CaCu_2O_8$ along the main symmetry lines of the body-centered tetragonal extended Brillouin zone. (Notation from Reference 11.)

character of these states, however, reveals important differences with respect to the $Bi_2Sr_2CaCu_2O_8$ case. While the Bi-O bands at E_F in $Bi_2Sr_2CaCu_2O_8$ originate mainly from the in-plane ppσ Bi-O hybrid, the Tl-O bands at E_F in Tl/2212 and Tl/2223 are mostly from oxygen p states hybridized (anti-bonding) with the Tl orbitals. In fact, the major Tl 6s bands are located at about 7 eV below E_F.

We have found that the 6s electrons of the Tl ions in $Tl_2Ba_2CaCu_2O_8$ are covalently bonded to the out-of-plane oxygens, O2 and O3; similarly, the Bi p orbitals in $Bi_2Sr_2CaCu_2O_8$ form weak covalent bonds with the in-plane O2 oxygens. This result is in contrast to the case of the $La_{2-x}M_xCuO_4$ and $YBa_2Cu_3O_7$ systems where the presence of strongly electro-positive 3+ ions (e.g., La^{3+}, Y^{3+}) is essential. We shall see that this has a significant effect on the electronic structure and may be relevent to understanding the superconducting mechanism.

One of the significant effects of the strong hybridization of the Tl s (d_{z^2}) with O2 and O3 p_z states (discussed above) on the pDOS structure of Tl/2212 and Tl/2223 is the existence of a gap between the non-bonding $p_{x,y}$ bands of O2 and O3 and the anti-bonding Tl s(d_{z^2}) - O2 p_z bands. (This gap, ~2.1 eV wide in Tl/2212, is reduced to $\lesssim$1 eV in Tl/2223 as a consequence of a Madelung shift of the non-bonding O2 states.) These systems are therefore seen to realize alternating metal/semiconductor superstructures, with the metal Fermi level slightly above the conduction band bottom of the semiconductor, a situation reminiscent of the Allender, Bray, and Bardeen [33] model for excitonic superconductivity (for a critical evaluation of a possible shortcoming of this model, see Ref. 34) which we will discuss later.

The calculated $N(E_F)$ for Tl/2212 and Tl/2223 are 2.82 states/eV-cell and 3.80 states/eV-cell, respectively. Thus, the additional CuO_2 sheet increases $N(E_F)$ by 1 state/eV-cell while the other components of $N(E_F)$ change by only 10-20%. Consistent with this is the fact that when we subtract the contribution from the Tl-O3-O2 bands, the $N(E_F)$ per Cu-atom is reduced to ~1.0 states/(eV-Cu atom), which is about the same as in [28] $Bi_2Sr_2CaCu_2O_8$.

The Fermi surfaces (FS) of $Tl_2Ba_2CaCu_2O_8$ (Tl/2212) are shown in Fig. 5 in an extended zone scheme. The electron pockets c and d centered around Γ and Z, respectively, are due to the Tl-O2-O3 bands. The Cu-O dpσ bands produce the two FS indicated by a and b in Fig. 5 (there is a third such surface for Tl/2223 lying between the two shown). These surfaces have a rounded-square shape centered around X. Fermi surface a especially shows striking nesting features along the (100) and (010) directions, with spanning vectors which are not commensurate. This high degree of FS nesting is expected to give rise to singularities in the generalized susceptibility, $\chi(\vec{q})$, of this highly 2D system, and may therefore have important consequences as possible electronically-driven instabilities (e.g., incommensurate charge density waves).

The simple FS of the 2D Cu-O bands in Tl/2212 shown in Fig. 5 as a and b should have a simple origin when looked at from the usual tight binding point of view. In a 2D square lattice, the simple tight-binding band is described by:

$$\varepsilon(\vec{k}) = \varepsilon_0 - 2t_1 (\cos k_x a + \cos k_y a) + 4t_2 \cos k_x a \cdot \cos k_y a \quad (1)$$

where t_1 represents the nearest neighbor (n.n.) interaction and t_2 the next-nearest-neighbor (n.n.n.) interaction. From a comparison of the tight-binding bands and the dpσ anti-bonding bands of Tl/2212 and Tl/2223, we showed that the Cu-O1 dpσ anti-bonding bands crossing E_F cannot be properly fitted with a n.n. only tight-binding model while they can be reasonably well described by including the n.n.n. (most likely to be O-O) interactions. We therefore expect that the correct Fermi surface can only be obtained from the fuller tight-binding treatment and not from a simple n.n. tight binding interaction. In fact, the inclusion of the n.n.n. term in Eq. 1 yields a FS which is substantially different from the FS of a simple tight-binding band with only n.n. interactions. As shown in Fig. 6, the square centered at X with perfect nesting along the (110) direction (for n.n. only) has been transformed dramatically (by adding n.n.n.) into a rounded square with strong nesting features along the (100) and (010) directions which closely resembles the actual FS of Tl/2212 (and Tl/2223).

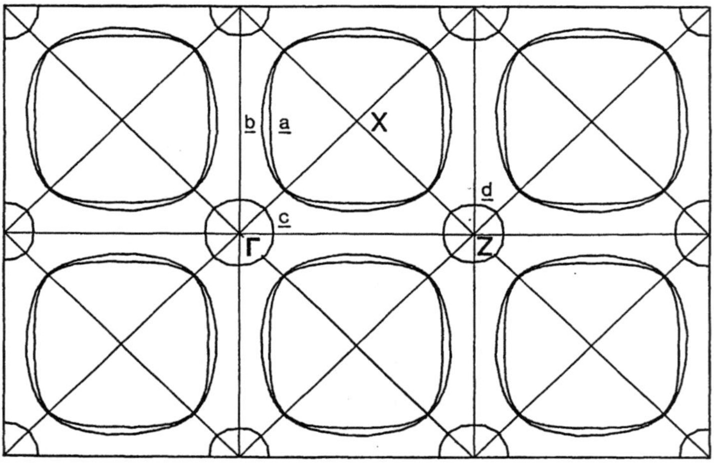

Figure 5 Fermi surfaces of $Tl_2Ba_2CaCu_2O_8$ in an extended zone scheme.

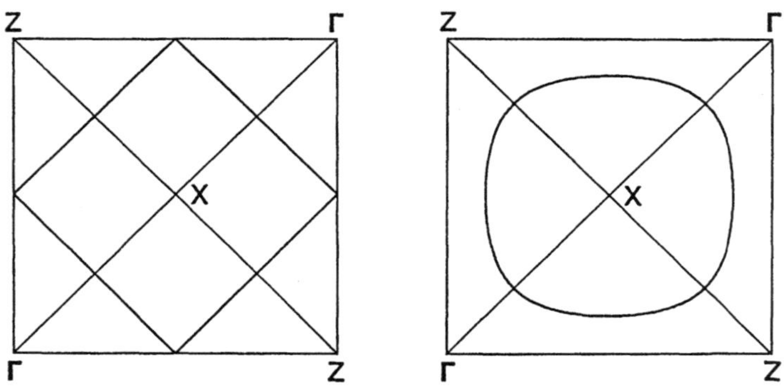

(a) (b)

Figure 6 Fermi surfaces of the tight-binding bands for (a) $t_2/t_1 = 0.0$ (with the n.n. interactions only) and (b) $t_2/t_1 = 0.45$ (with n.n.n. interactions included). (See text for details.)

Finally, it is important to note that the same result is also true for the $YBa_2Cu_3O_7$ system, [17] where the FS of the 2D dpσ bands at E_F (see Fig. 3) are rounded squares centered at S with nesting along (100) and (010) directions. This result implies that the commonly used tight-binding model Hamiltonian with only n.n. interactions is not sufficient to describe the anti-bonding bands crossing E_F in Tl/2212 as well as $YBa_2Cu_3O_7$ in that it yields incorrect results. This has important consequences for all such model Hamiltonian descriptions used for explaining the high T_c. Thus, for example, the inclusion of the n.n.n. interactions leads to the reduction of the effective on-site Coulomb repulsion due to the delocalization of the Wannier states.

Again, it is important to emphasize that the good agreement between LDF band theory predictions and experiment establishes strong evidence for the Fermi liquid nature of the metallic ground state of the high T_c Cu-oxide superconductors, and justifies the use of LDF band theory as an excellent starting point for describing the transition to the superconducting state of the high T_c Cu-oxide systems.

3. Excitonic Mechanism of High T_c Superconductivity

We have made crude estimates of the electron-phonon interaction in the Cu-oxide superconductors, $La_{2-x}M_xCuO_4$, $YBa_2Cu_3O_7$, $Bi_2Sr_2CaCu_2O_8$ and $Tl_2Ba_2CaCu_2O_8$, using the rigid muffin-tin approximation (RMTA) [35] to calculate the McMillan-Hopfield constant η and the electron-phonon coupling constant, λ. For all the Cu-oxide superconductors, the largest contributions to η come from the Cu and O ions in the CuO_2 planes, indicating the important role played by the "metallized" oxygens. As a crude approximation - and assuming the most favorable conditions, e.g., strong phonon softening $\theta_D \lesssim$ 100 K - we estimated the T_c of these systems by using the strong coupling formula of Allen and Dynes [36]. The highest calculated T_c is found to be 36 K. Even though the T_c for the $La_{2-x}M_xCuO_4$ is close to the values found in the RMTA calculations, it is unlikely that a purely electron-phonon interaction is responsible for its high T_c because these are most favorable (unrealistic) estimates and the corresponding λ values are much larger than the experimental values [12]. For the other systems ($YBa_2Cu_3O_7$, $Bi_2Sr_2CaCu_2O_8$, and $Tl_2Ba_2CaCu_2O_8$), the estimates of T_c are so far off (more than a factor of three) that despite the crudeness of the RMTA approach, they cast doubt on a purely electron-phonon explanation of the observed high T_c. These results suggest the possibility and importance of a non-phonon mechanism of high T_c superconductivity.

Many authors have discussed the excitonic mechanism [37-38] of superconductivity, in which the effective attractive interaction between conduction electrons originates from virtual excitations of excitons rather than phonons. The basic idea of the models proposed is that conduction electrons residing on the conducting filament (or plane) induce electronic transitions on nearby easily polarizable molecules (or complexes), which result in an effective attractive interaction between conduction electrons. As perhaps a striking realization of the excitonic mechanism of superconductivity, $YBa_2Cu_3O_{7-\delta}$ has two 2D conduction bands and additional highly polarizable 1D electronic structure between the two conduction planes.

We have previously discussed [17] the importance of the 1D feature in the electronic structure near E_F, pointing out the possible role played by charge transfer excitations ("excitons") of occupied (localized) Cu1-O dpπ orbitals into their empty (itinerant) Cu1-O dpσ anti-bonding partners. As shown schematically in Fig. 7, we can characterize the 1D electronic structure with

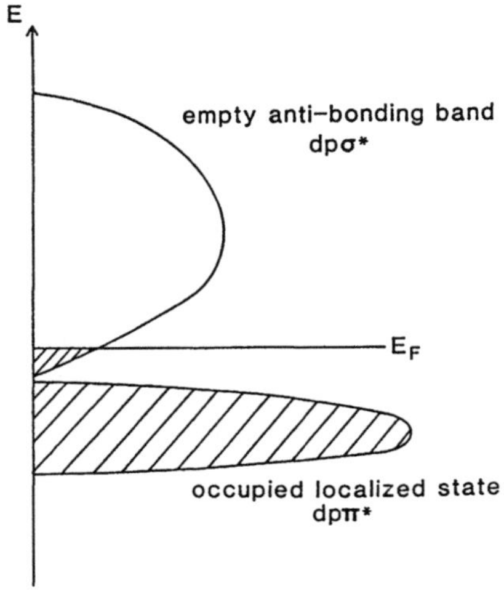

Figure 7 Schematic drawing of the 1D electronic structure in $YBa_2Cu_3O_7$.

two types of electronic states in it, one free-electron-like (the well-dispersed dpσ band) and the other localized (the almost flat dpπ state). When the localized hole is created by the excitation, a strong attractive correlation between the hole and excited electron may lead to an electron-hole bound state ("exciton"). Hence, this excitation of the localized dpπ to the extended dpσ with the electron-hole correlation in the 1D electronic structure will give rise to a strong polarization in the 1D chains between two conduction planes and couple to the 2D conduction electrons, which carry most of the superconductivity.

In comparing the electronic states of the four oxide superconductors, a number of common features emerges which supports the excitonic model of superconductivity: In all materials, the 2D Cu-O dpσ bands dominate the electronic structure near E_F. These bands consist of anti-bonding combinations of Cu $d_{x^2-y^2}$ and in-plane O $p_{x,y}$ orbitals of the CuO_2 planes, which give rise to the strong two-dimensionality of the bands. The remarkable 2D nature of the electronic structure of La_2CuO_4 leads to a simple picture of the conductivity confined essentially to the metallic CuO_2 planes separated by ionic (insulating) planes of the rock-salt type La_2O_2 layers. We note that the slab (LaO)-(CuO_2)-(LaO) has the correct stoichiometry and is charge neutral, where the ionic $La^{3+}O^{2+}$ layers provide residual charge to the CuO_2 layers. Indeed, the (LaO)- (CuO_2)-(LaO) slab becomes a basic building block (with moderate modifications) for the other high T_c Cu-oxide superconductors.

We have seen that in the 90 K superconductor $YBa_2Cu_3O_{7-\delta}$, the building block was modified by introducing the oxygen deficient Y layer between the CuO_2 layers. The new building block for $YBa_2Cu_3O_{7-\delta}$ thus becomes (BaO)-(CuO_2)-(Y)-(CuO_2)-(BaO), where the middle three layers (CuO_2)-(Y)-(CuO_2) correspond to the single CuO_2 layers for La_2CuO_4. Similarly, in $Bi_2Sr_2CaCu_2O_8$, a common building block would be (SrO)-(CuO_2)-(Ca)-(CuO_2)-(SrO) and in $Tl_2Ba_2CaCu_2O_8$, the corresponding one becomes (BaO)-(CuO_2)-(Ca)-(CuO_2)-(BaO). Finally, the one in $Tl_2Ba_2Ca_2Cu_3O_{10}$ is a mere extension of the one in $Tl_2Ba_2CaCu_2O_8$, i.e., (BaO)-(CuO_2)-(Ca)-(CuO_2)-(Ca)-(CuO_2)-(BaO).

For all of these compounds, the La, Ba, Y, Sr, and Ca atoms are purely ionic and supply extra charges to each CuO_2 layer. In contrast to the strong ionic contribution of the La, Ba, Y, Sr, and Ca atoms, the 2D CuO_2 planes become metallic and give rise to the well dispersed Cu-O dpσ bands at E_F, which are essentially confined within each CuO_2 layer. These 2D Cu-O dpσ bands are essential for all the high T_c Cu-oxide superconductors. In addition, the common structural feature of the layered Cu-oxides superconductor suggests the (intercalated) layer structure as another essential element in the high T_c Cu-oxides. Once we regard the CuO_2 planes as major conduction layers and the Cu-O chains, Bi_2O_2, and Tl_2O_2 as intercalated semi-metallic or insulating layers, then the role of CuO chains, Bi_2O_2, and Tl_2O_2 layers must be to enhance superconductivity.

We have discussed some details of the additional electronic structure induced by these intercalated layers. What all of these electronic structures due to the intercalated layers have in common, is almost empty bands having strong covalent (anti-bonding) character. Furthermore, we also find the existence of occupied localized flat bands or non-bonding bands connected to the anti-bonding bands above E_F. This local electronic structure arising from the intercalated layers can be viewed simply as shown diagramatically in Fig. 7.

As discussed above, in $YBa_2Cu_3O_7$, we have proposed charge transfer excitations of occupied (localized) dpπ states, to empty dpσ bands as a representation of the interband interactions. In $Tl_2Ba_2CaCu_2O_8$ (and similarly in $Bi_2Sr_2CaCu_2O_8$) it becomes clear that the interband interactions between the non-bonding O p states and the almost empty Tl-O sp(dpσ) bands will lead to virtual excitations which couple to the conduction electrons in the CuO_2 planes and may play an important role in their high T_c superconductivity. Indeed, we can consider the role of CuO chains, Bi_2O_2, and Tl_2O_2 layers as providing the low lying charge excitations which couple the conduction electrons in the 2D CuO_2 planes.

We are in the process of quantifying this picture of charge transfer excitations. Such an approach requires detailed calculations of the full dielectric tensor $\varepsilon(\vec{Q}, \vec{Q}')$, including the (important) Umklapp processes, using our band structure results as the starting point.

Acknowledgments

Work supported by NSF (through the Northwestern University Materials Research Center, Grant No. DMR85-20280) and the Office of Naval Research (Grant No. N00014-81-K-0438). We are grateful to NASA Ames and Kirkland Air Force Base personnel for help with the use of their Cray 2. We thank C.L. Fu, D.D. Koelling, T.J. Watson-Yang and J.H. Xu for collaboration on the early aspects of this work.

References

1. Bednorz, J. G.; Müller, K. A. Z. Phys. (1986) B64, 189.
2. Wu, M. K.; Ashburn, J. R.; Torng, C. J.; Hor, P. H.; Meng, R. L.; Gao, L.; Huang, Z. J.; Wang, Y. Q.; Chu, C. W. Phys. Rev. Lett. (1987) 58, 908.
3. Maeda, H.; Tanaka, Y.; Fukutomi, M.; Asano, T. Jpn. J. Appl. Phys. (1988) 27, in press.
4. Chu, C. W.; et al., Phys. Rev. Lett. (1988) 60, 941.
5. Sheng, Z. Z.; Hermann, A. M.; El Ali, A.; Almason, C.; Estrada, J.; Datta, T.; Matson, R. J. Phys. Rev. Lett. (1988) 60, 937.

6. Manuel, A. A.; Peter, M.; Walker, E. Europhys. Lett. (1987) 6, 61; Smedskaer, L.; et al., Physica C156, 269 (1988).
7. Allen, P. B.; Pickett, W. E.; Krakauer, H. Phys. Rev. (1987) B36, 3926.
8. Allen, P. B.; Pickett, W. E.; Krakaure, H. Phys. Rev. (1987) B37, 7482.
9. Jansen, H. J. F.; Freeman, A. J. Phys. Rev. (1984) B30, 561.
10. Wimmer, E.; et al., Phys. Rev. (1981) B24, 864.
11. Yu, J.; Freeman, A. J.; Xu, J. -H. Phys. Rev. Lett. (1987) 58, 1035.
12. Freeman, A. J.; Yu, J.; Fu, C. L. Phys. Rev. (1987) B36, 7111.
13. Xu, J. -H.; Watson-Yang, T. J.; Yu, J.; Freeman, A. J. Physics Lett. (1987) A120, 489.
14. Fu, C. L.; Freeman, A. J. Phys. Rev. (1987) B35, 8861.
15. Torrance, J. B.; Tokura, Y.; Nazzal, A. I.; Bezinge, A.; Huang, T. C.; Parkin, S. S. P. Phys. Rev. Lett. (1988) 61, 1127.
16. Massidda, S.; Yu, J.; Freeman, A. J.; Koelling, D. D. Physics Lett. (1987) 122, 198.
17. Yu, J.; Massidda, S.; Freeman, A. J.; Koelling, D. D. Physics Lett. (1987) 122, 203.
18. Anderson, P. W. Science (1987) 235, 1196.
19. Anderson, P W. Bull. Am. Phys. Soc. (1988) 33, 459.
20. Arko, A. J.; List, R. S.; Fisk, Z.; Cheong, S. -W.; Thompson, J. D.; O'Rourke, J. A. JMMM (1988) 75, L1.
21. Redinger, J.; Freeman, A. J.; Yu, J.; Massidda, S. Phys. Lett. (1987) 124A, 469.
22. Tozer, S. W.; Kleinsasser, A. W.; Penney, T.; Kaiser, D.; Holtzberg, F. Phys. Rev. Lett. (1987) 59, 1768.
23. Willis, J. O.; Fisk, Z.; Thompson, J. D.; Cheong, S. -W.; Aikin, R. M.; Smith, J. L.; Zirngiebl, E.. J. Magn. Matls., (1987) 67, L139.
24. Yu, J. and Freeman, A. J. (to be published).
25. Vaknin, D.; et al. Phys. Rev. Lett. (1987) 58, 2802.
26. Brewer, J. A.; et al. Phys. Rev. Lett. (1988) 60, 1073.
27. Terakura, K.; et al. Phys. Rev. (1984) B30, 4734.
28. Massidda, S.; Yu, J.; Freeman, A. J. Physica C:Superconductivity (1988) 152, 251.
29. Sunshine, S. A.; Siegrist, T.; Schneemeyer, L. F.; Murphy, D. W.; Cava, R. J.; Batlogg, B.; van Dover, R. B.; Fleming, R. M.; Glarum, S. H.; Nakahara, S.; Farrow, R.; Krajewski, J. J.; Zahurak, S. M.; Waszczah, J. V.; Marshall, J. H.; Marsh, P.; Rupp, Jr., L. W.; and Peck, W. F. Phys. Rev. B38, 893 (1988).
30. Hazen, R. M.; Finger, L. W.; Angel, R. J.; Prewitt, C. T.; Ross, N. L.; Hadidiacos, C. G.; Heaney, P. J.; Veblen, D. R.; Shen, Z. Z.; El Ali, A.; Hermann, A. M. Phys. Rev. Lett. (1988) 60, 1657.
31. Subramanian, M. A.; Torardi, C. C.; Calabrese, J. C.; Gopalakrishnan, J.; Morrissey, K. J.; Askew, T. R.; Flippen, R. B.; Chowdhry, U.; Sleight, A. W. Science (1988) 239, 1015.
32. Yu, J.; Massidda, S.; Freeman, A. J. Physica C:Superconductivity (1988) 152, 273.
33. Allender, D. W.; Bray, J. M.; Bardeen, J. Phys. Rev. (1973) B7, 1020.
34. Cohen, M. L.; Louie, S. G. Superconductivity in d- and f-band Metals; Douglass, D. H., Ed.; Plenum; New York, 1976.
35. Gaspari, G. D.; Gyoffry, B. L. Phys. Rev. Lett. (1972) 28, 801.
36. Allen, P. B.; Dynes, R. C. Phys. Rev. (1975) B12, 905.
37. Little, W. A. Phys. Rev. (1964) 134, A1416.
38. Ginzburg, V. L. JETP (1964) 46, 397.

Local-Spin-Density-Functional Approach to High-T_c Copper Oxides

Atsushi Oshiyama[1], *N. Shima*[2], *T. Nakayama*[3], *K. Shiraishi*[2], *and Hiroshi Kamimura*[2]

[1]NEC Fundamental Research Laboratories, Miyamae, Kawasaki 213, Japan
[2]Department of Physics, University of Tokyo, Bunkyo, Tokyo 113, Japan
[3]Department of Physics, Chiba University, Chiba 280, Japan

A state-of-the-art total-energy band-structure calculation within the local-spin-density-functional formalism has been performed for the high-T_c copper oxides, $(La_{1-x}Sr_x)_2CuO_4$. It is shown that the local-spin-density approximation does not produce a stable antiferromagnetic insulating state observed in La_2CuO_4, but that it still provides useful information on the electronic and atomic structures of the materials.

1. INTRODUCTION

Electronic structure calculation has been providing us with valuable information about physical properties of a variety of materials. In particular, a first-principles total-energy band-structure calculation within the local-density-functional formalism[1] enables us to obtain reliable results[2] not only for the electronic but also for the structural properties: e.g. energy bands, lattice constant, cohesive energy, bulk modulus and phonon structure.

The high-T_c copper oxides[3], since their discovery, have also been big targets from the state-of-the-art electronic structure calculation. MATTHEISS[4] and YU et al[5] have performed band-structure calculations for La_2CuO_4 at a very early stage of the high-T_c research, and have elucidated the roles of copper-oxygen planes in the electronic structure of the materials. Since then, a considerable amount of the state-of-the-art calculations has been done for the band structures[6], phonon structures[7] and electron-phonon coupling[8] of the high-T_c copper oxides.

Yet, the energy bands from the spin-unpolarized calculations[4-6] somewhat disagree with the photoemission data[9], and are inconsistent with the hole character[10] of the carriers in the materials. Importance of exchange-correlation interaction between electrons in the magnetic and superconducting behaviors of the materials, on the other hand, is suggested from some theoretical works[11] and from the neutron diffraction measurements[12].

It is thus of importance to perform the total-energy band-structure calculation for the high T_c materials within the local-spin-density functional formalism which is capable of taking the exchange-correlation effect into account partially in its parameter-free theoretical framework. The aim of this work is to obtain quantitative information about the electronic structure of the high-T_c materials from the parameter-free calculation on the one hand, and to examine validity and limitation of the local-spin-density approximation (LSDA) on the other hand. We here report the state-of-the-art pseudopotential total-energy band-structure calculations for $(La_{1-x}Sr_x)_2CuO_4$ within the LSDA.

Springer Series in Materials Science, Vol. 11
Mechanisms of High Temperature Superconductivity
Editors: H. Kamimura and A. Oshiyama

2. CALCULATION

The calculation has been done by the norm-conserving pseudopotential method within the LSDA using the gaussian-orbitals basis set. The effect of including core states into valence states on the spin-polarization is examined by preparing two sets of the pseudopotentials constructed from the first principles.

2.1 Local Spin Density Approximation (LSDA)

Total valence energy E of interacting electron system under crystal potential V_c from ion cores is expressed as a functional of valence electron spin density $n_+(\mathbf{r})$ and $n_-(\mathbf{r})$ ($n = n_+ + n_-$)[1],

$$E = T + \int V_c n(r)\,dr + \int\int \frac{n(\mathbf{r})\,n(\mathbf{r}')}{|\mathbf{r}-\mathbf{r}'|}\,d\mathbf{r}\,d\mathbf{r}' + E_{XC}[n_+(\mathbf{r}), n_-(\mathbf{r})] \qquad (1)$$

in atomic units, where T and E_{XC} are the kinetic energy and exchange-correlation energy, respectively. Minimization of (1) with respect to the spin density leads to the Euler equation (Kohn-Sham equation):

$$\left\{-\frac{\nabla^2}{2} + V_c + \int \frac{n(\mathbf{r}')}{|\mathbf{r}-\mathbf{r}'|}\,d\mathbf{r}' + \mu_{XC}^{\sigma}[n_+, n_-]\right\}\Psi_{i\sigma} = \varepsilon_{i\sigma}\Psi_{i\sigma} \qquad (2)$$

and

$$n_\sigma(\mathbf{r}) = \sum_i |\Psi_{i\sigma}(\mathbf{r})|^2 \qquad (3)$$

where $\mu_{XC}{}^{\sigma}$ is the exchange-correlation potential obtained from the functional derivative of E_{XC} with n_σ. Equations (2) and (3) constitute the coupled equations which are solved selfconsistently by using the localized-orbitals basis-set used in this paper.

As for the exchange-correlation energy, we adopt here the LSDA,

$$E_{XC} = \int n(\mathbf{r})\,\varepsilon_{XC}(n_+(\mathbf{r}), n_-(\mathbf{r}))\ . \qquad (4)$$

Here ε_{XC} is the exchange-correlation energy of the electron gas with uniform electron spin densities, n_+ and n_-. We use the parametrized form[13] fitted to the results by Ceperley and Alder[14].

2.2 Normconserving Pseudopotentials

The crystal potential V_c is expressed as a sum of the non-local norm-conserving atomic (ionic) pseudopotentials placed at each atomic site. The atomic (ionic) pseudopotential is constructed from first-principles in the following way [15]. First, all-electron atomic (ionic) Dirac equation is solved, and the eigenvalues and the eigenfunctions of the valence states are obtained. Next, a pseudopotential for each valence state is constructed numerically so as to reproduce the eigenvalue and the eigenfunction. Finally, the pseudopotential obtained numerically for each valence state is fitted to the analytic form for the application to band-structure calculation.

We are allowed to make a choice in constructing the pseudopotentials: classification of each electron state to valence or core state. The core states are assumed to be frozen in a solid, and are not treated selfconsistently in the band-structure calculation. There is no ambiguity in the classification of the electron states of oxygen, lanthanum and strontium atoms. The (2s, 2p, 3d) of O, (5d, 6s, 6p) of La and (5s, 5p, 4d) of Sr are treated as the valence states. For copper, however, careful classification is required. Since electron configuration of a Cu atom is $(3d)^{10}(4s)^1$, a natural choice of the valence state for Cu is (3d, 4s, 4p). We construct the Cu pseudopotential according to this choice, and the resulting set of pseudopotentials are denoted as the set-1 pseudopotentials hereafter. In that case, the 3s and 3p state of Cu are regarded as the core states. The eigenfunctions of the 3s and 3p, however, have considerable amplitude outside the peak of the Cu 3d orbital. Therefore, the effect of inclusion of the 3s and 3p into the valence states on the final band-structure should be examined. We have thus constructed the second set of the pseudopotentials (the set-2 pseudopotentials hereafter) in which Cu (3s, 3p, 3d) is treated as the valence state.

2.3 Gaussian Orbitals Basis Set

In order to solve the coupled equations, (2) and (3), we introduce a gaussian-orbitals basis set[2]. The exponents of each Gaussian orbital are determined by fitting to the pseudo-valence orbital numerically obtained in constructing the pseudopotential. It is found that 2 or 3 exponents are enough to reproduce each valence orbital.

In general, the plane-wave basis set is superior to the localized-orbitals basis set in the sense that a systematic increase of the number of plane-waves enables us to obtain exact solution of (2) and (3). The gaussian-orbitals basis set, however, has been found to produce accurate values, comparable with the plane-wave-basis-set results, for the ground state quantities, as long as the exponents are determined carefully[2].

3. RESULTS AND DISCUSSION

Now we report calculated results for the total valence energy, the energy-band structure and distribution of electron spin density. The results[16] from the set-1 pseudopotentials are presented first, and then the set-2 results[17,18] follow.

3.1 Spin-polarized Solution with the Set-1 Pseudopotentials

The first question we can address is which states, a paramagnetic (PM) or an antiferromagnetic (AFM) states, the LSDA produce as a normal ground state for $(La_{1-x}Sr_x)_2CuO_4$. Figure 1 shows a tetragonal unit cell of $(La_{1-x}Sr_x)_2CuO_4$ which is double in size of the usual unit cell. In the PM state, two kinds of Cu, Cu(a) and Cu(b), are equivalent to each other. But in the AFM state investigated here, the spins around Cu(a) are opposite to the spins around Cu(b) (two-sublattices configuration).

To assure the same accuracy in the calculated total energy, we use the unit cell shown in Fig. 1 for both the PM and the AFM configurations. The calculated total valence energy using the set-1 pseudopotentials is shown in Table 1. It is found that for La_2CuO_4 the total energy for the AFM state is lower than that for the PM state by 0.1 eV per unit cell. The energy gain comes from the exchange-correlation energy. The energy bands for the AFM state of La_2CuO_4 are shown in Fig. 2. The metallic Cu $d(x^2-y^2)$ - O p(x,y) band in the case of the PM state [4,5] splits into two bands with an energy gap. That is the AFM insulator. This splitting is a consequence from doubling of the total spin-dependent crystal potential due to induced spin density. The AFM state found in this calculation is triggered by the Fermi-surface nesting of the original PM band structure.

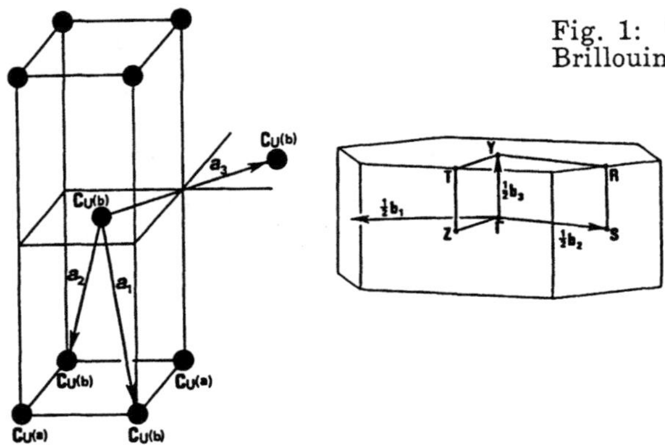

Fig. 1: Tetragonal unit cell and Brillouin zone of $(La_{1-x}Sr_x)_2CuO_4$

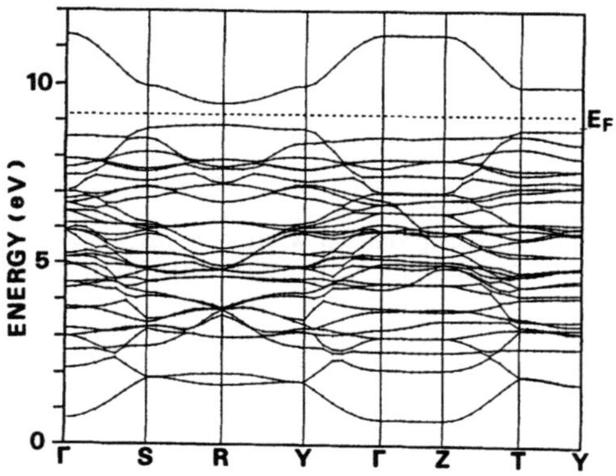

Fig. 2: Calculated energy-band structure for the AFM state of La_2CuO_4 using the set-1 pseudopotentials

Table 1: Calculated total valence energy (Ry) per unit cell for the PM and the AFM $(La_{1-x}Sr_x)_2CuO_4$ using the set-1 and set-2 pseudopotentials. Virtual crystal approximation has been used for the case of $x \neq 0.0$.

	set-1			set-2
x	0.0	0.1	0.5	0.0
E_{PM}	−468.16	−467.43	−463.86	−1049.52
E_{AFM}	−468.17	−467.43	−463.87	−1049.52

Figure 3 shows the calculated spin density on the Cu-O plane of La_2CuO_4. We notice that the spin is well localized at Cu site, although the hybridization between Cu and O orbitals results in the small induced spin at O sites. The spins are primarily carried by Cu $d(x^2-y^2)$ orbitals and its integrated value around Cu is 0.43 μ_B which is comparable with the experimental value[19].

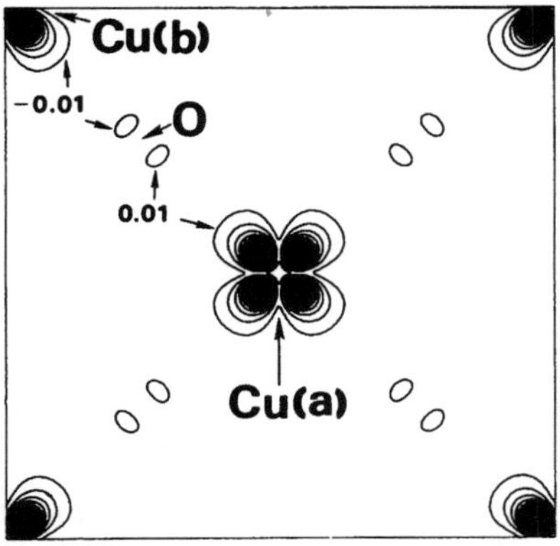

Fig. 3: Contour plot of the spin density of the AFM La_2CuO_4 on the Cu-O plane with the contour spacing of 0.028 electrons/(a.u.)^3 using the set-1 pseudopotentials

Upon Sr doping holes are created in the lower $d(x^2-y^2)$ - p(x,y) band and give rise to the positive Hall coefficient. The condition for the Fermi-surface nesting is, however, destroyed to some degree by Sr doping so that the AFM state is no more stable energetically. For $x=0.1$ the total energies of the PM and the AFM states which are obtained by the virtual crystal approximation combined with the LSD scheme are comparable to each other (Table 1).

In the case of heavy Sr doping, the AFM state becomes stable again (Table 1). The energy bands for both PM and AFM configurations for $LaSrCuO_4$ are shown in Fig. 4. We can see occurrence of two types of bands near the Fermi level in Fig. 4. In addition to the $d(x^2-y^2)$-p(x,y) band, $d(z^2)$-$O_2p(z)$-$O_1p(x,y)$ band, where O_1 and O_2 denote the in-plane and out-of-plane oxygen atoms, respectively, crosses the Fermi level in the PM energy bands. Our local-density-functional calculation reveals that the crossing of $d(z^2)$-$O_2p(z)$-$O_1p(x,y)$ bands with the Fermi level occurs at larger x values than 0.3 in $(La_{1-x}Sr_x)_2CuO_4$. Nesting conditions for the two conduction bands in the PM configuration of $LaSrCuO_4$ are satisfied, and the AFM configuration is stabilized.

The occurrence of the two types of bands upon heavy Sr doping is a consequence of the structural change around the Cu atom. Figure 5 shows the

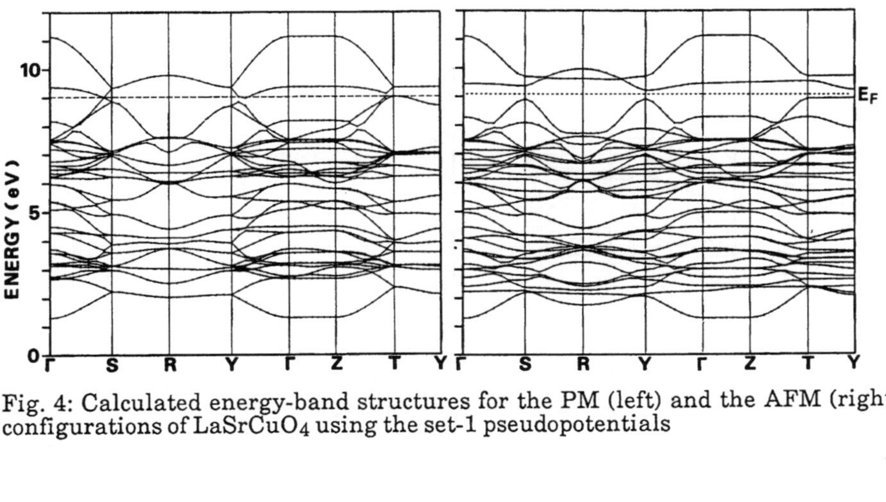

Fig. 4: Calculated energy-band structures for the PM (left) and the AFM (right) configurations of $LaSrCuO_4$ using the set-1 pseudopotentials

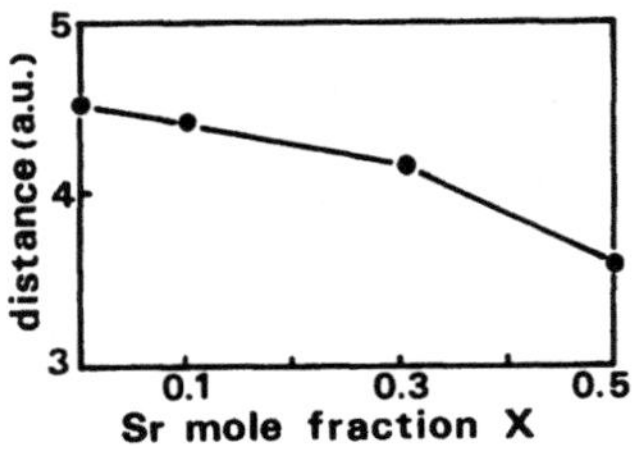

Fig. 5: Calculated distance from a Cu atom to an O atom situated outside of the Cu-O plane in $(La_{1-x}Sr_x)_2CuO_4$

calculated distance from a Cu atom to an O atom which sits outside the Cu-O plane. This is obtained from the total-energy minimization within the local density approximation. (The in-plane distance between Cu and O is fixed to the observed value.) The calculated distance for La_2CuO_4 is almost identical to the experimental value[20]. Further, it is noted that the calculated Cu-O distance decreases rather sharply around the Sr content, $x=0.3$[21]. As a consequence, the octahedron consisting of one Cu and six O atoms changes its shape from the elongated form to the regular form. Thus, the splitting between $d(x^2-y^2)$ and $d(z^2)$ orbitals disappears, and the two different types of energy bands appear around the Fermi level.

3.2 Vanishing Spin-polarization with the Set-2 Pseudopotentials

The results presented so far come from the set-1 pseudopotentials in which Cu 3s and 3p are regarded as core states. The 3s and 3p orbitals, however, possess considerable amplitudes outside the peak of the Cu 3d valence orbital. This could cause errors in estimating the exchange-correlation effect on the magnetic properties of the material[22]. We have thus prepared the set-2 pseudopotentials in which the 3s and 3p states are included in the valence states.

The calculated total valence energies for the PM and the AFM configurations of La_2CuO_4 are shown in Table 1. The values for the two configurations are identical to each other within the numerical error. Further, in sufficiently converged situation (i.e. mean square difference between the input and the output total crystal potentials is less than 10^{-10} in atomic units), the magnitude of the spins in the AFM configuration is less than 0.05 μ_B. The energy bands in the AFM configuration is shown in Fig. 6. The obtained energy bands are very

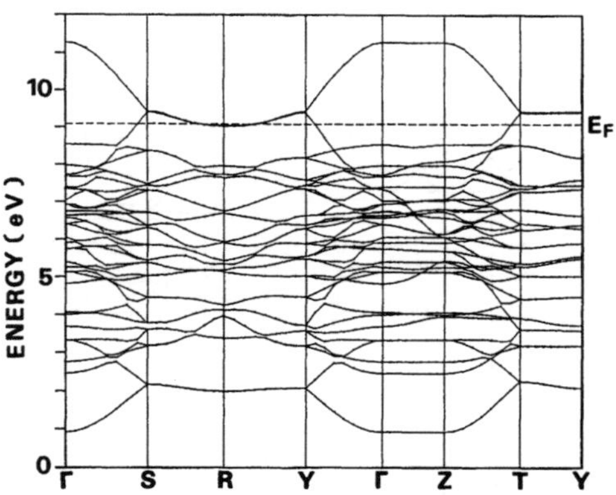

Fig. 6: Calculated energy bands for the AFM configuration of La_2CuO_4 using the set-2 pseudopotentials

similar to those in the PM configuration. In fact, the splitting of the metallic bands at the zone boundary is very small so that the bands are semimetallic. At this point, we conclude that the LSDA does not produce the stable insulating antiferromagnetic state for La_2CuO_4.

The reason why the spin has vanished in the set-2 pseudopotential calculation is rather simple. In the LSDA, the exchange-contribution to the potential-difference for up and down spins, $\mu_X{}^+ - \mu_X{}^-$, is written as

$$\mu_X^+(\mathrm{r}) - \mu_X^-(\mathrm{r}) = -4\left(\frac{3}{4\pi}\right)^{1/3}\left[n_+(\mathrm{r})^{1/3} - n_-(\mathrm{r})^{1/3}\right] . \tag{5}$$

The functional form $n^{1/3}$ is concave downward. In the set-2 pseudopotentials the integrated number of valence electrons around Cu^{2+} is about 17, whereas in the set-1 pseudopotentials it is about 9. Thus, even though the magnitude of $n_+(\mathrm{r}) - n_-(\mathrm{r})$ is comparable in the calculations with the two sets of the pseudopotentials, the potential-difference (5) is smaller in the set-2 calculation than that in the set-1 calculation. Namely, when the total valence-electron density increases, the role of the spin-dependent part of the crystal potential is diminished. This is the reason why the spin from the set-1 calculation has vanished in the self-consistent procedure with the set-2 pseudopotentials.

Occurrence of the stable AFM state in La_2CuO_4 depends on the delicate balance of the total energy. If the exchange-correlation energy is taken into account beyond LSDA, the insulating AFM state might be realized in La_2CuO_4.

In summary, we have performed a state-of-the-art total-energy band-structure calculation for $(La_{1-x}Sr_x)_2CuO_4$. It is found that the LSDA approximation does not produce a stable AFM insulating state for La_2CuO_4, but that it still provides us with reliable information about the structural properties of $(La_{1-x}Sr_x)_2CuO_4$.

REFERENCES

Present address: NTT Basic Research Laboratories, Musashino, Tokyo 180, Japan.

1. P. Hohenberg and W. Kohn: Phys. Rev. 136, B864 (1965), W. Kohn and L.J. Sham: Phys. Rev. 140, A1133 (1965).
2. See, for example, A. Oshiyama and M. Saito: J. Phys. Soc. Jpn. 56, 2104 (1987), Phys. Rev. B36, 6156 (1987) and references therein.
3. J.G. Bednortz and K.A. Müller, Z. Phys. B64, 189 (1986); M.K. Wu et al: Phys. Rev. Lett. 58, 908 (1987).
4. L.F. Mattheiss: Phys. Rev. Lett. 58, 1028 (1987).
5. J. Yu, A.J. Freeman and J.H. Xu: Phys. Rev. Lett. 58, 1035 (1987).
6. R.V. Kasowski and W.Y. Hsu: Phys. Rev. B36, 7248 (1987), L.F. Mattheis and D.R. Hamann: Solid State Commun. 63, 395 (1987); S. Massidda et al: Phys. Lett. 122, 198 (1987); M.S. Hybertsen and L.F. Mattheiss: Phys. Rev. Lett. 60, 1661 (1988); H. Krakauer and W.E. Pickett: Phys. Rev. Lett. 60, 1665 (1988).
7. R.E. Cohen et al: Phys. Rev. Lett. 60, 817 (1988).
8. W.E. Pickett et al: Phys. Rev. B35, 7252 (1987).
9. A. Fujimori et al: Phys. Rev. B35, 8814 (1987); A. Fujimori, E. Takayama-Muromachi and Y. Uchida: Solid State Commun. 63, 1009 (1987); A. Bianconi et al: Solid State Commun. 63, 1009 (1987); T. Takahashi et al: Phys. Rev. B36, 5686 (1987).
10. M.F. Hundley et al: Phys. Rev. B35, 8800 (1987); N.P. Ong et al: Phys. Rev. B35, 8807 (1987).

11. For example, P.W. Anderson: Science 235, 1196 (1987); H. Kamimura, S. Matsuno and R. Saito: Solid State Commun. 67, 363 (1988); J.E. Hirsch et al: Phys. Rev. Lett. 60, 1668 (1988); R.B. Laughlin: Phys. Rev. Lett. 60, 2677 (1988).
12. G. Shirane et al: Phys. Rev. Lett. 59, 1613 (1987).
13. J. Perdew and A. Zunger: Phys. Rev. B23, 5048 (1981).
14. D.M. Ceperley and B.J. Alder: Phys. Rev. Lett. 45, 566 (1980).
15. D.R. Hamann, M. Schlüter and C. Chiang: Phys. Rev. Lett. 43, 1494 (1979); G.B. Bachelet, D.R. Hamann and M. Schlüter: Phys. Rev. B26, 4199 (1982).
16. K. Shiraishi et al: Solid State Commun. 66, 629 (1988); Oshiyama et al: Physica C153-155, 1235 (1988).
17. H. Kamimura et al: Proc. 19th Int. Conf. Physics on Sermiconductors (Warsaw, 1988).
18. All-electron LSD band-structure calculations have been also reported. See J. Kübler et al: Physica C153-155 1237 (1988).
19. D. Vaknin et al: Phys. Rev. Lett. 58, 2802 (1987).
20. J.B. Boyce et al: Phys. Rev. B35, 7203 (1987).
21. K. Shiraishi et al: Jpn. J. Appl. Phys. 26, Suppl. 26-3, 987 (1987).
22. S.G. Louie et al: Phys. Rev. B26, 1738 (1982).

Part III

Experimental Approach

Quasielastic and Inelastic Spin Fluctuations in Superconducting $La_{2-x}Sr_xCuO_4$

R.J. Birgeneau [1,2], *Y. Endoh* [3], *Y. Hidaka* [4], *K. Kakurai* [3], *M.A. Kastner* [2], *T. Murakami* [4], *G. Shirane* [1], *T.R. Thurston* [2], *and K. Yamada* [1,3]

[1] Physics Department, Brookhaven National Laboratory, Upton, NY 11973, USA
[2] Department of Physics, Massachusetts Institute of Technology, Cambridge, MA 02139, USA
[3] Department of Physics, Tohoku University, Sendai 980, Japan
[4] NTT Opto-Electronics Laboratories, NTT Corporation, Tokai, Ibaraki, 319–11, Japan

We review the results of recent neutron scattering studies of the spin fluctuations in samples of $La_{1.89}Sr_{0.11}CuO_4$ which are ~ 80% superconducting with T_c = 10 K. The structure factor, $S(\vec{Q})$, reflects three dimensional modulated spin correlations with an in-plane correlation length of order 18 ± 6 Å. The fluctuations evolve with temperature from being predominantly dynamic at high temperatures to mainly quasielastic ($|\Delta E| < 0.5$ meV) at low temperatures. No significant differences are observed in the normal and superconducting states.

I. Introduction

A variety of experiments has indicated that the CuO_2 lamellar superconducting materials exhibit novel but complicated magnetic effects [1,2,3]. Recent neutron experiments in $La_{2-x}Sr_xCuO_4$ with $0.02 \le x \le 0.14$ show that as x increases the Cu^{++} moment is preserved but the basic antiferromagnetic state becomes progressively shorter ranged [1]. These experiments, however, are not definitive vis-a-vis the nature of the magnetism in the superconducting state since the samples studied there exhibit a Meissner fraction of at best 15%. Recently, however, two of us (Y.H. and T.M.) have made significant progress in the growth of single crystal $La_{2-x}Sr_xCuO_4$ of high crystalline perfection and with a large Meissner fraction (~ 80%) [4].

In this paper we review recent neutron scattering studies of samples of $La_{2-x}Sr_xCuO_4$ with T_c = 10 K and an 80% Meissner fraction [5]. The high quality of the samples has allowed a much more thorough study of the spin correlations than was possible previously. A number of new results have therefore been found. First, the static structure factor $S(\vec{Q})$ exhibits a complicated three dimensional incommensurate structure with a characteristic two dimensional

Springer Series in Materials Science, Vol. 11
Mechanisms of High Temperature Superconductivity
Editors: H. Kamimura and A. Oshiyama

(2D) correlation length of order 18 ± 6 Å. The low energy ($|\Delta E| < 0.5$ meV) part of $S(\vec{Q})$ exhibits pronounced three dimensional correlations at all temperatures (5 K to 350 K). The response function evolves with temperature from being predominantly inelastic ($|\Delta E| > 0.5$ meV) at high temperatures to mainly quasielastic ($|\Delta E| < 0.5$ meV) at low temperatures, $T \leq 50$ K. The integrated intensity is, however, preserved; further it is close to that observed under identical spectrometer conditions for pure La_2CuO_4 [6].

II. The Energy-integrating Experiments

The experiments were carried out on the H7 and H4M triple-axis spectrometers at the Brookhaven High Flux Beam Reactor. As will be discussed below, the experiments proved to be rather difficult and thus required a novel approach to data collection. Specifically, it was discovered early in the experiments that there was a striking thermal evolution in the distribution in energy of the scattering so that it was essential to separate the quasielastic ($|\Delta E| < 0.5$ meV) and integrated inelastic ($|\Delta E| > 0.5$ meV) contributions to $S(\vec{Q})$. Data were obtained primarily using neutrons with energy $E_i = 14.7$ meV and collimations 40′-40′-40′-80′; the monochromator and analyzer were pyrolytic graphite PG (002) and the incoming beam passed through two PG filters in order to eliminate completely any $\lambda/2$ neutrons. The spectrometer was set up in the triple-axis mode and all scans were carried out twice, first detecting neutrons scattered off the PG analyzer so that $|\Delta E| < 0.5$ meV and second detecting neutrons passing straight through the analyzer. The effective reflectivity of the analyzer was measured to be 78% so that by subtracting 29% of the first scan from the second scan, one obtained precisely in the latter scan the intensity integrated over all energies with $|\Delta E| > 0.5$ meV since the absorption by the analyzer is negligible. These latter data turn out to be particularly clean, with little contamination scattering.

As noted above two crystals of $La_{2-x}Sr_xCuO_4$ were studied with labels NTT-30 and NTT-35; the samples were ~ 2 x 2 x 0.2 cm^3 in volume with the thin direction along the orthorhombic $\vec{b}$, perpendicular to the CuO_2 planes ($\vec{a}$ - $\vec{c}$). For both crystals the tetragonal-orthorhombic structural phase transition occurred at 265 ± 10 K; from Fig. 3 of Ref. [1] this implies $x = 0.11 \pm 0.02$, close to, but slightly less than, the chemical analysis value of $x = 0.14 \pm 0.02$. Our new crystal growth technique is described elsewhere [4]; as discussed there, this technique produces large single crystals with an 80% Meissner fraction. As a check, the Meissner fraction was measured via both the zero-field and field cooled susceptibility in fields of order 10 Oe on a piece 6 mm x 6 mm x 2 mm broken off of NTT-35; the sample was found to be a least 80% superconducting with $T_c = 10$ K. These data will be discussed in more detail in Ref. [7]. We also confirmed directly using a neutron depolarization technique that both samples become superconducting at 10 K. We suspect that this low T_c is due to oxygen vacancies, and, indeed these presumed vacancies may play an important role in the spin fluctuation spectrum.

Closely similar results were obtained in both samples; however, our data for NTT-30 are much more extensive so we review primarily those results in this paper. The sample was mounted with the orthorhombic $\vec{a}$ (or $\vec{c}$ because of twinning) axis vertical (see Refs. [1,6] for diagrams of the real space lattice) first in a closed cycle refrigerator ($12 \text{ K} \leq T \leq 350$ K) and second in a pumped helium cryostat ($1.9 \text{ K} \leq T \leq 20$ K). The sample was masked very carefully in order to minimize parasitic peaks from the sample container, multiple scattering etc. Fortunately, the crystals themselves are of very high quality so that there is little or no contamination from powder lines, flux inclusions or unreacted material.

Representative scans across the 2D magnetic ridge are shown in Figs. 1 and 2. As discussed extensively in Ref. [1] and [6] these scans are along the direction (h, h − 0.45, 0) which for $E_i = 14.7$ meV has the feature that the outgoing neutron direction $\vec{k}_f/|k_f|$ is along $-\vec{b}^*$, that is, it is always perpendicular to the CuO_2 planes. The two-axis scan then

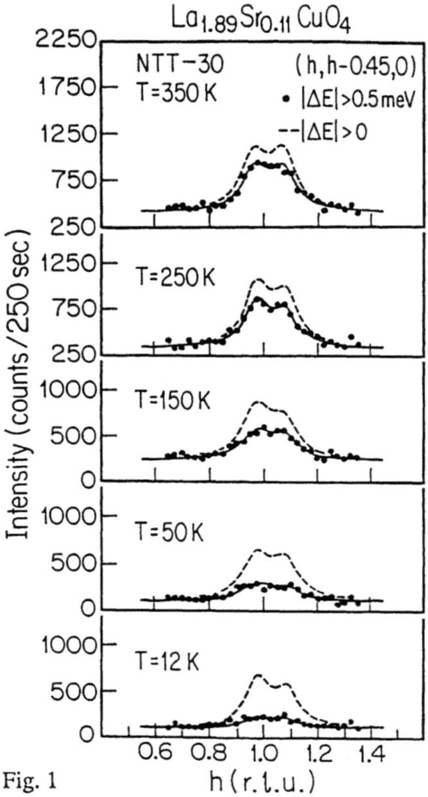

Fig. 1

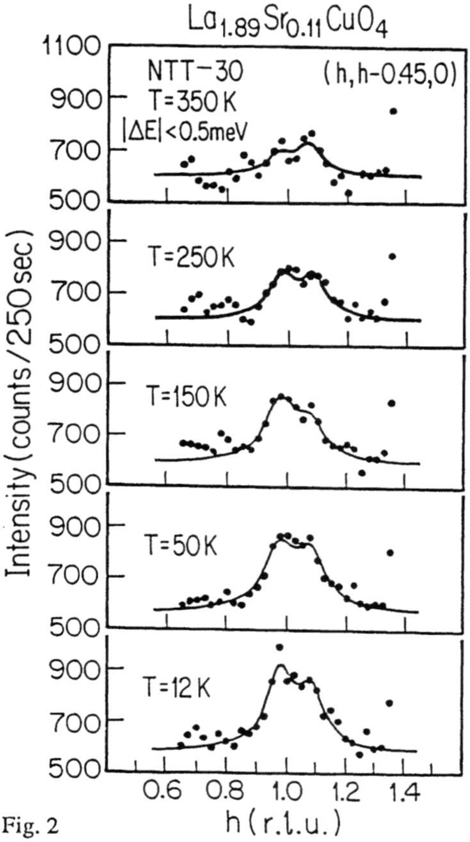

Fig. 2

Fig. 1. Integrated inelastic ($|\Delta E| > 0.5$ [meV]) scattering for scans across the magnetic ridge along (h, h – 0.45, 0); $E_i = 14.7$ [meV] and the collimator configuration is 40′-40′-40′-80′. The solid lines are the results of fits to two displaced 2D Lorentzians as discussed in the text. The dashed lines are the result of the best fits to the total scattering, elastic plus inelastic.

Fig. 2. Three-axis ($|\Delta E| < 0.5$ [meV]) scans across the magnetic ridge along (h, h – 0.45, 0); $E_i = 14.7$ [meV] and the collimator configuration is 40′-40′-40′-80′. The solid lines are the results of fits to two displaced 2D Lorentzians with width and positions held fixed at the values determined from fits to the total cross section, elastic plus inelastic.

automatically integrates over energy without varying the two dimensional (2D) momentum transfer $\vec{Q}_{2D}$. For the total scattering the energy integration range is from $\sim -kT$ (neutron energy gain) to $+E_i$ (energy loss). This is illustrated in the uppermost panel of Fig. 3 which shows an integrated inelastic scan (with $|\Delta E| > 0.5$ meV) with the 2D momentum transfer $\vec{Q}_{2D}$ held fixed at $1.05\,\vec{a}^*$ and $\vec{Q}_\perp = k\vec{b}^*$ varied; the scattering geometry is illustrated at the top of Fig. 3. This scan shows a well-defined peak at $\vec{Q}_\perp \simeq 0.6\,\vec{b}^*$ at which point $\vec{k}_f \parallel -\vec{b}^*$ for $E_i = 14.7$ meV. This verifies that there is a substantial inelastic 2D cross section at 350 K. As we shall discuss later, this is confirmed by direct inelastic measurements.

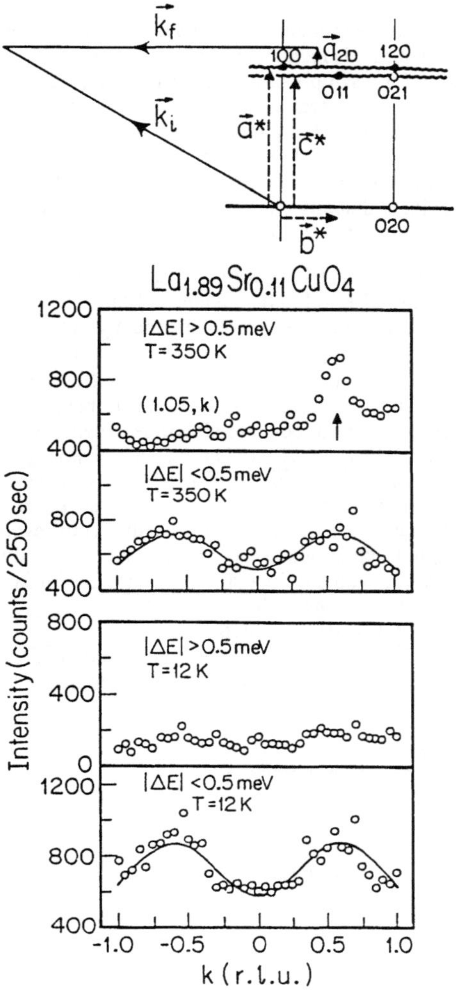

Fig. 3. Top: Superposition of $\vec{a}^*$ - $\vec{b}^*$ and $\vec{c}^*$ - $\vec{b}^*$ reciprocal lattice planes together with a representative scattering diagram for E = 14.7 [meV] neutrons. Bottom: Elastic ($|\Delta E| < 0.5$ [meV]) and integrated inelastic ($|\Delta E| > 0.5$ [meV]) scans ⊥ the CuO_2 sheets for $\vec{Q}_{2D} = 1.05\ \vec{a}^*$. The arrow gives the position at which the outgoing neutron wavevector $\vec{k}_f$ is ⊥ the CuO_2 sheets, that is, along $-\vec{b}^*$.

Figures 1 and 2 show the integrated inelastic ($|\Delta E| > 0.5$ meV), quasielastic, ($|\Delta E| < 0.5$ meV), and fitted total cross sections for the (h, h – 0.45, 0) scans across the 2D ridge at a sequence of temperatures. Two features are immediately evident. First the scattering is broad and flat-topped with some indication of a two-peaked structure. This incommensurate two-peaked structure was suggested in previous experiments [1] but was not established definitively. Second, the total cross section as measured in this particular cut through reciprocal space varies only weakly with temperature from 350 K to 12 K. However, the spin fluctuations change from being predominantly inelastic at 350 K to predominantly quasielastic at 12 K. We confirmed that the integrated intensity at room temperature is identical to within the errors (~ 20%) to the integrated 2D magnetic cross section for a sample of pure La_2CuO_4 ($T_N = 240$ K) measured under identical spectrometer conditions. Since the scattering near h = 1 comes predominantly from low energies, this implies that the full Cu^{++} moment or a significant part thereof is preserved in the superconducting samples. Figure 4 shows pure two-axis scans along (h, h – 0.4, 0) across the ridge at T = 20 K (normal) and T = 5 K (superconducting) together with a background scan along (h, –0.2, 0). It is evident that any change in the static structure factor, $S(\vec{Q})$, between the normal and superconducting states is below our limit of detectability.

Before discussing quantitatively the data in Figs. 1, 2 and 4 we consider further the qualitative features of the low energy spin fluctuations. The full Q-dependence of the quasielastic scattering is experimentally accessible albeit with considerable uncertainty in the background. As is evident in Fig. 2, the quasielastic scattering intensity increases gradually with decreasing temperature. Further the quasielastic line-shape for the (h, h – 0.45, 0) scan across the ridge is closely similar if not identical to that for the total cross section and thence $S(\vec{Q})$. The overall geometry in $\vec{Q}$-space, however, turns out to be quite elaborate. Figure 3 shows quasielastic scans at 12 K and 350 K in which $\vec{Q}_{2D}$ is held fixed at the peak position, $1.05\,\vec{a}^*$, and the momentum transfer ⊥ the CuO_2 planes, $\vec{Q}_\perp = k\vec{b}^*$, is varied. As is evident in Fig. 3, the quasielastic peak intensity exhibits a sinusoidal modulation ⊥ the CuO_2 sheets at both 12 K and 350 K. The period of the modulation is about 2 La_2CuO_4 unit cells. Thus, even at temperatures as high as 350 K the low energy spin fluctuations in $La_{1.89}Sr_{0.11}CuO_4$ are fully three dimensional in character. This contrasts markedly with the fluctuations in pure La_2CuO_4 which are essentially 2D above the Néel temperature [5]. Scans with $\vec{Q}_\perp \sim 0.5\,\vec{b}^*$ and $\vec{Q}_{2D}$ varied typically give either a flat-topped or a double-peaked structure with the incommensurability varying from $\sim 0.05\,\vec{a}^*$ to $\sim 0.2\,\vec{a}^*$ depending on the exact value of $\vec{Q}_\perp$. On the other hand, scans of $\vec{Q}_\perp$ at varying $\vec{Q}_{2D}$ all give the sinusoidal variation discussed above. Thus at low temperatures the Cu^{++} structure factor in superconducting

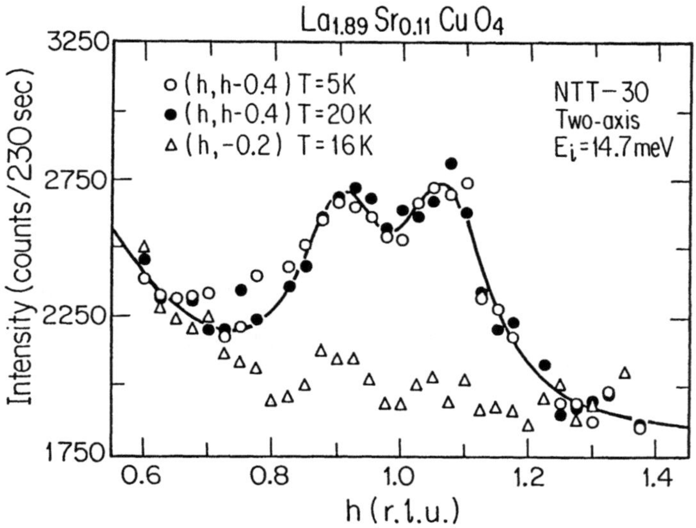

Fig. 4. Pure two-axis scans along (h, h – 0.4, 0) at T = 5 [K] and 20 [K] and along (h, –0.2, 0) at T = 16 [K]. The spectrometer had 1 PG filter and no analyzer. The solid line is the result of fits to two displaced 2D Lorentzians together with a background function determined from the (h, –0.2, 0) scan.

$La_{1.89}Sr_{0.11}CuO_4$ corresponds to a slowly fluctuating (< 10^{-11} sec) 3D modulated spin fluid. We note that recent μSR studies [2] on a sample prepared identically to ours indicate that the entire volume freezes magnetically below T ~ 4 K. Similar results in $YBa_2Cu_3O_{6.4}$ had been inferred earlier but could not be established definitively [2].

In order to compare these results with previous measurements [1] the data were fitted to several simple cross sections. The solid lines in Figs. 1, 2, and 4 are the results of fits to two displaced 2D Lorentzians. Clearly this simple model works well although it certainly is not unique. For the total scattering, the peak positions, intensities and width as well as the background were all varied. For the quasielastic and integrated inelastic components the peak positions and widths were fixed at the values determined from the fits to the total scattering and only the intensities and background were allowed to vary. The quasielastic and integrated inelastic components turn out to be well-described separately by the parameters characterizing the total cross section. The 2D instantaneous spin-spin correlation length is of order 18 ± 6 Å independent of temperature from 350 K to 5 K. The 2D incommensurability from these fits is of order 0.05 $Å^{-1}$ (Fig. 2) to 0.08 $Å^{-1}$ (Fig. 4) although as noted above, larger values are obtained from quasielastic scans ⊥ the rod so the exact value of the incommensurability should be treated cautiously. At 350 K the total cross section, which corresponds to an integral from ~ −kT to +14.7 meV, is predominantly (~ 75%) inelastic while at 12 K the $|\Delta E| < 0.5$ meV component accounts for ~ 75% of the observed scattering. We should note that all of the data reported in Ref. 1 are also well-described by the two-Lorentzian line shape. Further as shown in Fig. 5 the 2D correlation lengths so-deduced agree well with the average separation of the O^- holes as suggested in Ref. [1].

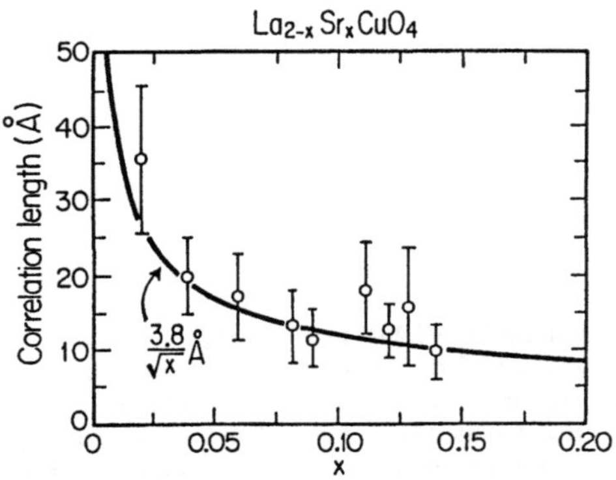

Fig. 5. Instantaneous spin correlation length vs temperature in $La_{2-x}Sr_xCuO_4$. The lengths are deduced from fits of two Lorentzians with identical widths symmetrically displaced about h = 1. The solid line is the function $3.8/\sqrt{x}$ [Å] which is just the average separation between the holes introduced by the Sr^{2+} doping.

III. Inelastic Measurements

As reported in more detail in Ref. [7], there are now some preliminary direct inelastic measurements of the spin excitations. Representative results at 6 meV are shown in Fig. 6. The lineshape for excitation creation closely mirrors the integrated response (Figs. 1, 3). Thus these represent excitations out of the slowly fluctuating modulated ground state. The excitation creation intensity is independent of temperature between 300 K and 5 K; this spans the region $kT >> \hbar\omega$ to $kT << \hbar\omega$. The excitation annihilation cross section is related to the creation process by the detailed balance factor $e^{-E/kT}$ as is required by time reversal symmetry. It has

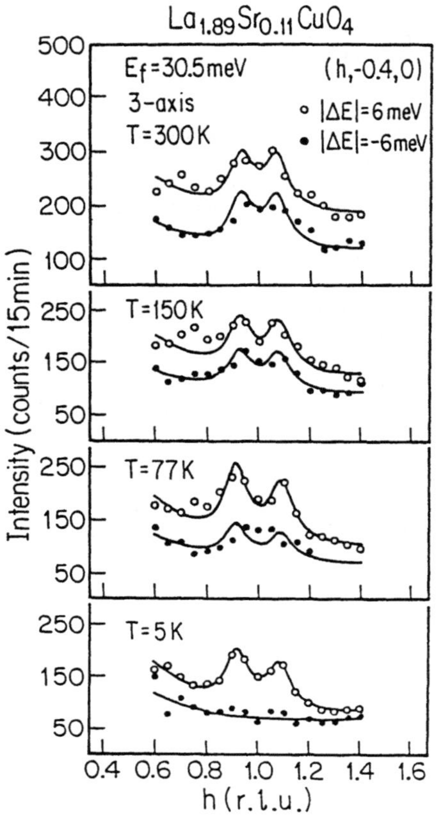

Fig. 6. Constant energy scans across the 2D rod along (h, −0.4,0) in NTT-35, $La_{1.89}Sr_{0.11}CuO_4$. The outgoing neutron energy was 30.5 meV and the collimation was 40′-40′-40′-80′. The lines for the +6 [meV] data are the results of fits to two symmetrically displaced Lorentzians. The lines for −6 [meV] are calculated from the +6 [meV] fits assuming detailed balance. The open circles represent excitation creation and the closed circles excitation annihilation scattering processes.

been explicitly verified [7] that there is no dispersion in the $\vec{b}$ direction at 6 meV, that is, the excitations are confined to the CuO_2 planes. This contrasts with the fluctuations with $|E| < 0.5$ meV which exhibit 3D sinusoidal correlations as shown in Fig. 3.

If the spin excitations were *bosons* as in pure La_2CuO_4 below T_N [1] then the intensity for E = +6 meV would have changed by a factor of 5 between 300 K and 5 K. On the other hand the temperature dependence of the excitation intensity is only consistent with *Fermi* statistics if the chemical potential is much larger than 25 meV. Thus the excitation statistics remain to be understood.

Limited measurements have also been performed [7] for energies varying between 4 meV and 18 meV. The lower limit of 4 meV is set by background considerations while above

18 meV phonon scattering processes dominate. The data at this point are incomplete but they do show that the excitation intensity depends only weakly on energy. Because of the broad distribution in energy the scattering at any given energy is quite weak. The count rate in Fig. 6 at 6 meV is 6 counts per minute; further, NTT-35 is a high quality single crystal 2 cm^3 in volume. It is not surprising, therefore, that similar inelastic neutron scattering experiments have not yet been successfully executed in other high temperature superconductors.

Finally, we consider the relationship between the dynamic (Fig. 6) and instantaneous spin correlations (Figs. 1-4). The integrated inelastic measurements cover the energy range from $-kT$ to $+E_i$, with the center 1 meV excluded. At 350 K this corresponds to ~ − 30 meV to + 14.7 meV. Further the energy gain part is amplified by the kinematical factor $k_f/k_i = (1 - \hbar\omega/E_i)^{1/2}$. At 12 K the energy integration range is from ~ 0.5 meV to 14.7 meV. Thus the diminution of the integrated inelastic scattering with decreasing temperature evident in Fig. 5 is simply a manifestation of detailed balance combined with the weak energy and temperature dependence of $S(\vec{Q},\omega)$ at $+\omega$. Heuristically, the thermal spin excitations seem to condense out yielding a slowly fluctuating, modulated, spin fluid response. Further, the slow part of $S(\vec{Q},\omega)$ is correlated three dimensionally.

From both the temperature dependence of the intensity and the dispersion, one may conclude that the inelastic scattering is consistent only with magnetic processes rather than lattice dynamical fluctuations. It has not, however, been rigorously proven that the quasielastic scattering shown in Figs. 2 and 3 is magnetic. However, the continuous trade-off in intensity evident in Fig. 1 between the dynamic and static fluctuations with decreasing temperature as well as the closely similar lineshapes is strongly suggestive. Certain theories such as that of Ref. [8] require a gap in the spin excitation spectrum in the superconducting state. These results seem to contradict these predictions. However, the low value of T_c of 10 K in our samples implies considerable disorder in the CuO_2 planes probably due to oxygen vacancies. It is possible that the quasielastic scattering originates from this disorder. Only future experiments on more perfect samples with T_c near 40 K can remove this ambiguity.

IV. Conclusions

Clearly, the magnetism in these samples of superconducting $La_{1.89}Sr_{0.11}CuO_4$ with $T_c = 10$ K is quite elaborate. The superconductivity occurs in the presence of a 3D-correlated, modulated, slowly fluctuating Cu^{++} spin fluid; the 2D correlation length is of order 18 ± 6 Å while the 3D correlations are sinusoidal in character with a period of ~ 2 La_2CuO_4 unit cells. It seems clear heuristically that this novel spin state is generated by the O^- holes [9,10,11] which also carry the supercurrent. So far we have not succeeded in observing a direct manifestation of the superconductivity in the spin fluctuations. The gradual freezing of the Cu^{++} spin fluid as the temperature is decreased from 350 K to 5 K is, in our view, one of the most remarkable features of these results. Of course, these experiments have also confirmed unambiguously that there is a substantial Cu^{++} moment in superconducting samples thence strengthening the case for a magnetic mechanism for the superconductivity in the lamellar CuO_2 superconductors.

We would like to thank our colleagues at MIT, Brookhaven National Laboratory, Tohoku University, and NTT for many helpful discussions of these results and R. B. Laughlin for invaluable critical comments. This work was supported by the U.S.-Japan Cooperative Neutron Scattering Program, and a Grant-In-Aid for Scientific Research from the Japanese Ministry of Education, Science and Culture. The work at MIT was supported by the National Science Foundation under Contract No. DMR85-01856 and No. DMR87-19217. Research at Brookhaven National Laboratory was supported by the Division of Materials Science, U.S. Department of Energy under Contract No. DE-AC02-CH00016.

V. References

1. R. J. Birgeneau et al.: Phys. Rev. B38, 6614 (1988) and references therein.
2. Y. J. Uemura et al.: Journal de Physique (Paris) Colloque (in press); Brewer et al.: Phys. Rev. Lett. 60, 1073 (1988).
3. J. M. Tranquada et al.: Phys. Rev. Lett. 60, 156 (1988).
4. Y. Hidaka et al.: (unpublished work).
5. R. J. Birgeneau et al.: Phys. Rev. B (submitted).
6. G. Shirane et al.: Phys. Rev. Lett. 59, 1613 (1987); Y. Endoh et al.: Phys. Rev. B37, 7443 (1988).
7. T. Thurston et al.: (unpublished work).
8. R. B. Laughlin: Science, 242, 525 (1988).
9. V. J. Emery: Phys. Rev. Lett. 58, 2794 (1987).
10. A. Aharony et al.: Phys. Rev. Lett. 60, 1330 (1988).
11. J. M. Tranquada et al.: Phys. Rev. B36, 5263 (1987).

Two Dimensional Quantum Spin Fluid – Progenitor of High Temperature Superconductivity –

Y. Endoh[1], R.J. Birgeneau[2,3], D.R. Gabee[2], Y. Hidaka[4], H.P. Jenssen[2], T. Murakami[4], M. Oda[4], P.J. Picone[2], G. Shirane[3], M. Suzuki[4], T.R. Thurston[2,3], and K. Yamada[1,3]

[1]Department of Physics, Tohoku University, Sendai 980, Japan
[2]Department of Physics, Massachusetts Institute of Technology, Cambridge, MA 02139, USA
[3]Physics Department, Brookhaven National Laboratory, Upton, NY 11973, USA
[4]NTT Opto Electronics Laboratories, NTT Corporation, Ibaraki, 319–11, Japan

Neutron scattering probed novel two dimensional antiferromagnetic spin correlations in pure La_2CuO_4 which is the mother compound of high temperature superconductors of doped $La_{2-x}M_xCuO_4$. Cu^{2+} moments are ordered instantaneously over very large areas in the CuO_2 plane but there is no measurable time averaged staggered moment. Since the dynamical behavior observed is much akin to that in the quantum antiferromagnetic chain, we concluded that spin fluctuations observed at finite temperatures are quantum spin liquid state like that in quantum antiferromagnetic chain at T = 0.

1. INTRODUCTION

It is widely accepted that the magnetism is the progenitor of the high temperature superconductivity at least in any CuO_2 lamellar compounds[1]. This has extensively been discussed in many theoretical models since Anderson[2] suggested that the quantum spin fluctuations in CuO_2 layers might be an important ingredient for the superconducting mechanism by the strong coupling between the electrical conductivity and the magnetism. Furthermore this discovery of high temperature superconductors led to renewal of an old but unsolved problem of thermodynamics in quantum spin systems described with the 2 dimensional (2D) Heisenberg antiferromagnetic Hamiltonian[3]. We have been elucidating static as well as dynamic spin fluctuations in both pure and doped $La_{2-x}Sr_xCuO_4$($0<x<0.15$) by utilizing large as well as high quality single crystals required for neutron scattering studies[4]. Following the preceding paper by R.J.Birgeneau[5] who describes the spin fluctuations in superconducting crystals with Sr doping, the present paper describes the unique feature of 2D quantum spin states in insulating La_2CuO_4. Therefore neutron scattering results are mainly cited from our previous publications and detailed results are found therein[4,5].

Although La_2CuO_4 belongs to the same family of K_2NiF_4, which was extensively studied by Birgeneau et al.[6] as the best known real system for a model of classical 2D

Heisenberg antiferromagnet, the crystal further transforms to the orthorhombic structure around 500K. This structural phase transition is driven by the freezing of the soft phonon mode[7] which introduces a slight tilt of CuO_6 octahedra around a principal axis in the tetragonal c plane. Then this distortion induces not only the antisymmetric exchange interaction[8] by breaking the inversion symmetry but the weak effective interplane interaction which should be nearly canceled in the tetragonal structure like K_2NiF_4. In fact La_2CuO_4 undergoes a magnetic phase transition to the 3D antiferromagnetic long range order (LRO) below relatively high Neel temperature (T_N) around 200K. Since the orthorhombic distortion depends on stochiometry of the oxygen content, T_N of the 3D LRO is dominated by the oxygen concentration[9] which differs sample to sample by the different crystal growing condition as well as the difference in further heat treatments. Nevertheless the basic magnetism in La_2CuO_4 is well modeled by an $S = 1/2$ nearest neighbor antiferromagnetic Heisenberg Hamiltonian on a square lattice with the large exchange constant J/k_B of about 1200K. The above conclusion was recently confirmed by Chakaravarty et al.[3] who made the analysis of the experimental results[4] of spin correlations in La_2CuO_4 on the basis of a quantum nonlinear sigma model. Another important conclusion derived by this theory is that the parameters fitted to La_2CuO_4 experiments suggest the existence of LRO at $T = 0$ like the classical 2D Heisenberg antiferromagnet.[10]

The variable range hopping conduction[11] is consistent with the good localization of magnetic moments unlike the earlier results of the band calculation. Thus we believe that Cu^{2+} ions carry $S = 1/2$ full moments, although the observed moment is much smaller due to the quantum spin reduction. It should be pointed out that the other members in the same family of La_2TMO_4, where TM ions are hence Ni^2 carries S=1[12], and Co^{2+} does S=3/2[13]. When we compare the 3d core level energy, the d level for Cu^{2+} may be the lowest in this family and the mixing must be largest due to the smallest energy difference between d level and conduction band. Therefore the localization of magnetic moments concluded by magnetic studies suggests that the electronic structure of Cu^{2+} is considered to be such that d gamma levels split by the Jahn Teller coupling locate near the Fermi level and a hole should go into the Cu orbital of $d_{x^2-y^2}$. These conclusions are along the same lines as the Hubbard model taking into account the strong electron correlation effects[14].

2. STATIC MAGNETIC PROPERTIES

Neutron scattering studies together with susceptibility and resistivity measurements provided the above mentioned conclusion that the basic magnetism is well modeled by the $S = 1/2$, 2D Heisenberg antiferromagnetic Hamiltonian[15]. It did not take a long time until the 3D spin structure was established since the magnetic susceptibility had been suggested the antiferromagnetic LRO in La_2CuO_4[16]. However the temperature dependence of the magnetic susceptibility (χ), in particular the appearance of a sharp peak near T_N in χ_b, where b

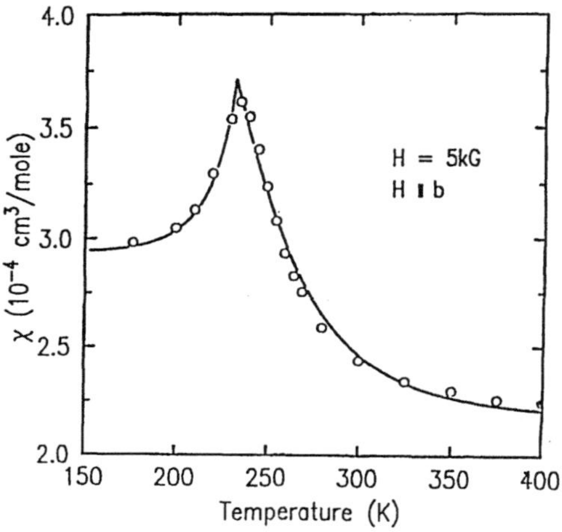

Fig.1. Magnetic susceptibility of b axis component from the single crystal of La_2CuO_4. Circles are calculated and solid line shows the experimental data.

specifies a component perpendicular to CuO_2 plane along orthorhombic b axis, has long been puzzling. The orthorhombic crystal structure generated by the rotation of CuO_6 octahedra allows an antisymmetric interaction of Dzyaloshinski-Moriya interaction which induces a coupling between the uniform and staggered magnetizations. The mean field calculation can fit beautifully to both temperature dependence of magnetic susceptibility of the b axis component and magnetization curves in the process of applying external field along the b axis[8]. Then Kastner et al.[17] confirmed spin canting exclusively by neutron diffraction experiments. They observed the magnetic phase transition from noncollinear to canted state at the critical field where the magnetization jumps. The fact is that half of the spins having an antiparallel component to the applied field along the b axis flip to be parallel to those of the other half. Therefore it is now clear that the spins are not exactly in CuO_2 planes nor in the plane perpendicular to the b axis as was orginally thought, but cant out of the planes as shown in Fig.2

Now we understand the reason why La_2CuO_4 becomes the 3D antiferromagnet with rather high T_N and also reasonably well understand how T_N varies by different heat treatments. Either the effective interplaner interaction (J') plays a dominant role or the 2D correlation length (ξ_{2D}) does, since T_N is proportional to $\xi_{2D}{}^2 J'$. ξ_{2D} seems not to be significantly affected by this heat treatment near T_N. We speculate that J' determines T_N due to the fact that oxygenation reduces orthorhombic distortion as well as making the b axis longer [18]. Another possible source is the creation of some spin disorder by the oxygenation. Shirane et al. showed that spins in La_2CuO_4 are strongly correlated in a wide temperature range over 300 degrees due to the large nearest neighbor Cu^{2+}-Cu^{2+} exchange interaction J in the CuO_2 sheets. By using this large value of

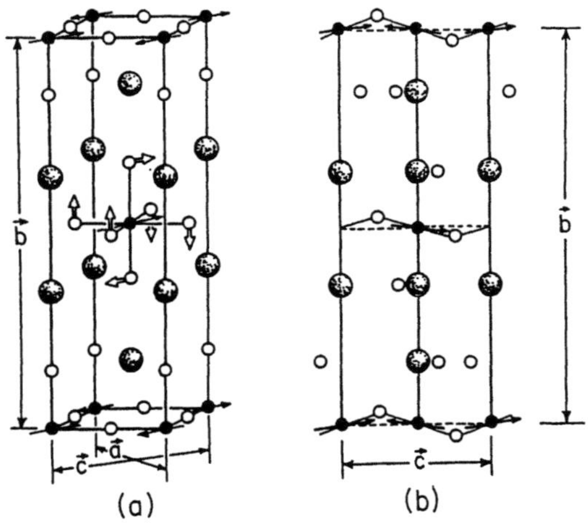

Fig.2. Schematic drawing of the spin structure of La_2CuO_4. (a) arrows from open circles (oxygen atoms) indicate the displacement of oxygen ions and the closed circles with arrow indicate Cu moment.(b) Projection on bc plane.

J/k_B about 1400K, we should observe a macroscopic correlation length even at room temperature based on the Heisenberg model without any anomalous quantum effect. In fact ξ_{2D} was observed to be about 400Å near T_N and furthermore quasielastic scattering seems not to be strongly divergent towards T_N. These important facts can be interpreted as the result of suppression of low energy, long distance spin fluctuations possibly by the quantum effect of S=1/2. The calculations of Chakravarty et al.[3] which are based on a classical model incorporating the quantum renormalization show a similar temperature evolution of spin correlations to those observed. According to their analysis of neutron scattering data of spin correlations, the parameters which give the best fit to the experimental data suggest the existence of 2D long range order (LRO) at T=0 like a pure classical 2D Heisenberg antiferromagnet. In other words spin correlations in a 2D quantum Heisenberg antiferromagnet are qualitatively not different from those of the classical 2D antiferromagnet but are quantitatively renormalized. Then the short correlation length around room temperature is considered in this context even it is about 1000 times shorter.

In order to thoroughly understand the quantum effect in 2D Heisenberg antiferromagnetism, we have also performed neutron scattering studies from La_2CoO_4[13]. This is a sister crystal in the same family and the main difference is that Co^{2+} moment is S=3/2 and the LRO structure is not La_2CuO_4 but like La_2NiO_4[12]. The static spin correlations in this crystal are just as expected when temperature approaches T_N. Quasielastic scattering of 2D nature above T_N tends to diverge critically. Furthermore the dimensional crossover effect is seen in the critical exponent of the order parameter just as seen in many other quasi 2D antiferromagnets. Therefore the big contrast in the spin correlations near above T_N is to be prevailed by the quantum effect. The quantum fluctuations suppress the development of spin correlations, which must be seen more clearly in spin dynamics.

3. DYNAMIC SPIN FLUCTUATIONS-QUANTUM SPIN FLUID STATE

Finally we describe a unique feature of spin dynamics in La_2CuO_4 near and above T_N. The experimental results may be characterized by the dynamical proporties of the 2D, S=1/2 Heisenberg antiferromagnet as discussed in preceding sections. It should be noted, however, that spin fluctuations at low temperatures well below T_N are approximated by the spin wave theory including additional interactions generating the 3D LRO[19].

As shown in Fig.3 where both two axis and three axis scans near T_N are illustrated, the very sharp flat topped scattering across the 2D magnetic rod are detected in energy integrated two axis scans[4]. But hardly any signal occurs in three axis scans in the same mode fixed the energy transfer to be zero. In other words, well developed 2D antiferromagnetic spin order is instantaneously visible in a picture taken with a very fast shutter speed but such an order is washed out in a time averaged picture. Therefore we straightforwardly come to a conclusion for spin fluctuations in La_2CuO_4 that all the spins

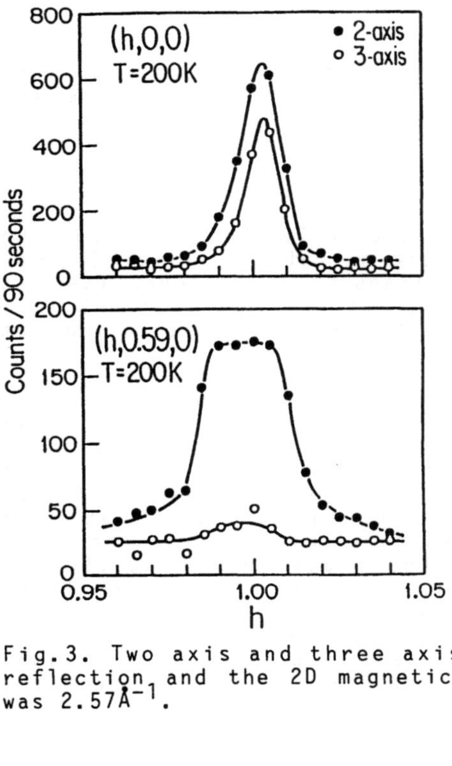

Fig.3. Two axis and three axis scans across the 3D magnetic reflection and the 2D magnetic rod. The incoming wave vector was 2.57Å^{-1}.

coherently fluctuate with the large characteristic velocity. We call this state as the quantum spin fluid or QSF. On the contrary quasielastic scattering from La_2CoO_4 near T_N is condensed to a response in the E=0 scattering[13].

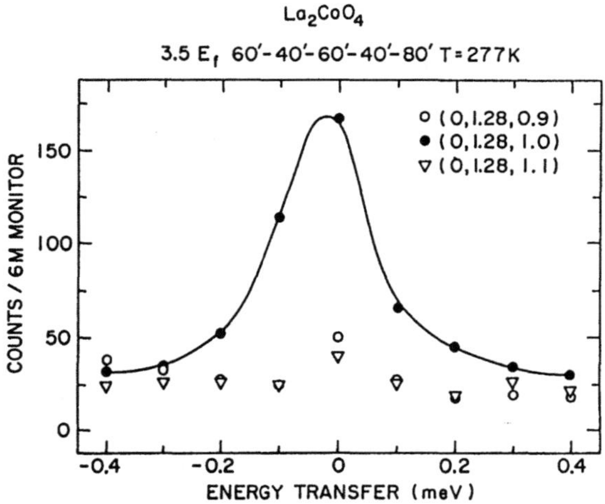

Fig.4. Energy scans around 2D magnetic rod in the constant Q mode in La_2CoO_4, where Q is presented in reciprocal unit.

The energy width of spin fluctuation can hardly be seen even with the high resolution measurements. Nevertheless the remarkable contrast in the features of spin fluctuations between two sister materials are considered to be predominated by the 2D, S=1/2 Heisenberg character.

The large inelasticity is more clearly seen in Fig.5. The data are for scans in which inplane momentum transfer Q_{2D} was varied at several fixed energy transfer conditions. Even at 12meV transfer the peak is centered at $Q_{2D}=a^*(c^*)$ or h=1 in a reciprocal unit. In other words, the magnetic excitations at h=1+q are not resolved.

When we analyze the data applying the standard spin wave theory convolving the resolution function, we estimate the large spin wave velocity of about 0.6eVÅ. This corresponds to the nearest neighbor exchange constant J/k_B to be 1400K, which is consistent to the estimate from the two magnon Raman scattering measurements[20]. It should be noted that both estimates may include ambiguities because we do not precisely know how the quantum effect takes place in the scattering spectrum. For instance the Raman spectrum becomes very broad and the energy position at which the scattering peak occurs is substantially renormalized by changing spin quantum. Nevertheless the large exchange interaction which is the origin of the large inelasticity is an important conclusion from the

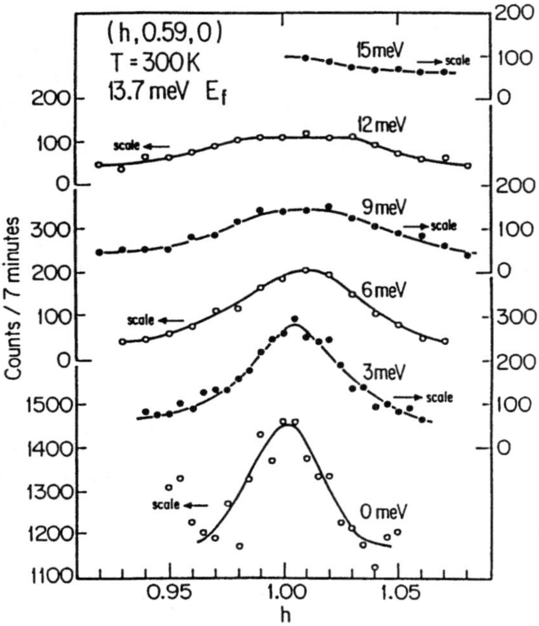

Fig.5. Inelastic constant energy scans across the 2D rod at T=300K. The outgoing neutron energy was fixed at 13.7meV.

inelastic neutron scattering. This conclusion was obtained by the theoretical analysis of spin correlations by Chakravarty et al.. This value is later compared with that of the strongly coupled Hubbard model including a large transfer of a hole between d orbitals and neighboring p orbitals. As a result the large super exchange energy is obtained and therefore the strong magnetic coupling of d-p orbitals can be an essential ingredient for any spin mediated theory for the high Tc superconducting mechanism[2].

The integrated intensity varies gradually with increasing energy, which clearly disagrees with the calculated intensity profile for the spin wave scattering. Since spin waves obey Bose statistics, the calculated intensity simulated the similar scans to the experiment should diverge as the energy transfer is decreased. From limited scans we make a conjecture that the dynamical response function from La_2CuO_4 has a sharp increase in intensity at the lower boundary and long tail towards higher energies. Then this response function reminded us the spin excitations from CPC at very low temperature [21]. Since the CPC is the best known real materials for the model of quantum Heisenberg linear chain, we can make another conjecture that the spin dynamics in the 2D, S=1/2 Heisenberg antiferromagnet is much more akin to those in the quantum chain. Therefore it is not so strange that statistics for spin excitations in this material does not obey Bose statistics since those in the quantum chain form a band at a given q from the lowest boundary of $\pi J \sin(qa)$ to the highest of $2\pi J \sin(1/2qa)$, and quasiparticles in the band are Fermions. Since the assumption of sharp excitations at higher temperatures without any considerations of energy broadening may be incorrect, much more extensive studies remain to be done. Although we know at present that the small Sr doping destroys the Neel state of La_2CuO_4 completely and changes to 2D, QSL state with much more smaller correlation range as will be described by Birgeneau, we have a little knowledge about spin fluctuations. It is also clear that spin fluctuations in the Sr doped material play a key role in the superconducting mechanism.

ACKNOWLEDGMENTS

One of authors (Y.E.) thanks K.Kakurai and M.Matsuda for their valuable contributions and discussions. The work was supported by the US-Japan Cooperative Neutron Scattering Program and a Grant-In-Aid from the Japanese Ministry of Education, Science and Culture. The work at Brookhaven was supported by the Division of Materials Science, US DOE under contract DEAC 0276 CH 00016. The research at MIT was supported by US-NSF Grants Nos. DMR 85-01856,87-19217 and 84-15336.

REFERENCES

[1] J.G.Bednorz and K.A.Muller, Z.Phys. B64, 189(1986)
[2] P.W.Anderson, Science 235,1196 (1987)
[3] S.Chakravarty, B.I.Halperin and D.R.Nelson, Phys.Rev.Lett.60,1057
[4] G.Shirane et al., Phys.Rev.Lett. 59, 1613(1987)
Y.Endoh et al., Phys.Rev.B37,7443 (1987)
[5] R.J.Birgeneau et al., Phys.Rev.B in press
[6] R.J.Birgeneau, J.Skalyo Jr. and G.Shirane, J.Appl.Phys.41
[7] R.J.Birgeneau et al., Phys.Rev.Lett.59, 1329(1987)
[8] T.Thio et al., Phys.Rev.B38, 905(1988)
[9] K.Yamada et al., Solid State Commu.64, 753(1987)
[10] S.H.Shenker and J.Tobochnik, Phys.Rev.B.22, 4462(1980)
[11] M.Oda et al., Solid State Commu. in press
[12] G.Aeppli and D.J.Buttrey, Phys.Rev.Lett.61 203(1988)
[13] K.Yamada et al., Phys.Rev.B in press
[14] M.Imada, J.Phys.Soc.Jpn. 56, 3793(1987)
[15] D.Vaknin et al., Phys.Rev.Lett.58 2802(1987)
[16] T.Fujita et al., Jpn.J.Appl.Phys.26 L402(1987)
[17] M.A.Kastner et al., Phys.Rev.B in press
[18] K.Yamada et al., Jpn.J.Appl.Phys. 24(1988)
[19] C.J.Peters et al., Phys.Rev.(1988)
[20] K.B.Lyons et al., Phys.Rev.B37 2393(1988)
[21] Y.Endoh et al., Phys.Rev.Lett. 26 718(1973)

Nuclear Resonance Studies of $YBa_2Cu_3O_{7-\delta}$

R.E. Walstedt and W.W. Warren, jr.

AT&T Bell Laboratories, Murray Hill, New Jersey, NJ 07974, USA

I. Introduction

The copper oxide high-T_c superconductors [1-3] have proven to be a rich and challenging arena of investigation for many condensed matter experimentalists, and particularly so for nuclear resonance investigators. In spite of a long, successful history of NMR and NQR studies of metallic solids, surprising and unprecedented behavior found with these techniques continues to fascinate workers in the field of high-T_c superconductors. Recent work on optical Raman scattering [4] and neutron diffraction [5] studies has made it clear that magnetism associated with the copper sites plays a major role in the properties of these systems. As we shall see, nuclear resonance provides an excellent local probe of both the static and dynamic magnetic properties. In this paper we shall present and discuss some recent NMR and NQR results obtained on the family of compounds $YBa_2Cu_3O_{7-\delta}$(YBCO). This system continues to attract a great deal of attention not only because it was the first liquid nitrogen superconductor, but also because it can be synthesized in both ceramic and single crystal form with the high quality and highly controlled microscopic properties necessary for fundamental research.

The structure [6] of YBCO contains pairs of quadratic layers of Cu(2) sites having five-fold oxygen coordination. These "conducting" layers are interlaced with individual planes of Cu(1) sites whose coordination varies from 2-fold (δ=1) to 4-fold (δ=0), depending on oxygen content [7]. The Cu(2) planes in these materials contain nominally Cu^{2+} ions which form, with the O^{2-} ligands, a half-filled $d_{x^2-y^2}$ band for $\delta = 1$ which, however, does not conduct because of strong correlation energies on the copper sites. Such a Mott-Hubbard insulator is frequently antiferromagnetic as found for YBCO [8] and only conducts (and indeed superconducts) when extra holes are doped into the Cu-O planes. For fully oxygenated material, there is one excess hole per formula unit, which appears to be distributed over both the plane and chain regions of the structure [9].

The nuclear species in YBCO studied thus far are $^{63,65}Cu$ [10-17] and ^{89}Y [18,19], with some very recent data on ^{17}O [20]. Our primary focus in this paper is on the copper NMR/NQR studies. Both copper isotopes have nuclear spin $I = 3/2$, with an abundance ratio $N_{63}:N_{65} = 69\%:31\%$ and with slightly different nuclear gyromagnetic ratios γ and quadrupolar moments Q. In fully

Springer Series in Materials Science, Vol. 11
Mechanisms of High Temperature Superconductivity
Editors: H. Kamimura and A. Oshiyama
Springer-Verlag Berlin Heidelberg © 1989

oxygenated ($\delta = 0$) material, the two copper sites give rise to ^{63}Cu NQR lines at 22.05 MHz (Cu(1)) and 31.5 MHz (Cu(2))[10]. Early evidence on which NQR line belonged to which site was inconclusive, leading to controversy and misidentification [21]. Strong evidence supporting the assignment given above has now been presented [16,17,22,23,24]. The electric field gradient (EFG) tensor components V_{xx}, V_{yy} and V_{zz} for the two copper sites have been determined to high accuracy [16,17]. They exhibit a remarkable contrast, where the asymmetry parameter $\eta = (V_{xx} - V_{yy})/V_{zz}$ is very nearly zero for the Cu[2) site [22] and almost exactly unity for the Cu(1) site [17]. Since the site symmetry is in both cases orthorhombic [6], these circumstances appear to be coincidental. They probably arise because the EFG tensors have two major and opposing contributions, a dominant one ($\sim$65 MHz) from the open Cu d-shell and a partially canceling contribution from the surrounding lattice. The corresponding ^{63,65}Cu high-field NMR spectrum has well-defined and clearly understood features [22]. NMR studies on single crystals [17] and oriented powders [25] have yielded precise data on the anisotropic shifts at both Cu sites.

In this paper we review and discuss the NMR shift and spin-lattice relaxation studies on $YBa_2Cu_3O_7$, which have revealed a greatly enhanced, highly differentiated relaxation process for the two Cu sites, with peculiar anisotropies and temperature dependences. This body of information provides useful guidance and a powerful constraint for any model theory of the fermion dynamics of YBCO. In Sec. II we discuss the existing body of relaxation and NMR shift data. The latter have been used to analyze the normal state susceptibility into its component parts [26,27]. In Sec. III these results are discussed in terms of available theoretical models.

II. Nuclear Spin-Lattice Relaxation and NMR Frequency Shifts in $YBa_2Cu_3O_{7-\delta}$: Experimental Results

A. Normal State Relaxation

The bulk of the T_1 measurements reported so far have been taken using the zero-field NQR technique [12-14,16], since this gives a single exponential decay and clearly distinguishes the two sites. T_1 data (NQR) are shown in Fig. 1, for $4K \leq T \leq 300K$, where we see comparable rates for the two copper sites in the normal state. The normal state temperature dependences for the two sites are, however, strikingly different, with neither site exhibiting simple Korringa behavior [28-30], $T_1^{-1} \propto T$. The Cu(2) (planar) site relaxation appears to be leveling off at higher temperatures, while the Cu(1) (chain) site rate is nearly linear in T, but actually turns upward near room temperature.

The remarkable feature of the copper relaxation process is its strength, which is two orders of magnitude greater *for both sites* than reasonable estimates [27] based on the Fermi surface density of states derived from the spin susceptibility. This large enhancement effect is undoubtedly connected with the unusual temperature dependences shown by the data. We defer any

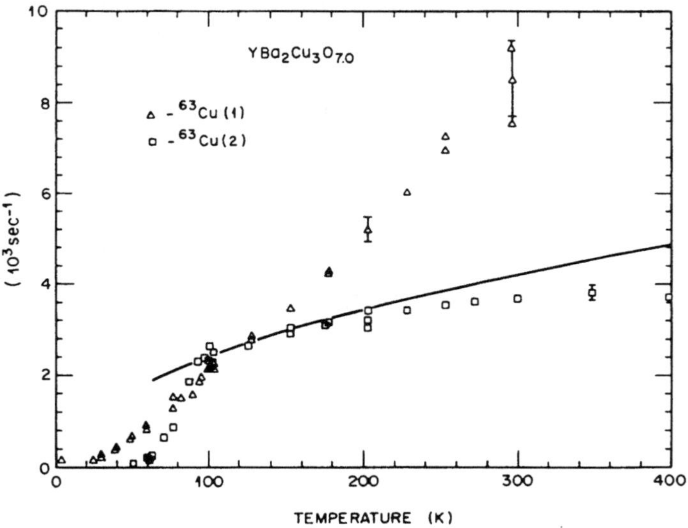

(1) NQR T_1 data for ^{63}Cu is plotted vs. temperature T for $4K \leq T \leq 400K$ for both Cu sites in $YBa_2Cu_3O_7$. The solid line is a model curve $T_1^{-1} \propto T^{1/2}$ as discussed in Sec. IIIA.

detailed discussion of enhancement mechanisms to Sec. III, but only note here the unusual fact that while the relaxation is greatly enhanced, the spin susceptibility is very nearly equal to its band theory (i.e., noninteracting electron) value (see Sec. IIC).

The foregoing discussion refers to zero-field NQR measurements of T_1, where (for Cu(2) sites) the axis of quantization lies along the c-axis. T_1 measurements employing NMR techniques on powder samples show [27] that there is a very substantial anisotropy to the Cu(2)-site T_1 process. Results for $100K \leq T \leq 300K$ at two angles θ of field $\vec{H}_0$ relative to the c-axis are plotted in Fig. 2, along with NQR ($\theta = 0$) data [31]. These results show an angular variation of the form $T_1^{-1} = a\sin^2\theta + b\cos^2\theta$, with a/b ~ 3.4. No variation of a/b with temperature is resolved by these data, suggesting that the anisotropy is simply a variation of hyperfine coupling coefficient. We consider it unlikely that there would be a significant anisotropy of the underlying spin fluctuations. As discussed in Sec. III, it is not possible to account for the T_1 anisotropy within the confines of the conventional enhanced Korringa model [30].

B. Nuclear Relaxation in the Superconducting State

Perhaps even more remarkable than the normal state behavior is the variation of T_1 below T_c. Here, again, the zero-field NQR measurements give the clearest characterization of the copper-site behavior [12-14,32]. As shown in Fig. 3, the planar sites undergo a dramatic freezeout of spin-lattice

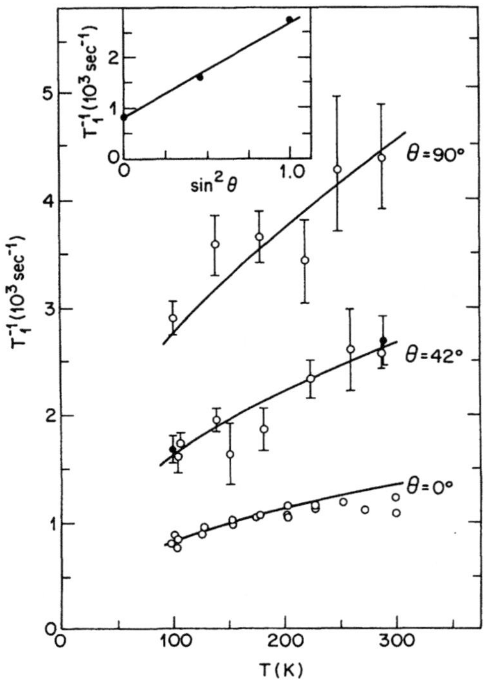

(2) NMR T_1 data from the high-field ($\theta = 42°$) and low field ($\theta = 90°$) singularities of the $^{63}Cu(2)$ powder pattern are plotted vs. temperature T, along with NQR ($\theta = 0$) T_1 data from Ref. 12. Here θ is the angle between the c-axis and the nuclear spin quantization axis. The solid lines are related by simple scaling factors.

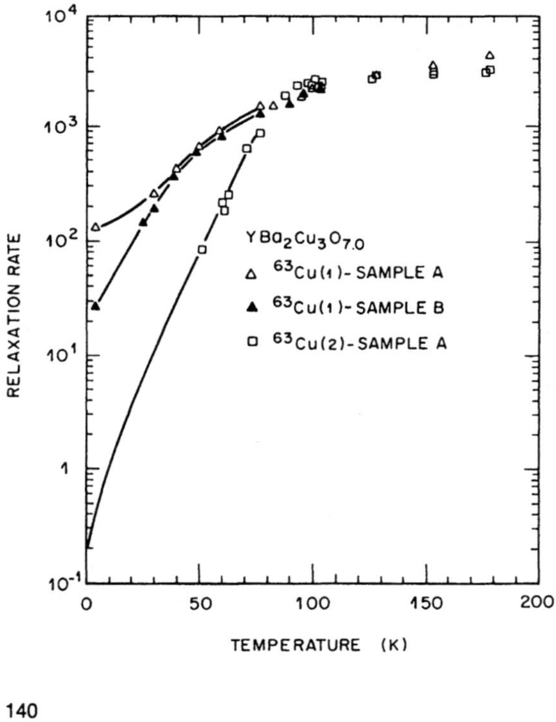

(3) ^{63}Cu NQR T_1 data for both Cu sites in a semilog plot for $T \leq T_c$. For Cu(1), the two samples shown give disparate and thus non-intrinsic behavior at low temperatures. The solid line for the Cu(2) shows behavior reported in Ref. 32.

relaxation below T_c, far steeper than permitted by the BCS weak-coupling model [33]. In stark contrast, the Cu(1) sites exhibit (Fig. 3) a much more gradual decline to a helium-temperature rate which varies between samples and is therefore in general non-intrinsic. The origin of the helium-temperature background rate is not understood at present. What seems most significant here is the dramatic collapse of the greatly enhanced Cu(2)-site rate owing to the pairing of superconducting carriers below T_c. It has been suggested that the Cu(2) sites are behaving like localized moments in the normal state [27,34]. If such moments are present at $T \sim 100K$, then the results in Fig. 3 show that they have effectively disappeared at helium temperatures. This is a severe constraint on any model theory involving localized moments. It suggests that the Cu(2) sites may in some respects be local-moment-like, but that the 3d holes are actually itinerant.

C. NMR Shift Results and Interpretation

We next turn to the Cu NMR shift results. In the normal state, both Cu sites present large positive, anisotropic shifts [25,35], which we summarize in Table I. Above T_c, the Cu shifts are very nearly independent of temperature. These shifts are thought to be dominated by the Van Vleck orbital contribution, because of the large orbital hyperfine field. A simple model calculation [11] shows that the orbital shift will be strongly anisotropic if the ground state of the single d-hole on the Cu^{2+} ion is $d_{x^2-y^2}$, as is widely assumed [36]. The only other important contribution to the Cu shift comes from the d-spin susceptibility. In a recent analysis of the shifts and susceptibility of YBCO [27], the powder average Cu shifts and susceptibility were related by the equation [37]

$$\overline{K}_1 + 2\overline{K}_2 = \beta \overline{\chi}_{VV} + \alpha_d \chi_p = 0.0250\,. \tag{1}$$

Here $\overline{\chi}_{VV}$ and χ_p are the powder average Van Vleck orbital and spin paramagnetic susceptibilities per mole formula unit, respectively, with β and

Table I. Analysis of the uniform susceptibility of $YBa_2Cu_3O_7$ (normal state) into spin paramagnetic and Van Vleck orbital contributions, using $\overline{K}_1 = 0.86\%$ and $\overline{K}_2 = 0.82\%$ for the powder average NMR shifts at Cu(1) and Cu(2), respectively. The orbital shift coefficient $\beta = K_{VV}/\chi_{VV}$ is taken to be 135 (emu/mole Cu)$^{-1}$ and results are given for d-spin shift coefficients for $\alpha_d = K_d/\chi_p = \pm 23$ (emu/mole Cu)$^{-1}$. Susceptibilities are in units of 10^{-4} emu/mole formula unit.

	$\overline{\chi}_{VV}$	χ_p	$\overline{K}_{VV1}$	$\overline{K}_{VV2}$	$K_{d1,2}$
$\alpha_d = -23$	2.23	2.22	1.03%	0.99%	0.17%
$\alpha_d = 23$	1.32	3.13	0.62%	0.58%	0.24%

α_d the corresponding shift coefficients proportional to the respective hyperfine fields per Bohr magneton [30]. To estimate $\beta = 2\mu_B <r^{-3}>/N_o$, where $<r^{-3}>$ corresponds to the Cu 3d wavefunction, we take $<r^{-3}> \sim 6$.a.u. [38], giving $\beta = 135$ [27]. The spin hyperfine coupling is more difficult to estimate. Assuming a typical, negative core-polarization hyperfine field for 3d metals [38] we estimate [27] $\alpha = -23$ (emu/mole cu)$^{-1}$. However, recent studies of the copper shifts below T_c [25] give evidence that α_d is actually *positive*. This work also leads to the conclusion that the c-axis hyperfine coupling parameter is rather less than that for a-b plane spin components, which is contrary to the observed T_1 anisotropy (Fig. 2). A more elaborate formulation of the problem will be required to accommodate these new results.

For our present purposes we consider the possibilities of both positive and negative α_d, with the results shown in Table I. This analysis is the result of combining Eq. (1) with the relation $\chi_{exp} = \chi_{VV} + \chi_p + \chi_{core}$, where the experimental powder average susceptibility is $\chi_{exp} = 2.7 \cdot 10^{-4}$ emu/mole f.u. [26,27] and we adopt the estimate $\chi_{core} = -1.75 \cdot 10^{-4}$ emu/mole f.u. for the core diamagnetism [26]. In the breakdown of shift in Table I, the powder average spin shift is assumed the same for both Cu sites. The broad conclusion from the results of Table I is that the spin and orbital paramagnetism are of comparable magnitude and that the spin susceptibility is not significantly larger than band theory estimates of Pauli paramagnetism. The effect of a positive value for α_d is to lower significantly the estimate of χ_{VV} and increase χ_p by $\sim 40\%$. This does not change our earlier conclusions [27] in any qualitative sense.

III. Model Theories of Nuclear Relaxation

A. Normal State

In typical transition metals the d-electron Korringa process of nuclear spin-lattice relaxation [29,30] consists of three terms,

$$T_1^{-1} = (T_1^{-1})_{dip} + (T_1^{-1})_{cp} + (T_1^{-1})_{orb}\,, \tag{2}$$

where the first two arise from d-spin hyperfine fluctuations through the dipolar and core-polarization coupling, respectively, and the third comes from d-orbital fluctuations. In one-electron theory each of the terms in Eq. (2) has the form

$$(T_1^{-1})_\alpha = 4\pi\hbar[\gamma H_\alpha^{hf}\eta_\alpha]^2 k_B T S_\alpha\,, \tag{3}$$

where H_α^{hf}, S_α and η_α are the hyperfine field, symmetry factor and density of states parameters appropriate to that specific term, respectively. As noted in Sec. IIA, the observed rate for both Cu sites is greatly enhanced over estimates based on Eq. (2) and (3) using the Cu-site density of states $\eta_d = 3.5.10^{11}$ (erg · atom · spin direction)$^{-1}$ derived from the susceptibility analysis described in Sec. IIC [27].

The enhancement of T_1^{-1} as a consequence of electron-electron interactions has been treated both in the limit of exchange couplings in an otherwise free electron gas [39] as well as with on-site correlation energies in a tight-binding (Hubbard) model [40). One general result is that *only* the dynamic spin susceptibilities appear to undergo q-dependent enhancement effects. Thus, the orbital process in the presence of interactions continues to be given by Eq. (3), which falls short of the experimental relaxation rate by well over an order of magnitude [27]. For the spin-hyperfine terms it is useful to generalize the relaxation rate (Eq. (3)) to [39]

$$(T_1)_{\alpha}^{-1} = (\gamma^2 k_B T/g^2\mu_B^2)(A_{\alpha}^2+B_{\alpha}^2)\sum_{\vec{q}}\chi_{\alpha}''(\vec{q},\omega_o)/\omega_o\,, \tag{4}$$

where A_α and B_α are the hyperfine field parameters along the two axes perpendicular to the quantization axis and $\chi_\alpha''(\vec{q},\omega_o)$ is the imaginary part of the appropriate dynamic susceptibility at $\vec{q}$ and ω_o, ω_o being the nuclear resonance frequency. In Eq. (4) there could be a peak in $\chi_\alpha''(\vec{q},\omega_o)$ at large $\vec{q}$ with negligible enhancement near $\vec{q}=0$, thus explaining the enhancement of T_1^{-1} when there is no significant susceptibility enhancement. This case arises when there are antiferromagnetic correlations [40], which seem appropriate for YBCO because of the antiferromagnetic phase which occurs for $\delta > 0.6$.

Since it gives qualitatively correct behavior, it is useful to consider the results of the foregoing (Hubbard) model of weak antiferromagnetism [40] in more detail. In the first place, this calculation assumes a single correlated band. Thus, it may not adequately represent a multi-component system such as YBCO, where e.g., many feel that conduction takes place on the oxygen lattice alone [41-43]. Nevertheless, if we take the model prediction for T_1,

$$T_1^{-1} \propto T(T-T_N)^{-1/2} \tag{5}$$

and let $T_N \rightarrow 0$ for the superconducting part of the phase diagram, we obtain at least a fair representation of the normal state T_1 data for $\delta=0$ ($T_1^{-1} \propto T^{1/2}$, see Fig. 1). For larger values of δ, i.e., nearer the antiferromagnetic phase boundary, one might expect the applicability of Eq. (5) to improve. That this is apparently *not* the case may be seen from relaxation data for one of the Cu(2) NQR lines [44] which occur in a $T_c = 60$K ($\delta = 0.3$) sample which we have subsequently studied in detail. The correspondence of this model curve with the data is very poor.

Another problem with this sort of model is its inability to generate the observed level of T_1 anisotropy (Fig. 2). With isotropic dynamics, we note that the hyperfine anisotropy in the conventional picture [30] is given entirely by the dipolar interaction, which is inadequate to explain the data [27]. There are two possible answers to this dilemma. First, it may simply be that the dynamics are anisotropic, with greater enhancement for fluctuation along the c-axis. Another possibility is that the non-interacting, cubic model of Ref. 30 may be inadequate to deal with the anisotropic relaxation in this system.

B. Model Theories of Nuclear Relaxation: Superconducting State

The standard reference point for nuclear relaxation at $T < T_c$ is, of course, the BCS theory [33] in which the asymptotic behavior for $T << T_c$ is $T_1^{-1} \propto e^{-\Delta/kT}$, where 2Δ is the superconducting energy gap, and where in clean systems with nearly isotropic gap parameters there is a peak in T_1^{-1} just below T_c. In YBCO the relaxation behavior is decidedly *not* "BCS-like", since neither of the two copper sites nor the ^{89}Y nuclei have been found to exhibit a peak in the relaxation rate below T_c [18,22], and T_1^{-1} for both the Cu(2) site [12-14,16] and the ^{89}Y [18] have been found to decline much more steeply just below T_c than the BCS prediction [45]. A relaxation peak has been reported for ^{17}O by one group [17]. In view of the fact that these data are taken in a high magnetic field and that there is an apparent conflict with the established behavior of the other nuclear resonances in YBCO, we consider it wise to wait for corroboratory evidence from other laboratories before coming to a conclusion on this point.

The steep decline in T_1^{-1} for the Cu(2) site is all the more remarkable in contrast with the behavior of the Cu(1) site, where a more gradual decline than the BCS prediction is observed, leveling off at low temperatures to what is still a remarkably large relaxation rate at helium temperatures. The low-temperature relaxation mechanism for this site, which is apparently not intrinsic (see Fig. 3), is not understood. The leveling off of T_1 for the Cu(2) site at helium temperatures [32] may be a transferred relaxation effect from the Cu(1) sites, for which T_1 is nearly two orders of magnitude shorter and which lie only a few angstroms away.

A serious theoretical attempt to understand the collapse of the Cu(2) site T_1 process below T_c has recently been put forward by Koyama and Tachiki [46]. In this BCS-like model, electron correlations are taken account of explicitly, leading to strong enhancement of T_1^{-1} in the normal state. Below T_c, temperature dependence of the available density of states causes this T_1 enhancement to collapse very abruptly. Unfortunately, the correlations modeled here are *ferromagnetic*, so that along with the T_1 enhancement there is a concomitant enhancement of the d-spin shift and susceptibility. This is not observed, as we have noted in Sec. IIC. Hopefully, similar treatment of antiferromagnetic correlations may yield a more realistic picture. Another major question remaining in this work is the behavior of the Cu(1) sites, for which there is equally large enhancement for $T > T_c$, but which is only moderately affected by the superconducting pairing. Can this result be explained within the same framework as the Cu(2)? The answer to this as well as to many other questions awaits further developments.

REFERENCES

1. J. G. Bednorz and K. A. Müller, Z. Phys. B **64**, 189 (1986).

2. R. J. Cava, B. Batlogg, R. B. van Dover, and E. A. Rietman, Phys. Rev. Lett. **58**, 408 (1987).

3. C. W. Chu, P. H. Hor, R. L. Meng, L. Gao, Z. J. Huang, and Y. Q. Wang, Phys. Rev. Lett. **58**, 405 (1987).

4. K. B. Lyons, P. A. Fleury, L. F. Schneemeyer, and J. V. Waszczak, Phys. Rev. Lett. **60**, 732 (1988).

5. R. J. Birgeneau, Y. Endoh, Y. Hidaka, K. Kakurai, M. A. Kastner, T. Murakami, G. Shirane, T. R. Thurston and K. Yamada, "Static and Dynamic Spin Fluctuations in Superconducting $La_{2-x}Sr_xCuO_4$," preprint.

6. T. Siegrist, S. Sunshine, D. W. Murphy, R. J. Cava, S. M. Zahurak, Phys. Rev. **35**, 7137 (1987); M. A. Beno, L. Soderholm, D. W. Capone, II, D. G. Hinks, J. D. Jorgensen, J. D. Grace, I. K. Schuller, C. U. Segre, and K. Zhang, Appl. Phys. Lett. **51**, 57 (1987).

7. F. Beech, S. Miraglia, A. Santoro, and R. S. Roth, Phys. Rev. B **35**, 8778 (1987). J. D. Jorgensen, M. A. Beno, D. G. Hinks, L. Soderholm, K. J. Violin, R. L. Hitterman, J. D. Grace, I. V. Schuller, C. U. Segre, K. Zhang, and M. S. Kleefisch, Phys. Rev. B **36**, 3608 (1987).

8. J. M. Tranquada et al., Phys. Rev. Lett. **60**, 156 (1988).

9. P. Bordet, C. Chaillout, J. J. Caponi, J. Chevenas, and M. Marezio, Nature **327**, 687 ((1987).

10. H. Lutgemeier and M. W. Pieper, Solid State Commun. **64**, 267 (1987).

11. R. E. Walstedt, W. W. Warren, Jr., R. F. Bell, G. F. Brennert, J. P. Remeika, R. J. Cava, and E. A. Rietman, Phys. Rev. B **36**, 5727 (1987).

12. W. W. Warren, Jr., R. E. Walstedt, G. F. Brennert, G. P. Espinosa, and J. P. Remeika, Phys. Rev. Lett. **59**, 1860 (1987).

13. M. Mali, D. Brinkmann, L. Pauli, J. Roos, H. Zimmermann, and J. Hulliger, Phys Lett. **124**, 112 (1987).

14. Y. Kitaoka et al., J. Phys. Soc. Japan **57**, 29 (1988).

15. I. Furo, A. Janossy, L. Mihaly, P. Banki, I. Pocsik, I. Bakonyi, I. Heinmaa, E. Joon, and E. Lippmaa, Phys. Rev. B **36**, 5690 (1987).

16. T. Shimizu, H. Yasuoka, T. Imai, T. Tsuda, T. Takabatake, Y. Nakazawa, and M. Ishikawa, J. Phys. Soc. Japan **57**, 2494 (1988).

17. C. H. Pennington, D. J. Durand, D. B. Zax, C. P. Slichter, J. P. Rice, and D. M. Ginsberg, Phys. Rev. B **37**, 7944 (1988).

18. J. T. Markert, T. W. Noh, J. E. Russek, and R. M. Cotts, Solid State Commun. **63**, 847 (1987).

19. H. Alloul, P. Mendels, G. Collin, and P. Monod, Phys. Rev. Lett. **61**, 746 (1988).

20. Y. Kitaoka, K. Ishida, K. Asayama, H. Katayama-Yoshida, Y. Okabe, and T. Takahas, preprint.

21. The site assignment in Refs. 11, 12 and 15 is, for example, incorrect.

22. R. E. Walstedt, W. W. Warren, Jr., R. Tycko, R. F. Bell, G. F. Brennert, R. J. Cava, L. Schneemeyer, and J. Waszczak, Phys. Rev. B *38*, 9303 (1988).

23. P. C. Hammel, M. Takigawa, R. H. Heffner and Z. Fisk, Phys. Rev. B **38**, 2832 (1988).

24. Y. Kitaoka, S. Hiramatsu, Y. Kohari, K. Ishida, T. Kondo, H. Shibai, K. Asayama, H. Takagi, S. Uchida, H. Iwabachi, and S. Tanaka, Physica C **153-155**, 83 (1988).

25. M. Takigawa, P. C. Hammel, R. Heffner, and Z. Fisk, preprint.

26. A. Junod, A. Bezinge, and J. Muller, Physica C **152**, 50 (1988).

27. R. E. Walstedt, W. W. Warren, Jr., R. F. Bell, G. F. Brennert, G. P. Espinosa, R. J. Cava, L. F. Schneemeyer, and J. V. Waszczak, Phys. Rev. B *38*, 9299 (1988).

28. J. Korringa, Physica **16**, 601 (1950).

29. Y. Obata, J. Phys. Soc. Japan **18**, 1020 (1963).

30. Y. Yafet and V. Jaccarino, Phys. Rev. **133**, A1630 (1964).

31. A spot check confirms that to the accuracy of the data in Fig. 2, T_1 is independent of applied field.

32. T. Imai, T. Shimizu, T. Tsuda, H. Yasuoka, T. Takabatake, Y. Nakazawa and M. Ishikawa, J. Phys. Soc. Japan **57**, 1771 (1988).

33. L. C. Hebel and C. P. Slichter, Phys. Rev. **113**, 1504 (1962).

34. M. Horvatić, P. Ségransan, C. Berthier, Y. Berthier, and P. Butaud, preprint.

35. W. W. Warren, Jr., R. E. Walstedt, R. F. Bell, G. F. Brennert, R. J. Cava, G. P. Espinosa, and J. P. Remeika, Physica C **153-155**, 79 (1988).

36. L. F. Mattheiss and D. R. Hamann, Solid State Comm. **63**, 395 (1987); J. Yu, W. Massidda, A. J. Freeman, and D. D. Koeling, Phys. Lett. A **122**, 203 (1987).

37. The shift values here are taken from Ref. 25.

38. A. Abragam and B. Bleaney, *Electron Paramagnetic Resonance of Transition Ions*, Claredon Press, (Oxford, 1970), p. 455.

39. T. Moriya, J. Phys. Soc. Japan **18**, 516 (1963).

40. T. Moriya and K. Ueda, Solid State Comm. **15**, 169 (1974), and references therein.

41. V. J. Emery, Phys. Rev. Lett. **58**, 2794 (1987).

42. Y. Guo, J.-M. Langlois, and W. A. Goddard III, Science **239**, 896 (1988).

43. R. J. Birgeneau, M. A. Kastner, and A. Aharony, Zeitschrift für Physik (in press).

44. W. W. Warren, Jr., R. E. Walstedt, G. F. Brennert, R. J. Cava, B. Batlogg, and L. W. Rupp, Phys. Rev. B (Rapid Communications) to be published.

45. Here we are assuming $2\Delta \sim 3.5kT_c$ (weak coupling).

46. T. Koyama and M. Tachiki (unpublished).

NMR in High T_c Oxide Superconductors

Y. Kitaoka[1], *K. Ishida*[1], *K. Fujiwara*[1], *Y. Kohori*[1], *K. Asayama*[1], *H. Katayama-Yoshida*[2], *Y. Okabe*[2], *and T. Takahashi*[2]

[1]Department of Material Physics, Faculty of Engineering Science, Osaka University, Toyonaka, Osaka 560, Japan
[2]Department of Physics, Tohoku University, Sendai 980, Japan

Since the discovery of the high T_c oxide superconductor[1)], many experimental and theoretical efforts have been made to elucidate the mechanism for the occurrence of the superconductivity. In these systems the important problem is to investigate which atomic sites are responsible for the superconductivity. NMR measurement which provides information on the electronic state at each atomic site is a suitable tool for this study. Measurements of nuclear spin-lattice relaxation time T_1 have been performed on Cu sites at CuO_2 plane and CuO chain[2)3)4)5)] and on Y site[6)]. Recently we have prepared the ^{17}O substituted $YBa_2Cu_3O_7$ and measured T_1 of ^{17}O.[7)]

In this paper we compare and discuss the result of ^{17}O [7)] with $^{63,65}Cu$ in CuO_2 plane and ^{89}Y [6)] in $YBa_2Cu_3O_7$, ^{63}Cu in $(La,Sr)_2CuO_4$ [8)] and ^{205}Tl in $Tl_2Ba_2Ca_2Cu_3O_{10}$ and $Tl_2Ba_2Ca_1Cu_2O_8$ [9)] which are shown in Fig. 1.

(1) Cu in CuO_2 plane in $YBa_2Cu_3O_7$

The measurements of T_1 of ^{63}Cu and/or ^{65}Cu in CuO_2 plane have been made by several groups[2)3)4)5)]. $1/T_1$ in the normal state is not linear in T but changes slowly. This indicates that the antiferromagnetic spin fluctuations play an important role[5)]. In the narrow region from 100 K to 90 K, T_1 begins to follow the relation of T_1T=const. In order to see the behavior in the normal state in a wider temperature range, we have measured T_1 of ^{63}Cu in the Zn substituted system, $YBa_2(Cu_{0.98}Zn_{0.02})O_7$, where T_c is

Springer Series in Materials Science, Vol. 11
Mechanisms of High Temperature Superconductivity
Editors: H. Kamimura and A. Oshiyama

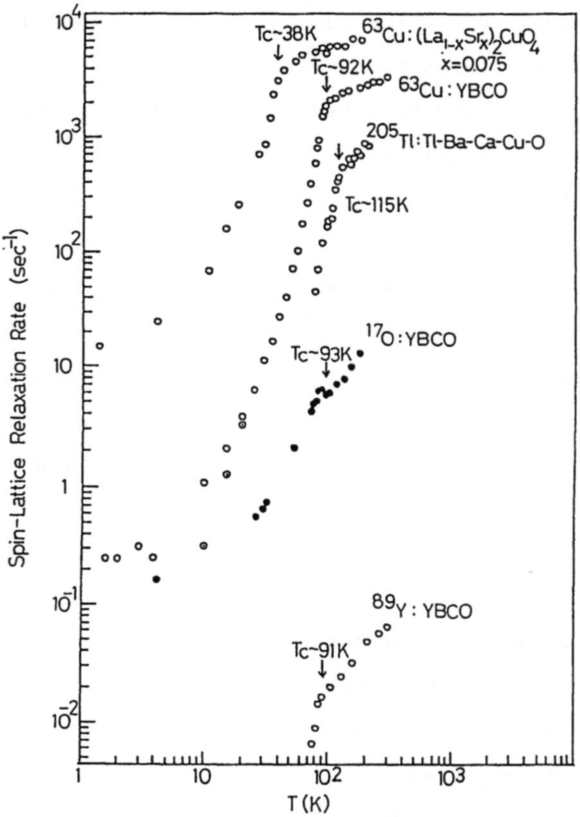

Fig. 1 Temperature dependence of the nuclear relaxation rates of ^{63}Cu in (La,Sr)CuO_4 (H=0), ^{63}Cu in CuO_2 plane (H=0), ^{89}Y (H=5T), ^{17}O (H=3T) in $YBa_2Cu_3O_7$ and ^{205}Tl in $Tl_2Ba_2Ca_2Cu_3O_{10}$ (H=1.2T). The mark ⊚ shows the rate of ^{63}Cu in YBCO due to magnetic interaction separated from the observed rate.[4)]

suppressed to 60 K.[10)] $1/T_1$ above 100 K is almost the same as in $YBa_2Cu_3O_7$, while below 100 K to 60 K $1/T_1$ follows the T_1T=const. law.[10)] In the case of 8% Zn substituted system, T_1 follows the relation of T_1T=const. down to 10 K. The nature of the magnetic moment of Cu is considered to be nearly localized but slightly itinerant.

In the superconduting state $1/T_1$ decreases rapidly without enhancement just below T_c. At low temperature the decrease of $1/T_1$ becomes slow. In our sample the quadrupole relaxation begins to appear below about 20 K which is confirmed by comparing T_1 of ^{63}Cu and ^{65}Cu.[4] The relaxation rate due to magnetic interaction is separated from the observed value as shown in the figure by ⊚ mark. $1/T_1$ due to magnetic interaction decreases nearly in proportion to T^3.

(2) ^{63}Cu in $(La,Sr)CuO_4$

Recently we have found the NQR signal of ^{63}Cu and ^{65}Cu in $(La_{0.925}Sr_{0.075})Cu_2O_4$ at 35.3 and 33.0 MHz.[8] The behavior of $1/T_1$ is quite similar to that in Y system. In the normal state $1/T_1$ is nearly proportional to T up to 60 K and then begins to show a saturation behavior with increasing temperature. At 4.2 K we have confirmed that the relaxation is governed by the quadrupole interaction.

(3) ^{89}Y in $YBa_2Cu_3O_7$

T_1 of ^{89}Y has been measured by Markert et al .[6] which is shown in the figure. $1/T_1$ in normal state follows T_1T=const. relation. Below T_c $1/T_1$ decreases rapidly which is quite similar to the behavior of ^{63}Cu in CuO_2 plane.

(4) ^{205}Tl in $Tl_2Ba_2Ca_2Cu_3O_{10}$ and $Tl_2Ba_2Ca_1Cu_2O_8$

$1/T_1$ of ^{205}Tl in $Tl_2Ba_2Ca_2Cu_3O_{10}$ is shown in the figure. As seen in the figure $1/T_1$ follows T_1T=const. above T_c, while $1/T_1$ in $Tl_2Ba_2Ca_1Cu_2O_8$ changes slowly[9]. Below T_c we have no enhancement just below T_c in both Tl systems, which is again similar to the behavior of Cu in CuO_2 plane.

(5) ^{17}O in $YBa_2Cu_3O_7$

The samples of ^{17}O substituted system is prepared as follows[7]. First $YBa_2Cu_3O_7$ was prepared by a solid state reaction of Y_2O_3, $BaCo_3$ and CuO all at least 99.99 % pure at 950°C in air for 24h. The gas exchange process consists of (1) removing ^{16}O from the sample at 700°C (2) keeping the pellet in $^{17}O_2$ atmosphere (51.244 at % ^{17}O) at 950°C for 24h, (3) cooling to 550°C at a rate of 50°C/h, (4) keeping at 550°C for 12h, and (5) cooling to room temperature at a rate of 50°C/h. There are three inequivalent O sites, CuO_2 plane, CuO chain and BaO layer. The measurement of Raman scattering showed that ^{17}O is replaced equally to these sites.[11)]

Figure 2(a) shows the NMR spectrum of ^{17}O in field of about 3T in $YBa_2Cu_3O_7$[7)]. The spectrum indicates a powder pattern of the first order electric quadrupole interaction with non zero asymmetry parameter, η. We first discuss which O sites the observed signal comes from. Figure 2(b) and (c) show, respectively, the spectra of ^{17}O in $YBa_2Cu_3O_{6.65}$ ($T_c \cong 60$ K) and $YBa_2Cu_3O_6$ (antiferromagnet) where the ^{17}O is replaced in almost the same procedure. As seen in the figure the signal intensity in $YBa_2Cu_3O_{6.65}$ is nearly the same as in $YBa_2Cu_3O_7$ in spite of the deficiency of oxygen at CuO chain. Furthermore, even in $YBa_2Cu_3O_6$ (Fig. 2(c)) having no Cu-O chain we have ^{17}O signal. Appreciable decrease in the integrated intensity was not observed compared with $YBa_2Cu_3O_7$.

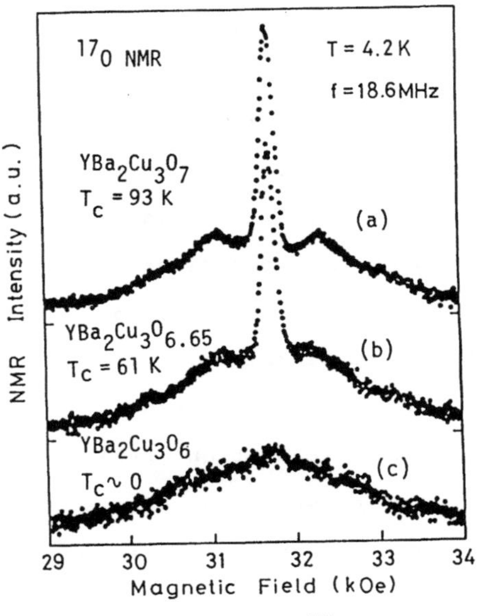

Fig. 2 NMR Spectra of ^{17}O in (a) $YBa_2Cu_3O_7$, (b) $YBa_2Cu_3O_{6.65}$ and (c) $YBa_2Cu_3O_6$.

(In the oxygen exchange process in $YBa_2Cu_3O_6$, the concentration of ^{17}O was about a half of those in O_7 and $O_{6.65}$ compounds which is the main reason for the slight decrease of the apparent integrated intensity in Fig. 2(c) compared with (a) and (b).

The disappearance of the central peak in $YBa_2Cu_3O_6$ is attributed to a broadening associated with the antiferromagnetic ordering. Even at the O site in CuO_2 plane which is a magnetically symmetrical position, a canting of the Cu magnetic moments due to the external field and/or an inhomogeneity in the sample produce a line broadening to smear the central line into the broad satellites. These results indicate that the contribution to ^{17}O spectrum from CuO chain is small. This is consistent with the fact that ^{17}O content in Cu-O chain is at most one-fourth of that in the CuO_2 plane even when ^{17}O is isotropically exchanged in all oxygen sites. Thus the signal observed is considered to come from CuO_2 and BaO layer although the contribution from the Cu-O chain cannot be excluded completely.

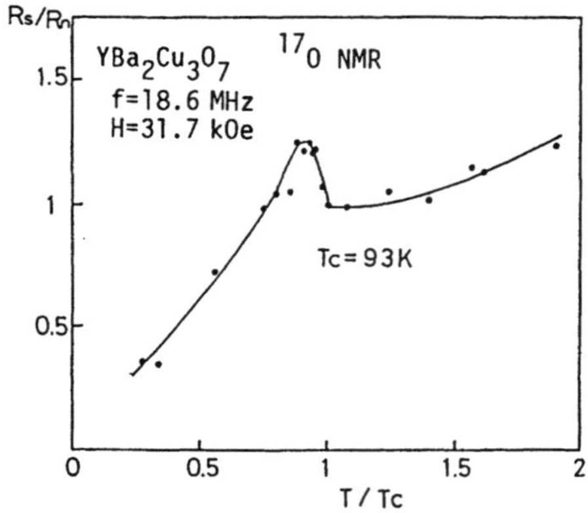

Fig. 3. Temperature dependence of the ratio of the nuclear relaxation rate in super and normal conducting state.

In Fig. 3 the ratio of the nuclear relaxation rate of ^{17}O in super and normal conducting state is plotted against temperature in order to see the relaxation behavior in Fig. 1 more clearly. The behavior of T_1 of ^{17}O is quite in contrast with those of $^{63,65}Cu$, ^{89}Y, and ^{205}Tl. Above T_c, $1/T_1$ of ^{17}O nearly follows the Korringa law up to 150 K, but deviates upward from a linear temperature dependence above 150 K. Just below T_c an enhancement appears which is characteristic for the BCS superconductor. At low temperature $1/T_1$ decreases slowly.

As is well known the presence of the coherence factor together with the divergence of the density of states of the quasiparticles at the gap edge both resulting from the s-wave pairing enhance $1/T_1$ just below T_c. In the p or d wave pairing the enhancement is small, sometimes disappears, owing to the absence of the coherence factor and the suppression of the density of state at the gap edge associated with the gap zeros on line or points at the Fermi surface. This anisotropy of the gap also produces a T power dependence of T_1 at low temperature in contrast to the exponential temperature dependence in s wave pairing. (The possibility of p-wave pairing is excluded in the Knight shift measurement.)[4)] We suggested the d-wave pairing from the relaxation behavior of Cu and Y.[4)]

However the appearence of the enhancement in ^{17}O relaxation suggests strongly s-wave pairing. The superconductivity is considered to be carried by the oxygen p holes Cooper pairing having s-symmetry. This result is consistent with the recent photo emission measurement[12)] which shows the doped holes to be predominantly oxygen 2p hole rather than Cu 3d-hole nearly localized at the Cu site. The relaxation of Cu is governed by the antiferromagnetic spin fluctuations considerably localized at the Cu site. The d-wave like behavior in Cu relaxation may be attributed to a decrease of the spin fluctuations below T_c assisted by a gapless effect due to spin fluctuation at Cu site.[13)] The strong spin fluctuations at the Cu site may induce a gapless state just below T_c to suppress the divergence of the density of states at the gap edge at the Cu site, while the antiferromagnetic fluctuations are more or less cancelled at the O site at CuO_2 plane to give an enhancement.

The relatively small enhancement $(1/T_{1s}/1/T_{1n})=1.3$ compared with those in conventional superconductors might be attributed to the gapless effect. The slow change of T_1 at low temperature may be attributed to some other relaxation mechanism such as spin diffusion to vortex cores. The relaxations of ^{89}Y and ^{205}Tl are considered to be governed by Cu fluctuations.

In conclusion, from the striking difference in the nuclear relaxation behavior between the oxygen and Cu sites, the high-T_c superconductor has been demonstrated to possess an unconventional nature although an uncertainty in site assignment still remains. The observation of an enhancement of $1/T_1$ just below T_c suggests an s-wave pairing in high-T_c superconductor. The different behavior in nuclear spin-lattice relaxation between oxygen and Cu may be explained in a scheme that the doped mobile holes are oxygen p-like, while the Cu d-holes are nearly localized with a freedom of spin fluctuation as derived by the photo-emission study.[5,6] The ^{17}O NMR study has presented that the role of oxygen is very important for high temperature superconductivity.

This work is partially supported by a grant in aid from Ministry of Education.

Reference

(1) J.G. Bednorz and K.A. Müller : Z. Phys. **B64** (1986).

(2) W.W. Warren et al. : Phys. Rev. Lett. **59** (1987) 1860.

(3) M. Mali et al. : Phys. Lett. **A124** (1987) 112.

(4) Y. Kitaoka et al. : J. Phys. Soc. Jpn. **57** (1988) 30.
Y. Kitaoka et al. : Physica **C153-155** (1988) 83.

(5) Imai et al.: J. Phys. Soc. Jpn. **57** (1988) 1771.

(6) Markert et al. : Solid State Commun. **63** (1987) 847.

(7) Y. Kitaoka et al. : submitted to Nature.
K. Ishida et al. : J. Phys. Soc. Jpn. **57** (1988) 2897.

(8) K. Ishida et al. : submitted to J. Phys. Soc. Jpn.

(9) K. Fujiwara et al. : in preparation
K. Fujiwara et al. : J. Phys. Soc. Jpn. **57** (1988) 2893.

(10)Y. Kohori et al. : J. Phys. Soc. Jpn. **57** (1988) 2905.

(11)H. Katayama-Yoshida et al. : Jap. J. Appl. Phys. Lett. **26** (1987) 2085.

C. Thomsen et al. : Solid State Commun. **65** (1988) 1139.

(12)T. Takahashi et al. : Phys. Rev. **B36** (1987) 5686.

(13)T. Koyama and T. Tachiki : submitted to Phys. Rev. B.

Charge Differentiation of Inequivalent Cu Sites of $YBa_2Cu_3O_y$ ($6.0 \leq y \leq 6.91$) Investigated by NQR and NMR

H. Yasuoka

Institute for Solid State Physics, University of Tokyo, Roppongi, Tokyo 106, Japan

Nuclear quadrupole resonance (NQR) and nuclear magnetic resonance (NMR) techniques have been utilized to characterize the local oxygen coordination of inequivalent Cu sites in $YBa_2Cu_3O_y$ ($6.0 \leq y \leq 6.91$). Essentially, four distinct NQR lines which correspond to 2, 3, 4 oxygen coordinated Cu(1) sites in the Cu-O chains and 5 oxygen coordinated Cu(2) sites in the Cu-O planes have been observed. The zero-field NQR frequencies of those are centered at about 30.1, 24.0, 22.0 and 31.5 MHz for ^{63}Cu, respectively. The electric filed gradient EFG which is mainly due to intra-atomic in origin depends only on the oxygen coordination, hence locally differentiated charged state of Cu atoms, in the Cu(1) sites. The charge differentiation is also found in the Cu(2) sites and the degree of it is found to be increased progressively with decreasing y from 6.91.

1. Introduction

The existence of oxides of intermediate or mixed Cu valence with transition from metallic to either high-T_c superconducting or antiferromagnetic insulating behavior adds an interesting new dimension to the problem of highly correlated electron system. Particularly, recent experimental developments in ternary and quaternary superconducting oxides, e.g. $(La_{1-x}Sr_x)_2CuO_{4-\delta}$, $YBa_2Cu_3O_{7-\delta}$ and $Tl_2Ba_2CaCu_2O_{8+\delta}$, have renewed considerable interest in the theoretical understanding of the mechanism of superconductivity. Although a number of theoretical models ranging from conventional phonon-mediated coupling to a variety of exotic mechanisms including the magnetic couplings have been explored, most of the theories are based on the unusual structural aspect of the Cu-O sublattices present in these compounds [1]. The tetragonal $La_{2-x}M_xCu_2O_4$ structure contains square-planar CuO_2 layers with weak coupling between them, while the orthorhombic $YBa_2Cu_3O_{7-\delta}$ structure consists of nearly square-planar CuO_2 layers (Cu(2) sites) with weakly bounded through Cu-O-Cu bonds along the c-axis to one dimensional CuO_3 chains (Cu(1) sites).

The most dramatic change in the transport and magnetic properties occurs in $YBa_2Cu_3O_y$ when the oxygen content y is varied. Namely, high temperature superconducting transition of about 90 K at y~ 7.0 decreases with decreasing y toward almost non-superconducting property at y~ 6.3. Further decreasing y, this system goes over to an antiferromagnetic ordered state. This transition seems to be accompanied by the metal to non-metal transition in a transport sense and the

Springer Series in Materials Science, Vol. 11
Mechanisms of High Temperature Superconductivity
Editors: H. Kamimura and A. Oshiyama

orthorhombic to tetragonal transition in a crystallographic sense. It is interesting to note that the superconducting phase exists exclusively in the orthorhombic phase. With changing y from 6.0 to 7.0, it is believed that starting with empty oxygen sites in the Cu(1) basal plane at y=6.0, excess oxygen atoms randomly (in most cases) fill sites between the Cu(1) atoms along the b-axes. This leads to a variation of the nominal Cu valence of the Cu(1) atoms (Cu^{1+} like valence state for y=6.0 to Cu^{3+} like for y=7.0), as well as an adjustment of the hole concentration of the Cu(2) planes. Presumably, the valence charge in the Cu atoms is associated locally with the oxygen coordination of those atoms.

Generally, in compounds with several metal-ion valence states, transfer of charge between sites of different valence can be of lower energy. In spite of this process, a number of mixed-valence transition-metal oxides have been found to undergo transitions to an insulating or a superconducting state. In order to examine such processes and transitions, it is desirable to use a probe which might resolve the charge and spin distributions of the inequivalent sites from both the static and dynamical points of view. In this paper, we describe our studies of the microscopic properties of $YBa_2CU_3O_y$, as a typical example, by means of Cu nuclear quadrupole resonance (NQR) and Cu nuclear magnetic resonance (NMR) techniques. These techniques offer an attractive way to investigate the microscopic properties since the response of inequivalent sites can, in principle, be resolved in the nuclear-resonance spectra and the dynamical properties of those can be studied by the nuclear magnetic relaxation at respective sites. It is also true that the variations in local behavior which might be expected, such as valence change in the Cu atoms, band paramagnetism, antiferromagnetism and superconductivity, all have quite characteristic signatures in nuclear-resonance experiments.

2. Sample Preparation and Experimental Procedures

It is well known that a number of physical quantities of $YBa_2Cu_3O_y$ strongly depends on the preparation procedure owing to the oxygen-nonstoichiometry and order-disorder transition. In this study, $YBa_2Cu_3O_y$ ($6.0 \leq y \leq 6.91$) with ordered Cu(1)-O planes (oxygen atoms are fractionally and randomly occupied only along the b-axis) was prepared by cooling slowly in either O_2 or N_2 gas at the Department of Chemistry, Kyoto University. Although details have been published elsewhere [2], we summarize briefly the procedure.

The starting material of $YB_2Cu_3O_{7-\delta}$ was prepared by the solid state reaction of Y_2O_3, $BaCO_3$ and CuO. The mixtures were pelletized, heated 900°C in air, ground and reheated. This process was repeated several times, following the final annealing for 5 hrs at 500 ℃ in air. The obtained material was confirmed to be a single phase with orthorhombic $YBa_2Cu_3O_{7-\delta}$. The samples with various oxygen content were prepared from the syntered (at 900 ℃) pellets as follows. Each of the pellets was precisely weighed and was firstly heated under air at 10 ℃ /min up to 800 ℃ and then cooled at the same rate at room temperature. In this run, the thermogravimetric analysis (TGA) traces show a distinct change in the curvature at 650 ℃, corresponding to the orthorhombic to tetragonal phase transition. The composition of the material at this temperature has been well established to be y=6.50 in equilibrium. The pellet was subsequently heated under oxygen or nitrogen (less than 0.001% O_2) gas at 10 ℃ /min up to the temperature range of 400-700 ℃, depending on the aimed value of y, and then cooled slowly at the rate of 2.5 ℃/min to room temperature. Oxygen content of the samples obtained by this treatment was determined by weight loss or gain, between the weight at 650 ℃ (y=6.50) and the final point in TGA. The accuracy of the microbalance was ± 0.1mg in weight, corresponding to ± 0.02 in y. The lattice parameters were significantly different from those of the quenched sample [2]. The characteristic features are noted as follows; (1) The a- and c-axes linearly increase with the same rate and the b-axis decreases with decreasing oxygen content. (2) The orthorhombic phase extends to about y=6.2 beyond y=6.5, and samples with $6.0 \le y \le 6.2$ show the tetragonal structure.

The nuclear resonance experiments have been performed by a spin-echo pulse sequence mode using conventional pulsed NMR spectrometers. For the frequency spectrum data taking, the super-heterodyne technique has been used for the signal detection and the amplitude of the time averaged spin-echo signal is recorded as a function of frequency at typically 0.1 MHz step. All the data shown in this paper were taken at 1.3 K. Since the absolute intensity measurements have not made in the present study, the amplitude of the spin-echo signal in the frequency spectra was normalized to the maximum value of it in the frequency range.

3. Results and Discussions

The superconducting and magnetic phase diagram in $YBa_2Cu_3O_y$ has been established by now from the neutron diffraction [3], μ SR [4] and NQR experiments [5]. With decreasing y from 7.0, the T_c of 90 K decreases, having some plateau around 60 K, to zero at y~ 6.3 where the orthorhombic to tetragonal crystallographical transition takes place. Further decreasing y, the antiferromagnetic long range order is developed and the Neel temperature T_N reached its maximum value of about 500 K at y=6.0. As is described in the previous section, when y is varied from 6.0 to 7.0 the excess oxygen in our sample believed to occupy fractionally and randomly at the sites in between the Cu(1) atoms along the b-axis. Then, we would expect three different oxygen coordinated Cu(1) sites which may have different valence states, hence NQR frequencies. We start from showing the results for y=6.0.

The zero-field spin-echo spectra in $YBa_2Cu_3O_{6.0}$ measured at 1.3 K are shown in Figs. 1 (a) and (b). Aside from the detailed discussion of the line broadening, the classification of the spectrum is rather straightforward from the frequencies for the resonance peaks [6]. First, the ratio of those in Fig. 1 (a) is found to be exactly the same as that of the nuclear quadrupole moment of Cu nuclei ($^{63}Q/^{65}Q=1.081$).

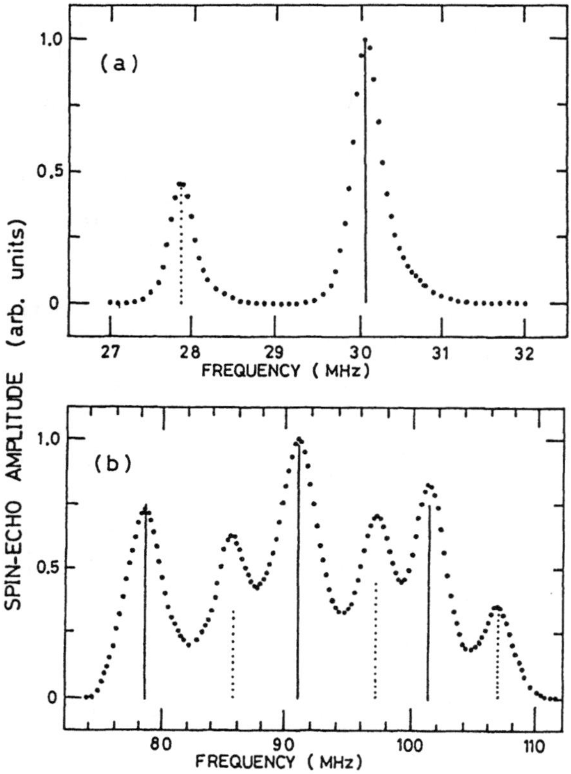

Fig. 1. ^{63}Cu and ^{65}Cu NQR (a) and AFNR (b) spin-echo spectra in $YBa_2Cu_3O_{6.0}$ at 1.3 K. Solid and dashed bars indicate the peak frequency and relative intensity for ^{63}Cu and ^{65}Cu nuclear-resonances, respectively.

Hence, these resonances are classified to $^{63}Cu/^{65}Cu$ NQR lines with $\nu_Q(^{63}Cu)$ = 30.11± 0.02 MHz and $\nu_Q(^{65}Cu)$ = 27.89± 0.02 MHz. Second, the six peaks observed in Fig. 1 (b) are identified to be due to the quadrupolar split Zeeman transitions for ^{63}Cu and ^{65}Cu in the so-called antiferromagnetic nuclear resonance (AFNR). From the detailed analysis of the line profile [6], the following conclusions have been obtained. (1) Since any magnetic broadening can not account for the relative line width of ^{63}Cu and ^{65}Cu NQR lines, the transferred hyperfine or the dipolar field is completely canceled out in this site. This fact immediately leads to a conclusion that the NQR signals are associated with the Cu(1) sites which have two oxygen coordination. (2) The quadrupolar split AFNR lines are assigned to the five oxygen coordinated Cu(2) sites where the broadening of lines are found to be magnetic in origin. (3) The antiferromagnetic Cu moments reside only on the Cu(2) sites and the direction of moments is perpendicular to the c-axis. This conclusion and the complete cancellation of the magnetic field at the Cu(1) sites are in general accord with the predicted spin structure by the neutron diffraction experiment, i.e. the ordered wave vector within a Cu(2)-O plane is (1/2, 1/2) and the planes are coupled antiferromagnetically along the c-axis [7].

Next we show the NQR spectra for samples with different values of y in Fig. 2. Since all the NQR signals have appeared as a twin structure with the intensity ratio of 1 : 0.45 which is due to the ratio of natural abundance of ^{63}Cu to ^{65}Cu, we only discuss the signal observed at higher frequency side (^{63}Cu) hereafter. It is immediately observed in Fig. 2 that a new signal has gradually appeared at 23.96 MHz with increasing oxygen content from y=6.0. Since the excess oxygen atoms are considered to be located only at the sites between the Cu(1) atoms along the b-axis, the new signal is naturally assigned to be due to the three oxygen coordinated Cu(1) sites. The similar NQR signal has previously been observed in the tetragonal phase by Lutgemeier [8] and Kitaoka et al. [5], and they have assigned it as the same as here. From the NMR studies in high fields, we have found that the asymmetric EFG parameter is nearly one [9]. Then the quadrupole frequency ν_Q is determined to be 20.75 MHz. The line width increases with increasing y, and the line broadening is mainly accounted for the inhomogeneous distributions of EFG. It should be noted, however, that there exists no evidence for magnetic broadening at both the $Cu(1)_2$ and $Cu(1)_3$ sites (subscripts denote the number of oxygen coordination). This clearly means that no ordered magnetic moments develop on the Cu(1) sites even at 1.3 K and no essential change in the antiferromagnetic spin structure takes place at the Cu(2) sites. Some of the neutron diffraction [10] and NQR experiments [11] concluded the existence of magnetic ordering at the Cu(1) sites that is inconsistent with the present results. However, the most recent neutron diffraction study by Tranquada et al. shows no magnetic order at Cu(1) sites [3]. We believe the discrepancies may come from difference in the sample quality. The latter evidence has been proved by the fact that no essential change in the AFNR spectrum has been observed. This indicates also that the magnitude of the antiferromagnetic moments is independent on the oxygen content.

For the samples with y>6.3, antiferromagnetic long range order disappears and superconductivity starts to develop. The crystal structure is transformed to orthorhombic. Therefore, we should expect signals of NQR for the Cu(2) sites. Those are actually observed at 31.45 MHz as a highest intensities, accompanied by several other signals. From the high field NMR studies of these samples, it is known that all the lines seen above 26 MHz have the asymmetric EFG parameter of essentially zero [9], which is expected only for either the $Cu(1)_2$ or the $Cu(2)_5$ sites. Since the line for $Cu(1)_2$ sites is seen around 30.5 MHz and the intensity is decreasing with increasing y due to the quick decrease of the probability of finding such sites, the rest of lines may be considered to be due to the $Cu(2)_5$ sites. It should also be noted in the spectrum with intermediate oxygen content that the frequency for the signal associated with the $Cu(1)_3$ sites stays almost unchanged.

More complicated but characteristic NQR spectra have been observed in the samples with much higher oxygen content. To explain this, it is better to start from the spectrum of y=6.91, since the complete site assignment has been made from the NMR studies using highly oriented powder sample [12] and single crystal [13]. The conclusion was as follows: The higher frequency line observed at 31.5 MHz is due to the

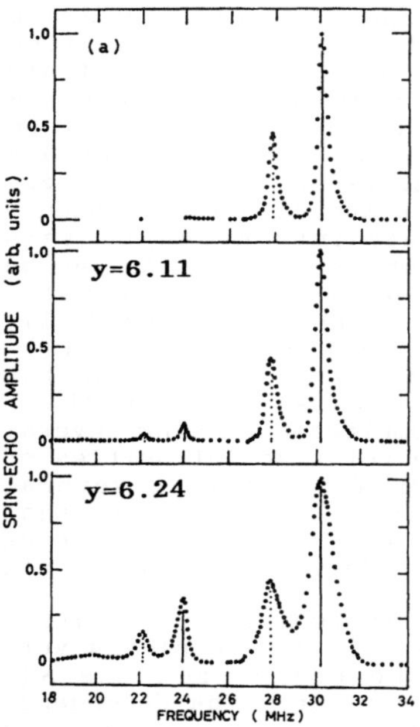

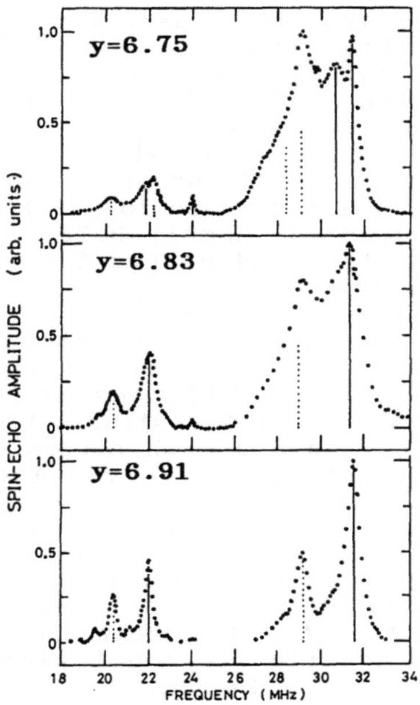

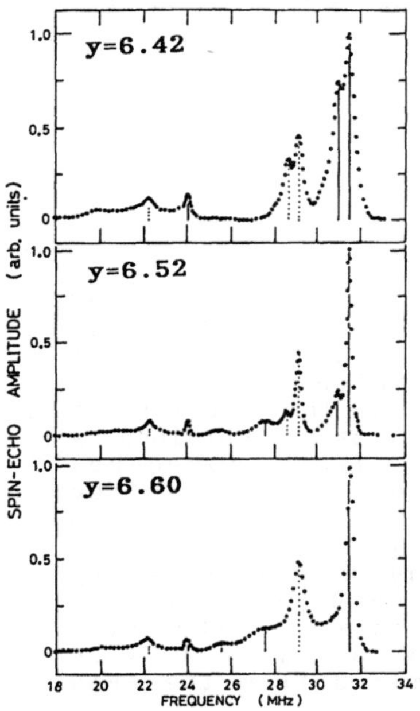

Fig. 2. $^{63}Cu/^{65}Cu$ NQR spin-echo frequency spectra in $YBa_2Cu_3O_y$ at 1.3 K. Solid and dashed bars, indicate the peak positions for ^{63}Cu and ^{65}Cu NQR in inequivalent Cu sites, respectively.

$Cu(2)_5$ sites with ν_Q=31.54± 0.05 MHz and η =0.01± 0.01. The lower line observed at 22.0 MHz is due to the $Cu(1)_4$ sites with ν_Q=19.35± 0.05 MHz and η =0.95± 0.02. The unexpectedly large value of η for $Cu(1)_4$ sites should be noted here. A theoretical calculation of the EFG tensors for this site predicts $\eta \sim 0.74$, where Cu 3d orbital contributions are included [14]. In order to explain the experimental findings, a lower valence state for O(4) sites in the Ba-O layers should be considered, namely holes are mainly introduced to this oxygen sites and nearly O^{-1} states are realized. As can be seen in Fig. 2, the oxygen content dependence for the Cu(1) signals is rather simple, i.e. the intensity for the $Cu(1)_4$ sites decrease with decreasing y while that of the $Cu(1)_3$ sites increases. This is qualitatively accounted for the change in probabilities of finding those sites with y. Contrary, the $Cu(2)_5$ signals become complicate, having several lines overlapped each others and the relative intensities of those lines depend on y. The reason for existing several lines at the $Cu(2)_5$ sites may be related to the fact that these sites are coupled to the Cu(1) sites having different oxygen coordinations, hence different Cu valence states, via the oxygen atoms in the Ba-O layers.

Although the full investigation including the relative intensities for different samples has not been completed yet, rather qualitative conclusions may be drawn from the NQR spectrum shown in Fig. 2. To see more clearly the dependence of each NQR frequency on the oxygen content, the peak frequency for the respective sites are plotted as a function of y in $YBa_2Cu_3O_y$ in Fig. 3. For the Cu(2) sites, we have used the decomposing procedure of the observed spectra to

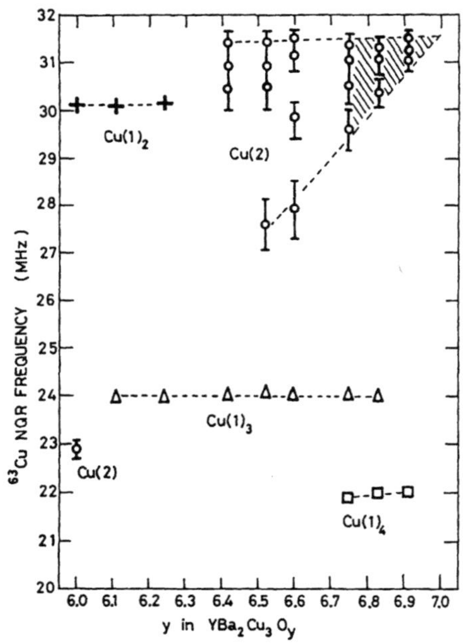

Fig. 3. Oxygen content dependence of the ^{63}Cu NQR frequencies for the inequivalent Cu sites in $YBa_2O_3O_y$. The data for the Cu(2) sites are obtained by decomposing procedure. Note that spectra in the samples with higher oxygen contend is distributed continuously as shown by slanting bars. The NQR frequencies for the Cu(2) sites at y=6.0 was determined from the AFNR spectrum.

obtain the peak frequencies for the different sites. A schematic illustration of the local oxygen coordination around the Cu(1) and C(2) sites for y=6.0, 6.5 and 7.0 with the NQR frequencies associated with ^{63}Cu is shown in Fig. 4. It is immediately apparent from Fig. 3 that the three distinct NQR frequencies for the Cu(1) sites are independent of y, although the relative y-dependent intensities are in general accord with the prediction based on the probability calculations.

The EFG is usually written as,

$$eq = eq_{ion} + eq_{el},$$
$$eq_{ion} = eq_{latt}(1-\gamma), \qquad (1)$$

where eq_{ion} and eq_{el} are the ionic and electronic contributions, respectively. The ionic term may be evaluated as the contributions from charges on the surrounding lattice points, eq_{latt}, and distorted core interactions which may be accounted for in terms of the familiar Sternheimer antishielding factor γ[15].

A simple point charge calculation predicts about 10% change in the NQR frequencies when y is varied, that is not consistent with the experimental findings. Therefor above fact yields a conclusion that main source of the EFG is intra-atomic in origin and it depends only on the charged state of Cu atoms which are well differentiated by the oxygen coordination nearby. We believe that the lattice contribution to the EFG is relatively small and contributes only the line broadening. Contrary to this, the NQR frequencies for the Cu(2) sites tends to spread toward the lower frequency side. Two interesting features are noticeable: (1) The frequency for the most varying line is seems to be joined eventually to that observed in AFNR at y=6.0, hence this line may comes from the Cu(2) sites with two oxygen coordinated Cu(1) sites along

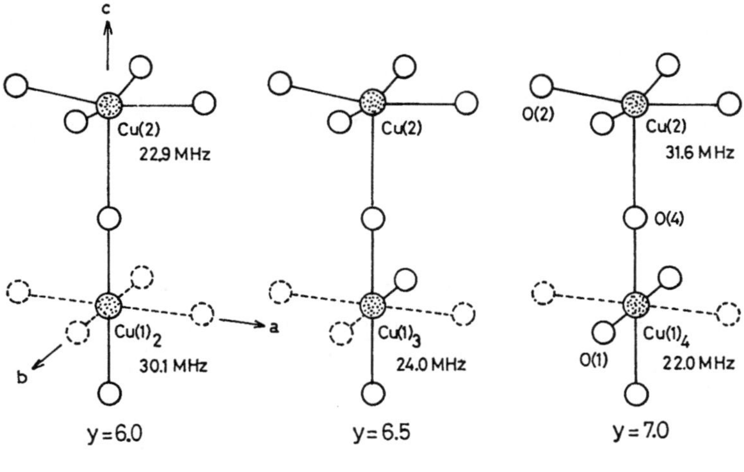

Fig. 4. Schematic illustration of the Cu sites with locally different oxygen coordinations and corresponding NQR frequencies for ^{63}Cu nuclei.

the c-axis. (2) All the split lines seems to be collapsed to a unique line at y=7.0 where we would expect the narrowest line width. Although the detailed assignment of the observed line is not given at present, it is nevertheless clear that the charge differentiation exists even in the Cu(2) layers and the degree of that is found to be increased with decreasing oxygen content, i.e. having more three or two oxygen coordinated Cu(1) sites. This general trend might be related to the decrease of T_c and play an important role for the understanding of the mechanism of high-T_c superconductivity in this system.

4. Concluding Remarks

Our measurements of oxygen concentration dependence of the NQR lines for inequivalent Cu sites in $YBa_2Cu_3O_y$ show the following characteristic features.

(1) The antiferromagnetic moments reside only on the Cu(2)-O plane sites and the direction of the moments is perpendicular to the c-axis.

(2) The electric field gradient EFG which is mainly due to intra-atomic in origin depends only on the oxygen coordination, hence on locally differentiated charged state of the Cu atoms, in the Cu(1) chain sites.

(3) The charge differentiation is also found in the Cu(2) plane sites and the degree of it is found to be increased progressively with decreasing y from 6.91, that might be related to the decrease of T_c with decreasing y in this system.

Acknowledgments

The author would like to acknowledge for the collaboration with T. Shimizu, T. Imai and S. Sasaki for resonance measurements, and Y. Ueda and K. Kosuge for sample preparation. Present research is supported by Special Project Research on High Temperature Oxide Superconductors of Grant-in-Aid for Scientific Research from the Ministry of Education, Science and Culture.

References

1. See for example papers in Novel Superconductivity, ed. S.A. Wolf and V.Z. Kresin (Plenum New York, 1987).
2. Y. Ueda and K. Kosuge, Physica, C156 (1988) 281.
3. J.M. Tranquada, A.H. Moudden, A.I. Goldman, P. Zolliker, D.E. Cox, G. Shirane, S.K. Sinha, D. Vaknin, D.C. Johnston, M.S. Alvarez and A.J. Jacobson, Phys. Rev. B.
4. N. Nishida et al., J. Phys. Soc. Jpn., 57 (1988) 597, and J.H. Brewer et al., Phys. Rev. Lett., 60 (1988) 1073.
5. Y. Kitaoka, K. Ishida, K. Asayama, H. Takagi, H. Iwabuchi and S. Uchida, Proc. Int. Conf. of Magnetism, Paris, 1988.
6 H. Yasuoka, T. Shimizu, Y. Ueda and K. Kosuge, J. Phys. Soc. Jpn., 57 (1988) 2659.

7. J.M. Tranquada, D.E. Cox, W. Kunnman, G. Shirane, M. Suenaga, P. Zolliker, D. Makim, S.K. Sinha, M.S. Alvarez, A.J. Jacobson and D.C. Johnston, Phys. Rev. Lett., 60 (1988) 56, and J. Rossat-Mignor, P. Barlet, M.J.G. Jurgens, J.Y. Henry and C. Vettier, Physica, C1152 (1988) 19.
8. H. Lutgemeier, to be published in Physica C153 (1988).
9. T. Shimizu, H. Yasuoka, Y. Ueda, K. Kosuge, in preparation.
10. H. Kadowaki, M. Nishi, Y. Yamada, H. Takeya, H. Takei, S. M. Shapiro and H. Shirane, Phys. Rev. B37 (1988).
11. Y. Kitaoka, S. Hiramatsu, K. Ishida, K. Asayama, H. Takagi, H. Iwabuchi, S. Uchida and S. Tanaka, J. Phys. Soc. Jpn., 57 (1988) 737.
12. T. Shimizu, H. Yasuoka, T. Imai, T. Tsuda, T. Takabatake, Y. Nakazawa and M. Ishikawa, J. Phys. Soc. Jpn., 57 (1988) 2494.
13. C. H. Pennigton, D.J.Durand, D. B. Zax, C. P. Slichter, J. P. Rice and D. M. Ginsberg, Phys. Rev., B37 (1988) 7944
14. F. J. Adrian, Phys. Rev., B38 (1988) 2426
15. R.M. Sternheimer, Phys. Rev., 95 (1954) 736.

Mössbauer Studies of High-T_c Oxides

Teruya Shinjo[1] *and Saburo Nasu*[2]

[1]Institute for Chemical Research, Kyoto University, Uji, Kyoto-fu 611, Japan
[2]Faculty of Engineering Sciences, Osaka University, Toyonaka 560, Japan

1. INTRODUCTION

Mössbauer spectroscopy has been known as one of the very useful experimental tools in fundamental solid state physics. Main parameters which we obtain from Mössbauer spectra are: (1) isomer shift, (2) quadrupole interaction, (3) magnetic hyperfine interaction, and (4) recoilless fraction. An electronic structure of the relevant atom is argued from the value of isomer shift and in usual ionic cases the valence state can be determined unambiguiously. If a magnetic hyperfine splitting is observed, it is very helpful for confirming the existence of magnetic order. From the temperature dependence of hyperfine field, the magnetic transition temperature is estimated.

For example, results on $CaFeO_3$ are shown in Figs. 1 and 2./1/ This is a rather old work by the present authors' group but is useful for readers who are not familiar with the Mössbauer spectroscopy to learn about an information from Mössbauer measurements. $CaFeO_3$ is a cubic perovskite compound containing Fe^{4+} ions whose interesting characteristic is a charge disproportionation as shown below.

At 300K, the spectrum is a sharp single line and the isomer shift (the peak position) is a typical one for Fe^{4+} state, $+0.07 mm.s^{-1}$, which is remarkably smaller than the values of Fe^{3+} or Fe^{2+}. Below about 280K, the peak begins to split. The peak separation becomes larger with the decreasing of temperature and reaches a constant value, $0.33 mm.s^{-1}$, at about 200K. The spectrum at 154K shown in Fig. 1 is a doublet with equal intensities. Such a pattern is usually interpreted as a result of quadrupole interaction. In this case, however, the reason is not a quadrupole interaction but is two valence states of equal amounts. This conclusion is drawn from the result at 4.2K. As also shown in Fig. 1, the spectrum at 4.2K is a superposition of two, magnetically-split 6-line patterns. The hyperfine fields and isomer shifts for the two portions are 41.5 and 27.8T, and 0.34 and $0.00 mm.s^{-1}$, respectively. The charge states are entirely different from each other and regarded as Fe^{3+} and Fe^{5+}. This charge disproportionation, $2Fe^{4+} = Fe^{3+} + Fe^{5+}$, was discovered by Takano/2/. From the temperature dependence of the hyperfine fields, the magnetic transition temperature (T_N) is determined to be 115K.

In Fig. 2, the difference of isomer shifts in $CaFeO_3$ is shown as a function of temperature, together with the hyperfine fields. Below about 280K, the charge disproportionation occurs and two charge states are respectively stable at least within the characteristic observation time of the Mössbauer effect(10^{-7}s). One may understand that the valence fluctuation in $CaFeO_3$ is dynamic at higher temperatures and becomes static at lower than a certain temperature. At the time when we had made these

Springer Series in Materials Science, Vol. 11
Mechanisms of High Temperature Superconductivity
Editors: H. Kamimura and A. Oshiyama
Springer-Verlag Berlin Heidelberg © 1989

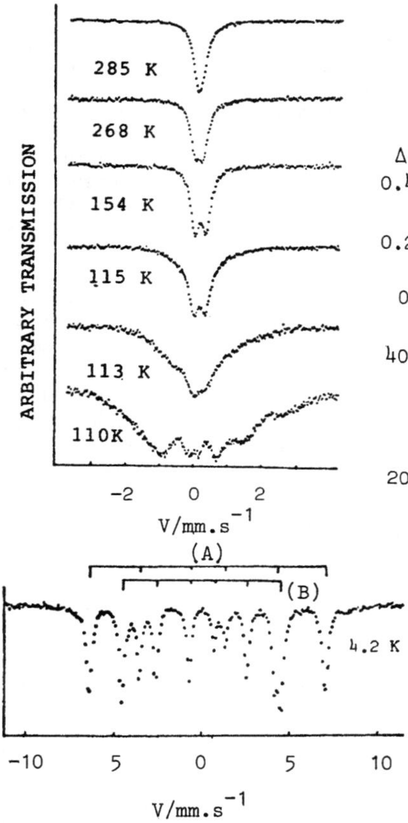

Fig.1. ^{57}Fe Mössbauer absorption spectra of $CaFeO_3$ at various temperatures /1/.

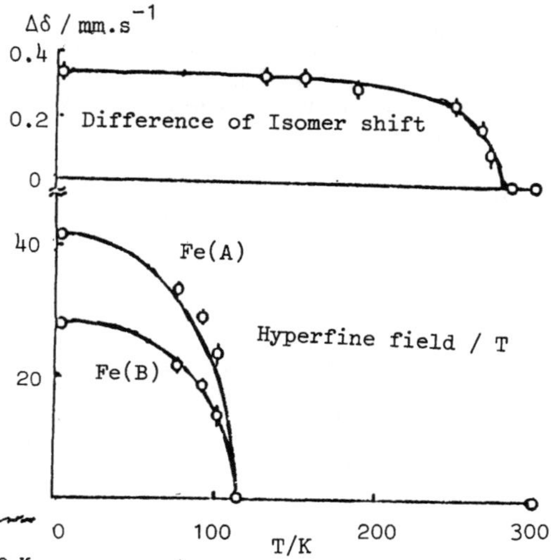

Fig.2. Difference of isomer shift in the two valence states in $CaFeO_3$ as a function of temperature (upper) and two hyperfine fields vs. temperature curves (lower) /1/.

measurements, none could imagine that such a compound has any relation with superconductivity. Since the discovery of high-Tc oxides, a great attention is being paid on magnetic perovskite compounds and therefore the results on $CaFeO_3$ have been recalled. Thus, the Mössbauer spectroscopy is very useful to study the valence state and the magnetic structure as far as they are static. For dynamical charge- or magnetic-fluctuations, on the other hand, we can get some information only if the relaxation time is near the Mössbauer characteristic time window.

2. MÖSSBAUER STUDIES WITH OTHER NUCLEI THAN Fe

Although the important elements in high Tc superconducting materials are Cu and O, unfortunately no Mössbauer isotope is available in these elements. Among rare earth (RE) elements, there are some Mössbauer isotopes and already several reports on RE ions in superconducting $REBa_2Cu_3O_7$ have been published. Eibschutz et al. measured ^{191}Eu Mössbauer spectra for

$EuBa_2Cu_3O_7$ and confirmed that the valence state of Eu is trivalent, whose isomer shift is almost the same as a non-conducting oxide such as Eu_2O_3/3/. The spectra were observed with changing temperature but no change was found at the superconducting-to-normal transition temperature(Tc), which is 94K in this sample(Fig.3). This result is reasonable since Eu ions are supposed not to participate in the superconductivity. However, it is to be noted that even if the relevant atom is involved in the superconductivity, a big change is not expected in the Mössbauer parameters. Accumulated results on superconducting Sn compounds indicate that a difference of electronic state between super and normal conductivity is in general too small to detect in such Mössbauer parameters as isomer shift or quadrupole interaction.

Similar results were obtained in ^{155}Gd measurements by Smit et al./4/ and ^{170}Yb measurements by Hodges et al./5/. The isomer shift values suggest that the conduction electron densities at the RE ions are almost zero and therefore it is thought that RE ions do not contribute to the conductivity. The measurements were also made at low temperatures and in both cases the occurrence of magnetic order was confirmed. For $GdBa_2Cu_3O_7$, T_N was estimated to be 2.5K and for $YbBa_2Cu_3O_7$, 0.35K(Figs.4 and 5). The magnetic ordering of RE ions at low temperature is regarded as another evidence that RE ions are not involved in the superconductivity. Therefore for the understanding of superconductivity, the role of RE ions may be less important. On the other hand, from a viewpoint of magnetism, the behaviors of RE ions in these compounds are of great interest. The RE sheets

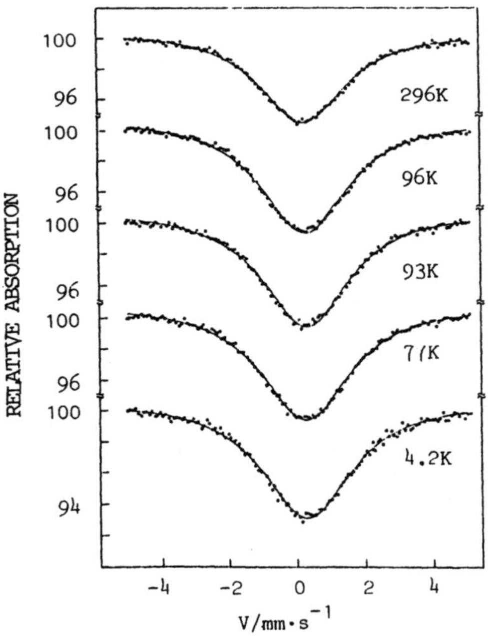

Fig.3. ^{151}Eu Mössbauer spectra of $EuBa_2Cu_3O_{7-\delta}$ as a function of temperature /3/.

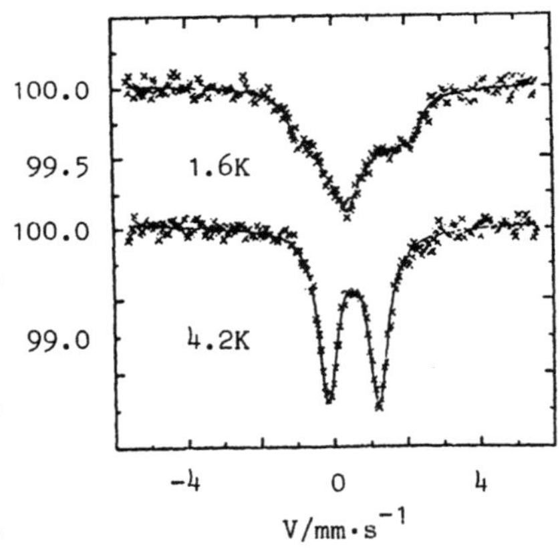

Fig.4. ^{155}Gd Mössbauer spectra of $GdBa_2Cu_3O_7$ at 1.6K and 4.2K /4/.

sandwiched by superconducting layers may be an ideal two-dimensional magnetic lattice. The dimensionality and the magnetic structure have been argued in Ref. 4.

As well as ^{57}Fe, ^{119}Sn is one of the most convenient nuclei for the Mössbauer effect. Measurements on Sn impurity in $YBa_2Cu_3O_7$ (Sn/Cu = 4%) were carried out by Yuen et al./6/. The authors suppose that Sn ions replace the Cu sites. However, the Mössbauer spectrum is a broad single line and an analysis of the line profile is very hard. The situation of the impurity Sn, for instance the site preference among the Cu sites, is not clear. Nevertheless, an interesting feature was found in the measurement of recoilless fraction as shown in Fig. 6. The recoilless fraction is estimated as the total area of the absorption spectrum and is proportional to $\exp[-\langle x^2\rangle]$, where $\langle x^2\rangle$ represents the mean square displacement of the relevant atom. The curve of the temperature dependence indicates that a lattice softening occurs below about 150K. In contrast,

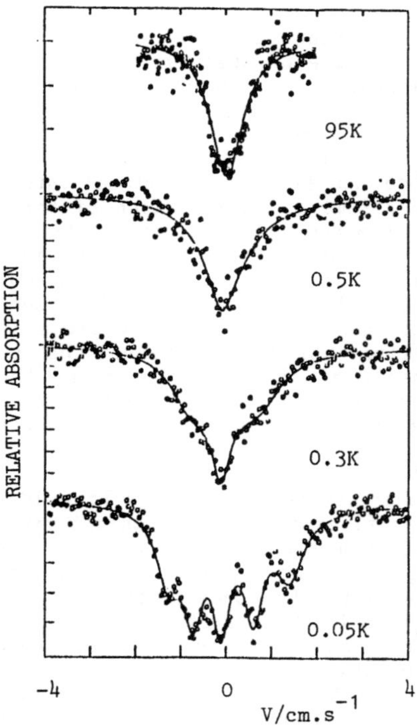

Fig.5. ^{170}*Yb Mössbauer spectra of* $YbBa_2Cu_3O_7$ *at low temperatures /5/.*

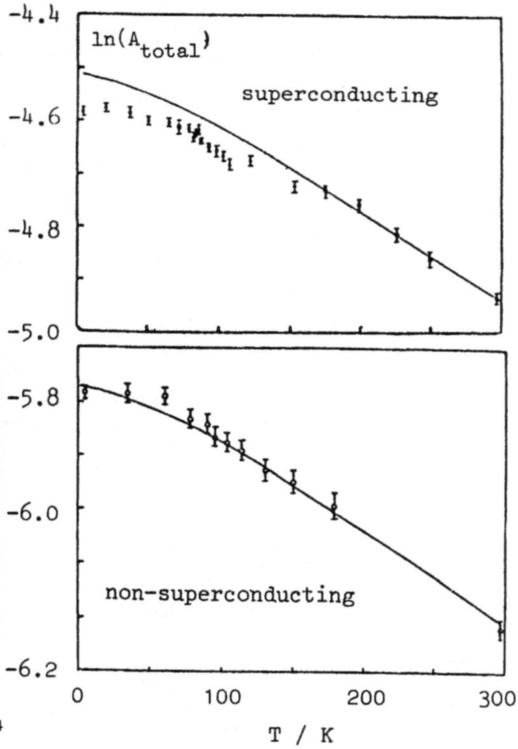

Fig.6. Recoilless fraction vs. temperature of ^{119}*Sn Mössbauer spectra of Sn-doped superconducting* $YBa_2Cu_3O_7$ *(upper) and of an oxygen-deficient, non-superconducting* $YBa_2Cu_3O_{7-x}$ *(lower). The curves are calculated ones using phonon spectra /6/.*

as also shown in the same figure, another sample, which is an oxygen-deficient non-superconducting $YBa_2Cu_3O_{7-x}$ doped with Sn, has no anomaly in the temperature dependent curve. Therefore, the softening is regarded to be correlated to the superconductivity. According to the authors' opinion, this softening cannot be considered as evidence in favor of the electron-phonon mechanism. A modification of local lattice vibration due to the occurrence of superconductivity is possible in other models. Since a quantitative determination of recoilless fraction is not easy, the reproducibility of this work has to be checked. Anyway the importance of Mössbauer spectroscopic studies on local lattice dynamics has been revealed and a further development is expected.

3. ^{57}Fe MÖSSBAUER STUDIES

Behaviors of impurity Fe in $REBa_2Cu_3O_7$ are of particular interest since Fe atoms occupy the Cu sites. Magnetic properties of the Cu-O system are considered to be crucial for understanding the mechanism of superconductivity. Through ^{57}Fe Mössbauer measurements, we expect to explore the magnetic properties of $YBa_2Cu_3O_7$.

Concerning Fe impurities in $REBa_2Cu_3O_7$, a great number of publications have already appeared. Figure 7 shows an example of spectrum for ^{57}Fe in $YBa_2Cu_3O_7$ (Fe/Cu=2%) measured at room temperature by the present authors' group /7/. Other investigators also get similar results and the essential features of the line profile are almost the same for all the results. However, the details of the line profiles, relative intensities for instance, depend on the investigators since the samples depend delicately on the conditions of preparation. If we assume a superposition of three doublet spectra, a fairly good fitting is obtained. The isomer shifts for the two major components are close to Fe^{4+} values but the remaining minor one is an Fe^{3+} value. As is well known, in the structure of $YBa_2Cu_3O_7$, there are two Cu sites, so-called "chain"(Cu-I) and "plane"(Cu-II). Then, one would naively attribute two major Mössbauer components to Fe atoms in the two Cu sites. However, the site assignment cannot be so straightforward. For the site assignment, it is useful to refer the results on the magnetic properties. Subsequently, the present authors reach the conclusion that Fe atoms preferentially occupy the Cu-chain sites/8/.

It has been regarded as one of the characteristics of high-Tc oxides that magnetic impurities do not strongly influence the superconductivity. The case of magnetic RE ions is easy to understand since they are not involved in the conduction bands. Although Fe atoms replace the Cu sites. the superconductivity is maintained even when the Fe concentration is 10%. Tc(R=0) of $YBa_2(Cu_{1-x}Fe_x)_3O_7$ was observed to be 78 and 34K, respectively for x=2 and 8%. The results from other groups are very similar, even quantitatively /9-11/. Mössbauer measurements at low temperature in the presence of a strong magnetic field is useful to check that Fe impurities really possess local magnetic moments in a superconducting matrix. Figure 8 shows the result for 2% Fe sample at 4.2K with and without an external field, 4.5T. Zero field spectrum has no significant change from that at 300K. On the other hand, by applying an external field, we obtain a magnetically-split spectrum, whose splitting is much larger than the external field. Namely each Fe ion has a paramagnetic moment and is significantly polarized by the external field. Although the spectrum is very broad, the average of the induced hyperfine field is estimated to be about 20T. A computer fitting is not applicable for this broad spectrum of

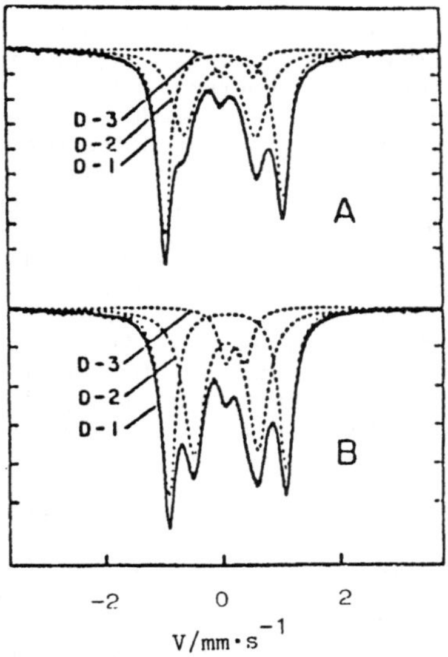

Fig.7. ^{57}*Fe Mössbauer spectra at 300K of Fe-doped* $YBa_2Cu_3O_7$*. Fe/Cu; 2%(A) and 8%(B).*

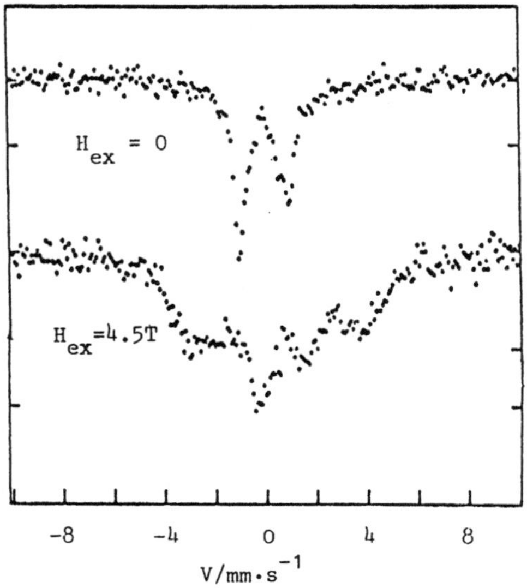

Fig.8. ^{57}*Fe Mössbauer spectra of Fe-doped* $YBa_2Cu_3O_7$ *(Fe/cu=2%), at 4.2K with and without an external field.*

powder sample: The situation is too complicated because there are three unequivalent sites and the effective hyperfine fields are not uniform in magnitude and almost at random in direction. If the paramagnetic spin polarization is assumed to be full and the direction of the hyperfine field is antiparallel to the external one, the hyperfine field corresponding to the local magnetic moment of an Fe impurity in $YBa_2CU_3O_7$ is obtained as 25T. Compared with such a value as 55T of Fe^{3+} ion, this is not very large but it is certain that each Fe atom has a sizeable local magnetic moment.

When the Fe concentration is more than 5%, a magnetic hyperfine structure appears in the spectra at 4.2K, without applying any external field. In Fig. 9, the Mössbauer spectra for a sample with 8%Fe measured in the temperature range from 25K to 0.1K are shown. From the disappearance of the hyperfine splitting, T_N is estimated to be 15K. A magnetic hyperfine structure can also be observed in case when a paramagnetic moment has a very long relaxation time. However this possibility is definitely excluded in this case, from the results of temperature dependence and also Fe-concentration dependence. This result indicates that a spatially stable magnetic order coexists with the superconductivity. With increase of Fe concentration, Tc has decreased but T_N shows an increase, which agrees well with the results by Tamaki et al./11/. By many investigators, spectra with similar profiles were obtained but computer analyses were not very successful /12/. The following assumptions, three hyperfine fields, three electric field gradients(efg) with axial symmetries and a unique angle between the magnetic fields and efg's axes, are not sufficient. The situation is certainly more complicated. Before developing further the discussion on the site assignment and the analysis of the spectrum, it seems worthwhile to refer the results on non-superconductiong, oxygen-deficient $YBa_2Cu_3O_{7-x}$.

An oxygen-deficient sample is prepared by quenching from 900°C, in contrast to a superconducting sample which is prepared by a slow cooling in an oxygen atmosphere. An oxygen-deficient $YBa_2Cu_3O_{7-x}$ has an isomorphic structure but does not exhibit superconductivity. Neutron diffraction studies confirmed that this compound is antiferromagnetic with a rather high T_N due to a strong magnetic coupling between Cu atoms in the plane sites /13/. Figure 10 shows the Mössbauer spectra for an oxygen-deficient sample with 8%Fe, as a function of temperature. At room temperature, the major fraction is non-magnetic and the Mössbauer parameters are almost the same as those of a superconducting sample in Fig. 7. However also observed is another fraction with a magnetic hyperfine splitting, whose valence state is Fe^{3+}. From the temperature dependence, T_N is determined to be 420K, which agrees well with the value of T_N from neutron diffraction studies. It is natural to attribute this fraction to the Fe atoms in the Cu-O plane sites. On the other hand, the major fraction shows a magnetic splitting only below about 100K. The average hyperfine field is estimated from the overall width and plotted as a function of temperature in Fig. 11, with the curve for the fraction of bigger hyperfine field. It has been revealed that there are two types of Fe having different magnetic properties. The bigger hyperfine field fraction is attributed to the Fe atoms substituting the Cu-plane sites, which participate in the antiferromagnetic order of the Cu-O plane. The other fraction with smaller hyperfine field should correspond to Fe atoms in the chain sites. Whose hyperfine field at zero temperature is again about 25T. The temperature dependence is unusual, deviating remarkably from a usual Brillouin-like curve. This behavior is accounted for if the exchange field at the chain sites is much smaller than that of the plane sites. This hypothesis is supported by the result from neutron diffraction study that the Cu atoms in the chain sites are almost non-magnetic. In summary, Fe atoms clearly reflect the magnetic behaviors of the two Cu sites in non-superconducting $YBa_2Cu_3O_{7-x}$: The Cu-O plane is strongly antiferromagnetic but the magnetic interactions in the Cu-O chain sites are very weak. Perhaps, the magnetic couplings in the chain sites can be enhanced by Fe impurities.

The spectrum at 4.2K for the oxygen-deficient $YBa_2Cu_3O_{7-x}$ is a very complicated one. However, if we subtract the 6-line fraction due to the plane sites from the total absorption spectrum, the resultant is found to be very similar with the spectrum for a superconducting sample at 4.2K in Fig. 9. The line profiles are practically the same. From this comparison, we can attribute the absorption in the superconducting sample to the Fe atoms in the Cu-chain sites. In Fig. 9, a fraction with larger hyperfine fields, corresponding to Fe atoms in the plane sites can be distinguished but the relative intensity is much weaker. Namely Fe impurities in $YBa_2Cu_3O_7$ preferentially occupy the chain sites. This seems to be the reason why Tc does not drop very rapidly with the substitution of Cu sites by magnetic Fe impurities. In the case of 8% Fe, a magnetic order exists at 15K, although Tc is 34K. Apparently a magnetic order and superconductivity coexist in the same crystal. However the present results suggest that the magnetic order is formed among Cu atoms and Fe impurities in the chain sites. The Fe concentration in the plane sites is not high enough to destroy the superconductivity. The result in Fig. 9 shows that Fe atoms in the plane sites also participate cooperatively in the magnetic order. Therefore a part of Cu atoms in the plane sites also might be involved in the magnetic order. However, it seems probable that the other part of Cu atoms involved in the superconductivity is not taking part in the static magnetic order.

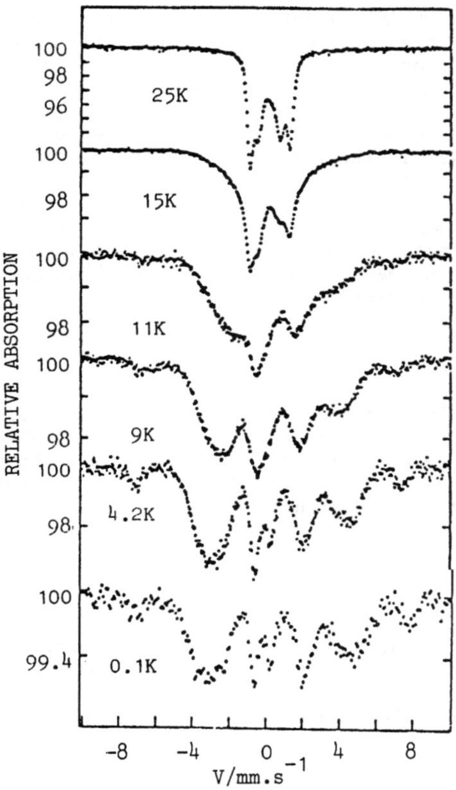

Fig. 9. ^{57}*Fe Mössbauer spectra of Fe-doped* $YBa_2Cu_3O_7$ *(Fe/Cu=8%), as a function of temperature.*

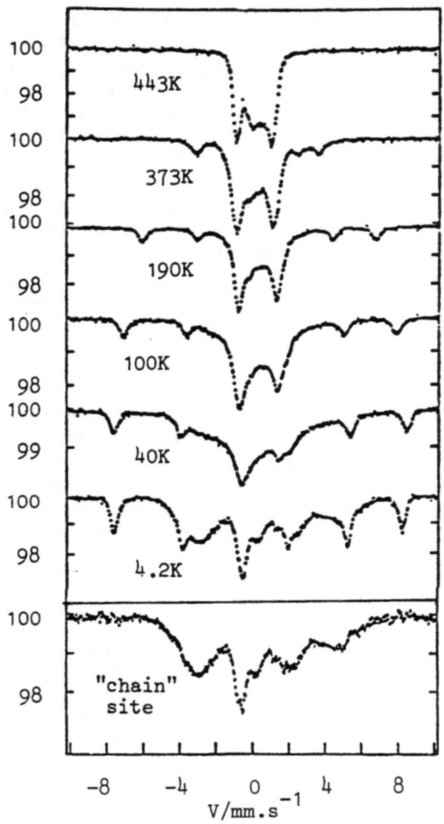

Fig. 10. ^{57}*Fe Mössbauer spectra of an oxygen-deficient (non-superconducting) Fe-doped* $YBa_2Cu_3O_{7-x}$ *as a function of temperature. "Chain" site in the figure is a synthesized spectrum by subtracting the 6-line part from the above spectrum at 4.2K, which corresponds to the Fe atoms in the chain site.*

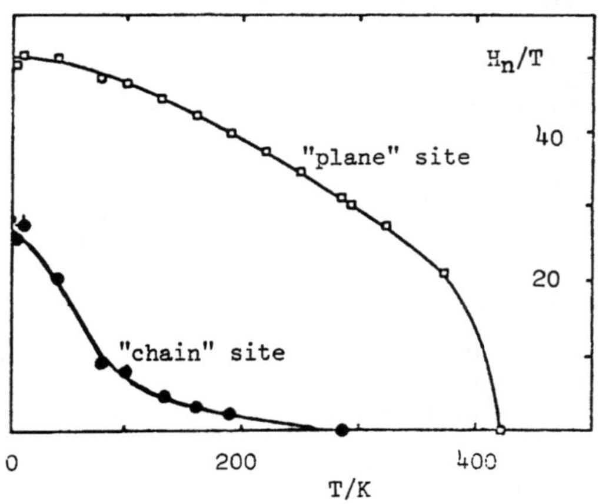

Fig. 11. Temperature dependences of the two Fe hyperfine fields in an oxygen-deficient (non-superconducting) $YBa_2Cu_3O_{7-x}$.

The 4.2K spectrum in Fig. 9 is too broad to get a unique result from a computer fitting and therefore the present authors conclude the site assignment by the visual comparison with the result on a non-superconducting sample. Pankhurst et al. attempted to analyze their very similar spectrum by assuming a broadening due to a dynamical magnetic fluctuation /14/. According to the present result, the degree of broadening is almost the same at 0.1K. If the reason of the broadening is a magnetic relaxation, the line width at very low temperature should become smaller. A more plausible model for such a broadening is a non-collinear spin arrangement. If the angle between the hyperfine field and the efg axis is not unique for all the sites, a considerable broadening may occur. If this is the case, the observed broadening in the Mössbauer spectra is an evidence of spin-glass type ordering in $YBa_2Cu_3O_7$.

4. SUMMARY

Fe impurities in a superconducting $YBa_2Cu_3O_7$ have local magnetic moments. If the Fe concentration is more than a few percent, a spatially stable magnetic order coexists with the superconductivity. However, Fe atoms preferentially occupy the Cu-chain sites. It seems that Cu atoms in the plane sites which carry the superconducting current are not involved in the magnetic order induced by the Fe impurities in the chain sites.

ACKNOWLEDGEMENTS

The authors thank H. Kitagawa, K. Shintaku and T. Kusuda for technical assistance and T. Kohara, Y. Oda, K. Asayama, F. E. Fujita, T. Takabatake and M. Ishikawa for discussions. This work was supported by a Grant-in-Aid for Scientific Research from the Ministry of Education, Science and Culture in Japan.

References

1. Shinjo, N. Hosoito, T. Takada, M. Takano and Y. Takeda: FERRITES, Proc. Intern. Conf. 1980, Japan, p.393.
2. M. Takano, N. Nakanishi, Y.Takeda, S. Naka and T. Takada: Mater. Res. Bull. 12, 923 (1977).
3. M. Eibschutz, D. W. Murphy, S. Sunshine, L. G. Van Uitert, S.M. Zahurak and W.H. Grodkiewicz: Phys. Rev. B 35, 8714 (1987). Eu measurements were also reported by A. K. Grover, S. K. Dhar, P. L. paulose, V. Nagarajan, E. V. Sampathkumaran and R. Nagarajan: Solid State. Commun. 63, 1003 (1987).
4. H. H. A. Smit, M. W. Dirken, R. C. Thiel and L. J. de Jongh: Solid State Commun. 64, 695 (1987). J. van den Berg, C. J. Van der Beek. P. H. Kes, J. A. Mydosh, G. J. Nieuwenhuys and L. J. de Jough ibid. 699.
5. J. A. Hodges, P. Imbert and G. Jehanno, Solid State Commun. 64, 1209 (1987).
6. T. Yuen, C. L. Lin, J. E. Crow, G. N. Myer. R. E. Salomon and P. Schlottmann: Phys. Rev. B 37, 3770 (1988).
7. S. Nasu, H. Kitagawa, Y. Oda, R. Kohara, T. Shinjo, K. Asayama and F. E. Fujita: Physica 148B,484 (1987).
8. T. Shinjo, S. Nasu, T. Kohara, T. Takabatake and M. Ishikawa: presented at the ICM'88 (Paris, 1988). To be published in J. de Phys.
9. G. Xiao, F. H. Streitz, A. Gavrin, Y. W. Du and C. L. Chien: Phys. Rev. B35, 8782 (1987).

10. I. Felner, I. Nowik and Y. Yeshurun: Phys. Rev. B 36, 3923 (1987).
11. T. Tamaki, T. Komai, A. Ito, Y. Maeno and T. Fujita: Solid State Commun. 65, 43 (1988).
12. Besides ours and Ref. 11, similar Mössbauer spectra were reported in the following papers; H. Tang et al., Phys. Rev. B 36, 4018 (1987), M. Takano et al., Jpn. J. Appl. Phys. 26, L1862 (1987), F. R. Bauminger et al., Solid State Phys. 65, 123 (1988), Q. A. Pankhurst et al., Phys. Letters 127, 231 (1988), E. B. Saitovitch et al. Phys. Rev. B 37, 7967 (1988), and S. Suharan et al., Solid state Commun. 67, 125 (1988). Some other papers might escape our attention.
13. J. M. Tranquada, D. E. Cox, W. Kunnmann, H. Moudden, G. Shirane, M. Suenaga, P. Zolliker, D. Vaknin, S. K. Sinha, M. S. Alvarez, A. J. Jacobsen and D. C. Johnston, Phys. Rev Lett. 60, 156 (1988). J. W. Lynn, W.-H. Li, H. A. Mook, B. C. Sales and Z. Fisk,: Phys. Rev. Lett. 60, 2781 (1988).
14. Q. A. Pankhurst, A. H. Morrish, M. Raudsepp and X. Z. Zhou: J. Phys. C; Solid State 21, L7 (1988).

Photoemission Studies of High-T_c Cu Oxides: Character of Doped Oxygen Holes and Pairing Mechanisms

A. Fujimori

Department of Physics, University of Tokyo, Bunkyo-ku, Tokyo 113, Japan

The electronic structure of Cu-oxide superconductors revealed by photoemisson spectroscopy combined with configuration-interaction cluster model analyses is reviewed. In particular, the results of recent cluster calculations suggest that under certain conditions the doped holes enter oxygen sites in the BaO (or LaO, SrO) planes as well as those in the CuO_2 planes and are weakly coupled to the Cu d^9 spins through ferromagnetic interaction. Implications of these results on various pairing mechanisms are discussed. Importance of long-range Coulomb interaction manifested in the photoemission spectra is pointed out.

1. Introduction

Since the discovery of high-T_c superconductivity in the La-Ba-Cu-O system [1], there has been a great deal of experimental and theoretical effort to elucidate the mechanism of superconductivity in the Cu-oxide systems. Although mechanisms proposed so far are so diverse, ranging from electron-phonon mechanisms to mechanisms of purely electronic origins, there is at least general agreement that the two-dimensional CuO_2 networks play an important role. The presence of one-dimensional Cu-O chains in $YBa_2Cu_3O_{7-\delta}$ would not be crucial since equally high T_c's have been observed in tetragonal samples (e.g., $YBa_2Cu_{3-x}Fe_xO_{7\pm\delta}$ [2]) where oxygen atoms in this layer are disordered and in the Bi-Sr-Ca-Cu-O and Tl-Ba-Ca-Cu-O systems where there are no Cu-O chains. I would like to note a common structural feature in the Cu-oxide superconductors, namely, the presence of adjacent CuO_2 planes <u>and</u> BaO, LaO, or SrO planes (referred to BaO planes, hereafter) for which oxygen atoms in the latter planes are located above the Cu atoms. Doped Nd_2CuO_4 which has CuO_2 planes but without such "BaO" planes does <u>not</u> become superconducting [3]. Also, some metallic Cu oxides having three-dimensional Cu-O networks such as $La_4BaCu_5O_{13}$ and $LaCuO_3$ are not superconducting [4]. It therefore seems that the presence of adjacent CuO_2 and BaO planes is a necessary condition for the occurrence of superconductivity in the Cu-oxide systems, although it is not a sufficient condition since metallic $La_2SrCu_2O_{6.2}$ which has this type of structure is not superconducting [4]. In this article, I will address the question of the role of oxygen in the BaO

Springer Series in Materials Science, Vol. 11
Mechanisms of High Temperature Superconductivity
Editors: H. Kamimura and A. Oshiyama
Springer-Verlag Berlin Heidelberg © 1989

planes based mainly on the results of photoemission spectroscopy and their analyses using the configuration-interaction cluster model.

The key information so far obtained by photoemission and other high-energy spectroscopies may be summarized as follows [5]: (i) Electron correlation is so strong for the Cu 3d electrons that they behave as nearly localized d^9 (Cu^{2+}) configurations; (ii) The Cu 3d and O 2p levels are close in energy in the ground state of the undoped sample and are therefore strongly hybridized with each other; (iii) Because of the strong electron correlation at the Cu sites and the proximity of the d and p levels, doped holes enter O 2p orbitals rather than Cu 3d. Namely, undoped Cu^{2+} oxides are charge-transfer insulators [6] rather than Mott-Hubbard insulators in the original sense, and their highest occupied states are O 2p-like.

In theoretical studies based on the extended Hubbard or periodic Anderson models which include both Cu 3d and O 2p states explicitly, it has usually been assumed that relevant physical processes occur within the CuO_2 planes. Of course, the p-d hybridization in the ground state and the superexchange coupling between the Cu spins (apart from the weaker interlayer coupling) take place predominantly within the plane due to the x^2-y^2 symmetry of the unoccupied orbital of the Cu^{2+} ion. However, it is not so obvious whether only the CuO_2 planes play important roles when the system is doped with extra holes. Recent polarized Cu L_3-edge x-ray absorption studies on single-crystal $YBa_2Cu_3O_{7-\delta}$ by BIANCONI et al. [7] have suggested that the doped holes have $3z^2-r^2$ symmetry with respect to the Cu site, i.e., the result is compatible with holes in the p_z orbitals of the BaO planes. The same conclusion has been drawn from the study of the $La_{1+x}Ba_{2-x}Cu_3O_{7\pm\delta}$ system by TAKAYAMA-MUROMACHI et al. [8], according to which a small amount of oxygen vacancies in the BaO planes are shown to reduce the T_C dramatically.

Various types of oxygen orbitals are shown in Fig. 1. The in-plane $p\sigma_{x,y}$ orbitals which have been most frequently employed in theoretical studies [9,10] are σ-bonded to the Cu $d_{x^2-y^2}$ orbitals and have the largest p-d overlaps. The magnetic coupling between these p holes and the Cu d^9 spins is antiferromagnetic and extremely strong. The $p\pi$ orbitals, on the other hand, are ferromagentically coupled to the Cu d^9 spins [11] and the coupling strength is weaker. The p_z orbtial of the apex oxygen ($p\sigma_z$ in Fig. 1) is also ferromagnetically coupled to the d^9. In the quantum mechanical treatment of the magnetic mechansim of superconductivity (on finite systems), the sign of the p hole-Cu spin magnetic coupling is crucial, particularly for strong magnetic coupling and large p-hole tansfer [12,13]. Also, reasonably strong coupling is shown to be required to yield high T_C's [12].

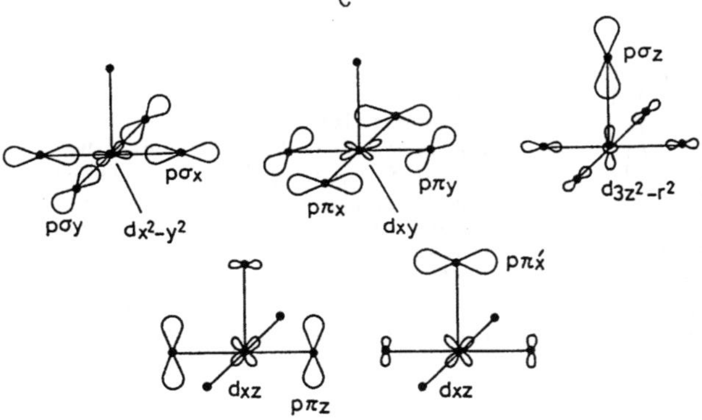

Fig. 1 Various types of oxygen p orbitals for the CuO_5 cluster. Also shown are Cu 3d orbitals which are hybridized with these p orbitals

2. Information Obtained from Photoemission Spectroscopy

Because of the strong intra-atomic correlation as compared to the interatomic interaction (band effects) for the Cu 3d states, single impurity models such as the cluster and the impurity Anderson models provide a good starting point to treat the Cu-oxide systems. As for the band-like O 2p states, its finite bandwidth is explicitly taken into account in the impurity Anderson model, whereas in the cluster model the O 2p band is treated as a set of molecular orbitals derived from the oxygen ligand orbitals. Nevertheless, these two models are virtually equivalent concering the ground states of insulators and also for their spectroscopic properties as far as gross spectral features are concerned [14]. The ground state of the $(CuO_5)^{-8}$ or $(CuO_6)^{-10}$ cluster representing the undoped Cu^{2+} compound is written as

$$\Psi_g = \alpha|d^9\rangle + \beta|d^{10}\underline{L}\rangle, \qquad (1)$$

where $\underline{L}$ denotes a hole in the ligand orbitals. The first term in (1) represents the purely ionic configuration with filled O 2p band, whereas the second term represents a p-to-d charge-trasfer state resulting from the p-d hybridization. The energy levels of this system (N-electron system) are characterized by the charge-transfer energy, $\Delta = E(d^{10}\underline{L}) - E(d^9)$, and the p-d hybridization, $T = \langle d^{10}\underline{L}|H|d^9\rangle$, and can be obtained by diagonalizing a 2 × 2 Hamiltonian for each symmetry. (The ground state has ^2E symmetry.)

The parameters, Δ and T, can be obtained from analyses of Cu core-level photoemission spectra as follows: The Cu core-hole state is given by

$$\Psi_f = \alpha_f'|\underline{c}d^9\rangle + \beta_f'|\underline{c}d^{10}\underline{L}\rangle, \qquad (2)$$

where $\underline{c}$ denotes a core hole. The energy difference between the two configurations, $E(\underline{c}d^{10}\underline{L}) - E(\underline{c}d^9)$, is now Δ − Q rather than Δ due to the Coulomb interaction between the core hole and the d electron, Q (> 0). Thus two peaks are observed in the spectrum corresponding to the $\underline{c}d^9$ and $\underline{c}d^{10}\underline{L}$ final-state configurations as shown in Fig. 2. Then it is possible to obtain Δ, T, and Q from the energy separation ΔE_{ms} and the relative intensities I_s/I_m of the two peaks following the procedure given by VAN DER LAAN et al. [15]. In Fig. 3, Δ and Q are plotted as functions of T

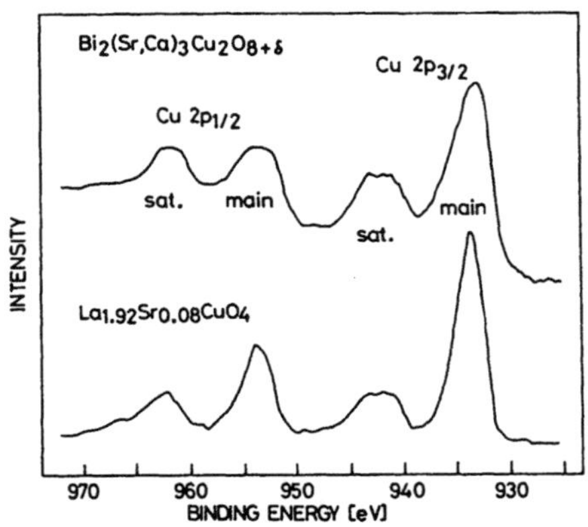

Fig. 2 Cu 2p core-level x-ray photoemission spectra of single-crystal $La_{2-x}Sr_xCuO_4$ and $Bi_2(Sr,Ca)_3Cu_2O_{8+\delta}$ [16]. Each of the Cu $2p_{3/2}$ and $2p_{1/2}$ spin-orbit components is further split into the main ($\underline{c}d^{10}\underline{L}$) and satellite ($\underline{c}d^9$) peaks. Overlapping Ca Auger emission has been subtracted from the $Bi_2(Sr,Ca)_3Cu_2O_{8+\delta}$ spectrum

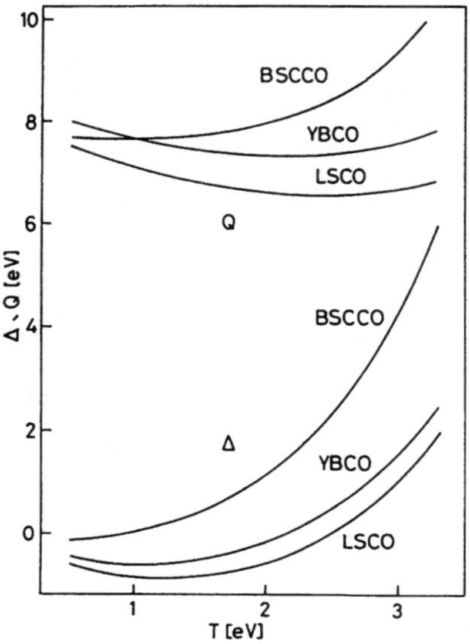

Fig. 3 Charge-transfer energy, Δ, and Cu 3d-Cu 2p Coulomb energy, Q, as functions of hybridization, T, for the observed ΔE_{ms} and I_s/I_m of $La_{2-x}Sr_xCuO_4$ (LSCO) [16], $YBa_2Cu_3O_{7-\delta}$ (YBCO) [17], and $Bi_2(Sr,Ca)_3Cu_2O_{8+\delta}$ (BSCCO) [16]

for the experimental ΔE_{ms} and I_s/I_m values of $La_{2-x}Sr_xCuO_4$ [16], $YBa_2Cu_3O_{7-\delta}$ [17], and $Bi_2(Sr,Ca)_3Cu_2O_{8+\delta}$ [16]. Since only two quantities are experimentally known for three unknown parameters, we make a further assumption that the intra-atomic Q does not vary strongly from one compound to another and that T varies only slightly according to the slight changes in the in-plane Cu-O distance. Parameter values thus estimated are listed in Table I where one finds that Δ is small as compared to T, which leads to considerably strong Cu-O covalency in the ground state (corresponding to a p-to-d charge transfer of 0.3-0.4 electron per Cu atom). Table I also indicates that Δ increases in going from $La_{2-x}Sr_xCuO_4$ to $YBa_2Cu_3O_{7-\delta}$ to $Bi_2(Sr,Ca)_3Cu_3O_{8+\delta}$.

The final state of valence-band photoemission is given by

$$\Psi_f = \alpha'_f|d^8\rangle + \beta'_f|d^9\underline{L}\rangle + \gamma'_f|d^{10}\underline{L}^2\rangle, \quad (3)$$

for each final-state symmetry [14]. Here, the energy differences between the configurations are parameterized as $E(d^9\underline{L}) - E(d^8) = \Delta - U$ and $E(d^{10}\underline{L}^2) - E(d^9\underline{L}) = \Delta$, where U is the Cu d on-site Coulomb energy. (Here, interatomic Coulomb energy between Cu and oxygen, V, has been neglected

Table I Relative intensities I_s/I_m and splittings ΔE_{ms} of the main (m) and satellite (s) peaks in the Cu $2p_{3/2}$ core-level photoemission spectra. Parameters, Δ, T, and Q evaluated from the I_s/I_m and ΔE_{ms} values are also given for each compound. Energies are in eV.

	I_s/I_m	ΔE_{ms}	Δ	T	Q
$Bi_2(Sr,Ca)_3Cu_2O_{8+\delta}$	0.60	7.9	2.0±0.5	2.4±0.2	8.1±0.2
$YBa_2Cu_3O_{7-\delta}$	0.33	8.5	0.7±0.2	2.6±0.2	7.3±0.1
$La_{1.92}Sr_{0.08}CuO_4$	0.25	8.2	0.4±0.3	2.7±0.2	6.5±0.1

for simplicity; The spectra are consistent with a small V of 0-1 eV.) Multiplet effects for the d^8 configuration are included using Racah parameters. Using essentially the same Δ and T values as those obtained from the core-level photoemission and adjusting U so as to give the best fits to experiment, the valence-band photoemission spectra of $La_{2-x}Sr_xCuO_4$ and $Bi_2(Sr,Ca)_3Cu_2O_{8+\delta}$ have been calculated as shown in Fig. 4. The resulting U is in the range of 6.5-7 eV, which is large enough to allow the impurity description of the d electronic states and makes one-electron band theory innapropriate.

A remarkable consequence of the above analyses is that the Coulomb interaction U is larger than the charge-transfer energy Δ. This results in $E(d^9\underline{L}) < E(d^8)$ for the N-1-electron system, namely, the doped carriers are holes at oxygen sites and not at Cu sites. This situation is schematically illustrated in Fig. 5 [6,15]. This conclusion is supported by the fact that any spectral features associated with Cu^{3+} (d^8) configuration have not been observed in the photoemission [5,16,17] and x-ray absorption studies [18].

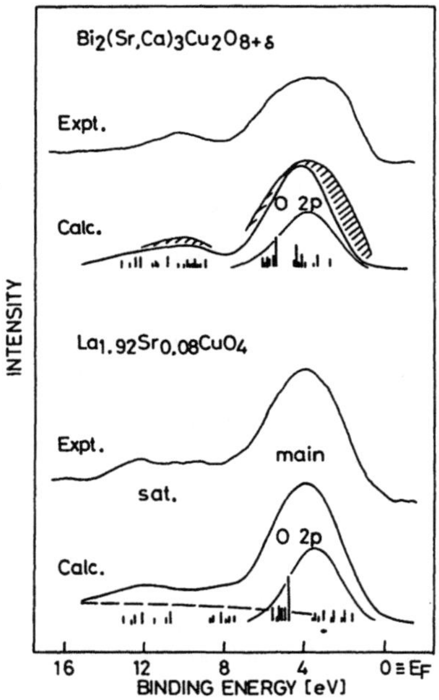

Fig. 4 Valence-band x-ray photoemission spectra of single-crystal $La_{2-x}Sr_xCuO_4$ and $Bi_2(Sr,Ca)_3Cu_2O_{8+\delta}$ compared with those calculated using the cluster model [16]. The spectra show satellites (d^8) at higher binding energies as well as the main band ($d^9\underline{L}$ and $d^{10}\underline{L}^2$) overlapping the O 2p band. The shaded area represents estimated Bi contributions. Shallow core levels such as La 5p and Sr 4p levels have been subtracted

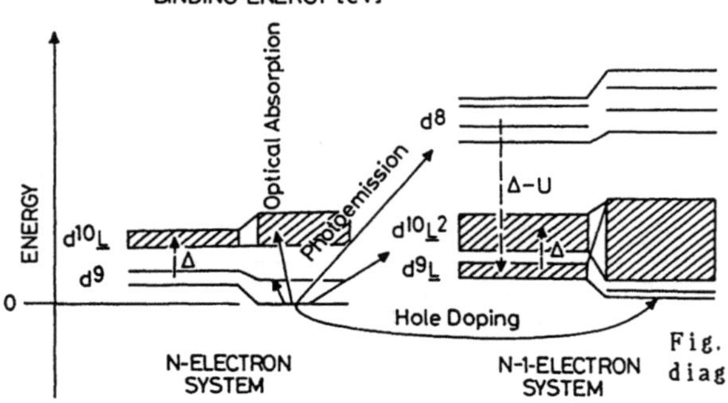

Fig. 5 Schematic energy-level diagram of the CuO_n cluster

3. Character of Doped Oxygen Holes

The next step to understand the electronic structure of high-T_C superconductors will be to determine the symmetry and site distribution of the extra holes, since the pairing mechanism depends on these, particularly in the case of magnetic fluctuation mediated superconductivity. Therefore, we have studied the lowest energy states of the doped (N-1-electron) CuO_5 and CuO_6 clusters [19]. Here, it should be noted that the cluster model describes the lowest energy states as correctly as the impurity Anderson model does if these states are split from the $d^9\underline{L}$ (+ $d^{10}\underline{L}^2$) continuum and form bound states. This is indeed the case for the Cu oxides due to the large p-d hybridization as shown by ESKES ans SAWATZKY [20].

Now we introduce, in addition to Δ, U, and T, a new parameter which describes the the p-orbital energy difference between the in-plane and apex oxygens, $\Delta\epsilon_p = \epsilon(p\sigma_z) - \epsilon(p\sigma_{x,y})$. In order to evaluate various p-d and p-p hybridization parameters in the elongated octahedral and pyramidal clusters, they have been assumed to scale with $R^{-3.5}$ and R^{-2}, respectively, where R is the atomic distance [21]. (p-p hybridization parameters within the CuO_2 plane have been taken form PARK et al. [22].) Thus the relative atomic distance of the out-of-plane Cu-O bond to the in-plane one, r, has been taken as another independent parameter. The new parameters, r and $\Delta\epsilon_p$, have already been included in the calculation of the valence-band photoemission spectra shown in Fig. 4, but the gross spectral features are largely determined by the in-plane quantities, Δ, U, and T, and not by r or $\Delta\epsilon_p$.

For a reasonable range of the parameters, only $^1A_1$- and $^3B_1$-symmetry states are found to become the ground state as shown in Fig. 6: In the $^1A_1$ state, the doped hole is in a molecular orbital of x^2-y^2 symmetry consisting only of in-plane oxygen $p\sigma_{x,y}$ orbitals and forms a singlet with the Cu d^9 spin; In the $^3B_1$ state, the hole is in a $3z^2-r^2$-symmetry molecular orbital consisting of the apex oxygen $p\sigma_z$ as well as the in-plane $p\sigma_{x,y}$ orbitals (Fig. 1) and is coupled to the Cu spin ferromagnetically. The singlet state is stabilized for large T, because the stabilization due to p-d hybridization is largest for the $d_{x^2-y^2}$-$p\sigma_{x,y}$ hybridization. The triplet state is stabilized for small r through increased p-d hybridization between the $d_{3z^2-r^2}$ and $p\sigma_z$ orbitals. This state is also stablized for large (positive) $\Delta\epsilon_p$ for the obvious reason that the hole in the triplet sate is mainly distributed on the apex oxygen.

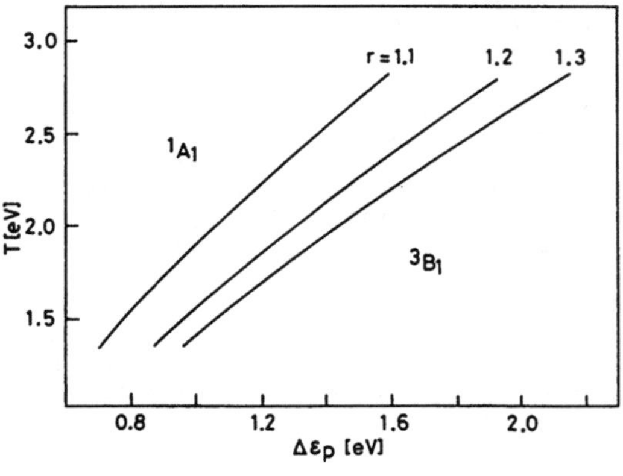

Fig. 6 Boundaries between the $^1A_1$ and $^3B_1$ ground states in the T-$\Delta\epsilon_p$ plane of the hole-doped CuO_5 cluster. U = 6.5 eV and Δ = 1.0 eV [19]

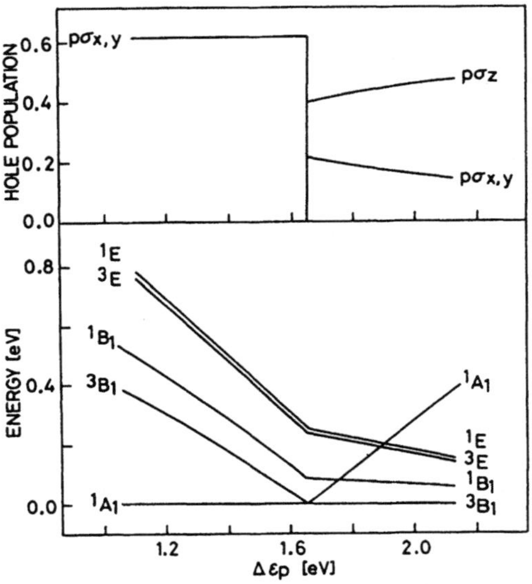

Fig. 7 Bottom: Energies of low-lying excited states. Top: Distribution of holes in different oxygen p orbitals. U = 6.5 eV, Δ = 1.0 eV, T = 2.34 eV, and r = 1.23. For these parameters, the triplet-singlet transition occurs at $\Delta\varepsilon_p \simeq 1.65$ eV as shown in the figure [19]

The magnetic coupling between the doped hole and the Cu spin, J_{pd}, can be seen from Fig. 7, where the energies of low-lying excited states of the doped cluster are plotted as functions of $\Delta\varepsilon_p$. There one can see that the $^3B_1$-$^1A_1$ splitting in the case of the $^1A_1$ ground state, which corresponds to an antiferromagnetic J_{pd}, varies considerably with $\Delta\varepsilon_p$ (and also with r) and can become huge ($J_{pd} \sim -0.2$-0.5 eV) for small $\Delta\varepsilon_p$ (or large r). The $^1B_1$-$^3B_1$ splitting in the case of the $^3B_1$ ground state corresponding to a ferromagnetic J_{pd}, on the other hand, is only weakly dependent on $\Delta\varepsilon_p$ and r and is relatively small ($J_{pd} \sim 0.07$ eV). It is also noticed from Fig. 7 that a ^3E state is only ~0.2-0.4 eV above the $^3B_1$ state. As the ^3E state is mainly derived from $p\pi_{x,y}$ orbitals of the apex oxygen ($p\pi'_x$ in Fig. 1), this state can become the ground state of the doped cluster if the $p\pi'_{x,y}$ orbital energy is higher than the $p\sigma_z$ only by 0.2-0.4 eV due to anisotropy at the apex oxygen site.

Thus we may conclude that the doped holes are partially distributed in the BaO planes and form local triplets with the Cu spins when $\Delta\varepsilon_p$ is larger than ~1.5 eV. $\Delta\varepsilon_p$ may be estimated from the mean energies of the oxygen p local densities of states (DOS) given by band-structure calculations. Thus we estimate $\Delta\varepsilon_p$ to be about 1 eV for La_2CuO_4 [22,23] and to be further increased by ~1 eV when adjacent La is replaced by Sr [23], resulting in a sufficiently large value to induce triplet states. As for $YBa_2Cu_3O_{7-\delta}$, $\Delta\varepsilon_p$ is about 1 eV for $7-\delta = 6$ and is expected to increase with $7-\delta$ because the oxygen p DOS of the BaO-plane is shifted toward the Fermi level (E_F) with increasing $7-\delta$ [22]. Further, the interlayer Cu-O distance or r decreases with increasing $7-\delta$ [24], favoring the formation of triplets. In this context, it is interesting to note that this Cu-O distance (and not the in-plane Cu-O distance) shows an anomalous decrease below T_c [25]. $\Delta\varepsilon_p$ is not necessarily large enough for $Bi_2Sr_2CaCu_2O_8$ [26], but it should be noted that its crystal structure and oxygen content have not yet been determined accurately enough to allow determination of the energy levels through band-structure calculations [16].

An effect which is not considered in the cluster study is the Coulomb correlation between oxygen p holes, which amounts to ~5 eV at the same oxygen site [27]. Due to this interaction, it would cost more energy than estimated above (possibly by ~1 eV) to accomodate an extra hole in the in-plane $p\sigma_{x,y}$ orbitals where a significant amount of holes (~0.2 hole per oxygen atom) exist already in the undoped sample due to the strong p-d covalency.

4. Implications on the Mechanism of Superconductivity

Magnetic pairing mechanisms have recently been investigated extensively by IMADA and co-workers [12]. They have investigated by exact diagonalization of finite systems which consist of two-dimensional antiferromagnetic Heisenberg spins (including 4 × 4 Cu sites, with superexchange $J_{dd} \sim -0.1$ eV) coupled to itinerant fermions (O 2p holes) through exchange J_{pd}. They have found attractive interactions in two regions of the parameter space. Namely, for strong antiferromagnetic spin-fermion coupling J_{pd} as compared to J_{dd} and for large fermion transfer t, i.e., for $t \sim |J_{pd}| >> |J_{dd}|$, an extended cloud of "singlet liquid" with relatively large binding energy ($\lesssim$0.1 eV) is formed, which they consider to be relevant for the high-T_C superconductivity. Another region is characterized by small t ($<< |J_{dd}|$) with J_{pd} being either antiferromagnetic or ferromagnetic, in which the binding energy is found much smaller. The same conclusion has been obtained for p_z holes in the BaO planes by NISHINO et al. [28]. SHIBA and OGATA [13] have also shown that the ferromagetic coupling does not lead to binding of two holes in the large transfer region.

The above theoretical studies suggest that, in the case of the triplet $^3B_1$ state, pairing interaction of magnetic origin is difficult to explain the observed high T_C's. On the other hand, HIDA [29] has studied the same model using the spin-wave approximation for the Cu spins and shown that it is possible to obtain a reasonably high-T_C's already in the weak coupling region irrespective of the sign of J_{pd}. He has shown that the Cooper pair extends to as large as about 10 lattice spacings and pointed out the limitation of finite-system studies. KAMIMURA et al. [30] have presented a spin-polaron mechanism in a system with strong ferromagnetic coupling between the localized Cu $d_{x^2-y^2}$ spins and doped $d_{3z^2-r^2}$ holes. The size of the Cooper pair in this type of mechanisms may not be small enough to be studied in the finite systems.

For the $^1A_1$ states in which carriers are confined within the CuO_2 planes, the increase in the charge-transfer energy Δ in going form $La_{2-x}Sr_xCuO_4$ to $YBa_2Cu_3O_{7-\delta}$ to $Bi_2(Sr,Ca)_3Cu_2O_{8+\delta}$ (and probably to Tl-Ba-Ca-Cu-O [31]) is consistent with the magnetic mechanism, because the increase in Δ suppresses charge fluctuations within the CuO_2 planes while enhancing magnetic fluctuations [32]. For the $^3B_1$ states, simple relations are not expected between the electronic structure parameters and the T_C since too many independent parameters are involved. At least we could not find significant differences in J_{pd} between these materials, hence no correlation between the T_C and J_{pd}. In the case of the ^3E states, J_{pd} (~ 0.01 eV) is probably too small for any magnetic mechanism to yield reasonably high T_C's.

In charge fluctuation mechanisms, the magnetic coupling J_{pd} is not a relevant quantity, and in this sense the formation of triplets is not inconsistent with the charge fluctuation mechanism. In the charge-transfer exciton mechanism first proposed by VARMA et al. [33] and later developped by HIRSCH et al. [34], too large interatomic p-d Coulomb interaction V > 5 eV has been assumed. On the other hand, TACHIKI and TAKAHASHI [35] have considered V of order of 1 eV, which is compatible with the photoemission results. Unfortunately, so far no clear evidence has been found for low-energy electronic excitations with $q = 0$ [36] or $q \neq 0$ [37] that have been predicted by these theories.

Another charge fluctuation mechamism proposed by TAKADA [38] emphasizes the role of long-range Coulomb interaction which is not effectively screened when the carrier density is low. This mechanism does not seem to seriously contradict with spectroscopic and other experimental results, and further might be able to explain the relatively high T_c's of $BaPb_{1-x}Bi_xO_3$ and $K_xBa_{1-x}BiO_3$ containing no Cu or other magnetic elements. The role of the BaO planes in this case could be to allow the doped holes to be more uniformly distributed in the crystal rather than to confine them in the two-dimensional CuO_2 planes. (Nd_2CuO_4, in which there are no "BaO" planes and the formation of $^1A_1$ states is expected when doped with holes, cannot be actually doped with holes [3].) It is noted that in the phase diagrams of Cu-oxide systems [39] the magnetism and superconductivity appear to exclude each other rather than to cooperate. Here I would like to point out that the photoemission spectra of the Bi-oxide [40] and Cu-oxide superconducotrs [5,16,17,32] have a common characteristic feature that the DOS at E_F is unusually low, compared not only with band theory [22,23] but also with impurity Anderson model calculations [20]. Thus the low DOS cannot be simply due to strong one-site Coulomb interaction since it is already included in the impurity models and is supposed to be much less important for the Bi oxides. Presumably, in analogy to the Coulomb gaps in disordered systems and gaps in Wigner crystals, the low DOS at E_F is due to long-range Coulomb interaction in low carrier-density metals [41] which has been neglected so far in most of theoretical studies.

I would like to gratefully acknowledge collaboration and discussions with E. Takayama-Muromachi, S. Takekawa, T. Takahashi, H. Katayama-Yoshida, and Y. Tokura.

References

1. G. Bednorz, K. A. Müller: Z. Phys. B64, 189 (1986)
2. Y. Maeno, M. Kato, Y. Aoki, T. Fujita: Jpn. J. Appl. Phys. 26, L1982 (1987)
3. Y. Tokura et al.: (to be published)
4. J. B. Torrance, Y. Tokura, A. I. Nazzal, A. Bezinge, T. C. Huang, S. S. P. Parkin: Phys. Rev. Lett. 61, 1127 (1987)
5. A. Fujimori, E. Takayama-Muromachi, Y. Uchida, B. Okai: Phys. Rev. B 35, 8814 (1987); Solid State Commun. 63, 857 (1987)
6. J. Zaanen, G. A. Sawatzky, J. W. Allen: Phys. Rev. Lett. 55, 418 (1985); S. Hüfner: Z. Phys. B61, 135 (1985)
7. A. Bianconi, M. de Santis, A. di Cicco, A. M. Flank, A. Fontaine, P. Lagarde, H. Katayama-Yoshida, A. Kotani, A. Marcelli: Phys. Rev. B38 (in press)
8. E. Takayama-Muromachi, Y. Uchida, A. Fujimori, K. Kato: Jpn. J. Appl. Phys. 27, L223 (1988)
9. V. J. Emery: Phys. Rev. Lett. 58, 2794 (1987)
10. M. Imada: J. Phys. Soc. Jpn. 57, 3128 (1988)
11. A. Aharony, R. J. Birgeneau, A. Coniglio, M. A. Kastner, H. E. Stanley: Phys. Rev. Lett. 60, 1330 (1988)

12. M. Imada, Y. Hatsugai, N. Nagaosa: J. Phys. Soc. Jpn. 57, 2901 (1988); (submitted)
13. H. Shiba, M. Ogata: In Proc. 6th Int. Conf. on Crystal-Field Effects and Heavy Fermion Physics, Fraknfurt, 1988 (to be published)
14. A. Fujimori: In Core-Level Spectroscopy in Condensed Systems, ed. by J. Kanamori, A. Kotani (Springer, Berlin, 1988) p. 99; A. Fujimori, F. Minami: Phys. Rev. B30, 957 (1984)
15. G. van der Laan, C. Westra, C. Haas, G. A. Sawatzky: Phys. Rev. B23, 4369 (1981)
16. A. Fujimori, S. Takekawa, E. Takayama-Muromachi, Y. Uchida, A. Ono, T. Takahashi, Y. Okabe, H. Katayama-Yoshida: Phys. Rev. B (in press)
17. H. M. Meyer III, D. M. Hill, T. J. Wagener, Y. Gao, J. H. Weaver, D. W. Capone II, K. C. Goretta: Phys. Rev. B (submitted)
18. J. M. Tranquada, S. M. Heald, A. R. Moodenbaugh: Phys. Rev. B36, 5263 (1987)
19. A. Fujimori: Phys. Rev. B (submitted)
20. W. A. Harrison: Electronic Structure and the Properties of Solids (Freeman, San Fransisco, 1980)
21. H. Eskes, G. A. Sawatzky: Phys. Rev. Lett. 61, 1415 (1988)
22. K.-T. Park, K. Terakura, T. Oguchi, A. Yanase, M. Ikeda: J. Phys. Soc. Jpn. (in press)
23. G. M. Stocks, W. M. Temmerman, S. Szotek, P. A. Stern: Supercond. Sci. Technol. 1, 57 (1988)
24. H. Asano, K. Takita, T. Ishigaki, H. Akinaga, H. Katoh, K. Masuda, F. Izumi, N. Watanabe: Jpn. J. Appl. Phys. 26, L1341 (1987)
25. T. Ishigaki, H. Asano, K. Takita, H. Kato, H. Akinaga, F. Izumi, N. Watanabe: Jpn. J. Appl. Phys. 26, L1681 (1987)
26. S. Massida, J. Yu, A. J. Freeman: Physica C152, 251 (1988)
27. D. van der Marel, J. van Elst, G. A. Sawatzly, D. Heitmann: Phys. Rev. B37, 5136 (1988)
28. T. Nishino, M. Kikuchi, J. Kanamori: Solid State Commun. (in press)
29. K. Hida: J. Phys. Soc. Jpn. 57, 1844 (1988); (in press)
30. H. Kamimura, S. Matsuno, R. Saito: Solid State Commun. 67, 363 (1988)
31. H. Katayama-Yoshida, A. J. Mascarenhas, Y. Okabe, T. Takahashi, T. Suzuki, J. I. Pankove, T. Ciszek, S. K. Deb. H. Adachi, K. Setsun, K. Horiuchi: (submitted)
32. Z.-X. Shen, J. W. Allen, J. J. Yeh, J.-S. Kang, W. Ellis, W. Spicer, I. Lindau, M. B. Maple, Y. D. Dalichaouch, M. S. Torikachvili, J. Z. Sun: Phys. Rev. B36, 8414 (1987)
33. C. M. Varma, S. Schmitt-Rink, E. Abrahams: Solid State Commun. 62, 681 (1987)
34. J. E. Hirsch, S. Tang, E. Loh, Jr., D. J. Scalapino: Phys. Rev. Lett. 60, 1618 (1988)
35. M. Tachiki, S. Takahashi: Phys. Rev. B38, 218 (1988)
36. S. Tajima, S. Uchida, H. Ishii, H. Takagi, S. Tanaka, U. Kawabe, H. Hasegawa, T. Aita, T. Ishibashi: Mod. Phys. Lett. B1, 353 (1988)
37. J. Fink, N. Nücker, H. Romberg, S. Nakai: In Proc. Int. Symp. on the Electronic Structure of High-T_c Superconductors, Rome, 1988 (to be published)
38. Y. Takada: Phys. Rev. B37, 155 (1988); (submitted)
39. Y. Kitaoka, K. Ishida, S. Hiramatsu, K. Asayama: J. Phys. Soc. Jpn. 57, 734 (1988)
40. H. Sakamoto, H. Namatame, T. Mori, K. Kitazawa, S. Tanaka, S. Suga: J. Phys. Soc. Jpn. 56, 365 (1987)
41. A. Fujimori, K. Kawakami, N. Tsuda: Phys. Rev. B38 (in press)

Experimental Approach to the Mechanism of High-T_c Superconductivity

H. Katayama-Yoshida, T. Takahashi, and Y. Okabe

Department of Physics, Faculty of Science, Tohoku University, Sendai 980, Japan

1. Introduction

One of the most fascinating problems in solid state physics is the mechanism of high-transition-temperature (high-Tc) superconductivity discovered in layered copper oxides. The knowledge of the electronic structure is a key step toward the theoretical understanding of the mechanism of high-Tc superconductivity. We determine the band structure of single crystal of $Bi_2Sr_2CaCu_2O_8$ system by angle-resolved resonant photoemission measurement with synchrotron radiation [1,2]. We give a direct evidence for the Fermi-liquid states with O-2p character at the Fermi level in section 2. The nuclear magnetic resonance (NMR) study is a powerful tool which allows us to differentiate the microscopic properties between the copper and oxygen sites in both normal and superconducting states. We performed NMR study of enriched-^{17}O in Y-Ba-Cu-O system by the collaboration with Prof. K. Asayama's group [presented in this symposium], and found BCS-like enhancement of the nuclear-spin-lattice relaxation rate ($1/T_1$) just below Tc [3,4]. We present an evidence that the high-Tc superconductivity is of s-wave type with O-2p hole Cooper pairing, while the Cu-3d holes are nearly localized with antiferromagnetic spin-fluctuation in section 3. To clarify the high-Tc mechanism, we also report the oxygen isotope effect for Y-Ba-Cu-O and Bi-Sr-Ca-Cu-O systems in section 4 [5,6]. Based upon the above experimental data related to the mechanism of superconductivity, we discuss possible electronic structure of high-Tc superconductors in section 5.

2. Evidence of the Fermi-liquid States with Oxygen 2p character in Bi-Sr-Ca-Cu-O System Studied by Angle-resolved Resonant Photoemission [1,2].

Figure 1 shows photoemission spectra in the Fermi level region obtained by long-time data acquisition. Figure 1 clearly shows that (i) there are two dispersive bands in the vicinity of the Fermi level and (ii) one of them (band A) crosses the Fermi level midway in the high-symmetry line ΓX. Due to the finite energy resolution, the peak position does not reach the Fermi level even when the band is closest to it. Band B gradually approaches the Fermi level until $\theta = 17.5^\circ$.

Figure 2 shows the band structure of $Bi_2Sr_2CaCu_2O_8$ determined from the angle-resolved photoemission. Experimental results obtained with $h\omega = 40$

Springer Series in Materials Science, Vol. 11
Mechanisms of High Temperature Superconductivity
Editors: H. Kamimura and A. Oshiyama
Springer-Verlag Berlin Heidelberg © 1989

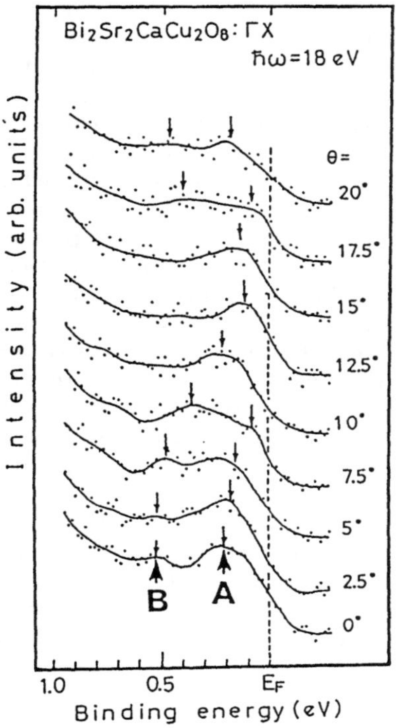

Fig.1: Angle-resolved photoemission spectra of $Bi_2Sr_2CaCu_2O_8$ in the vicinity of the Fermi level measured in the direction Γ X using hω = 18 eV.

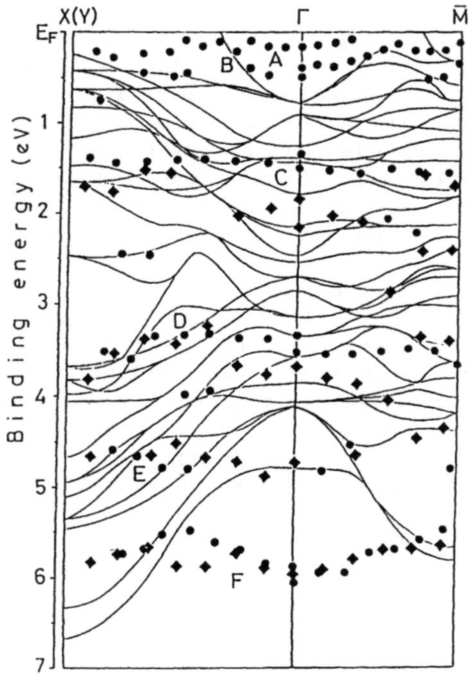

Fig.2: Band structure of $Bi_2Sr_2CaCu_2O_8$ determined by the angle-resolved photo-emission with hω = 18 eV (circles) and 40 eV (squares). Filled and open symbols represent strong and weak structures in the spectra, respectively. Note that the experimental results with hω = 18 eV are shown only for the Γ X direction in the high-binding energy region. A band structure calculation (Ref.9) is shown by thin solid lines for comparison.

eV and for another high-symmetry direction Γ M are also included. Three band structure calculations [7-9] for $Bi_2Sr_2CaCu_2O_8$ are available at present and they are essentially the same. One of them [9] is shown in Fig. 2 for comparison. We compare the present experimental result with the band structure calculation. Band A seems to correspond to the calculated Cu3d-O2p antibonding states located at 0.7 eV at the Γ point. Comparison of the two bands gives a rough estimation of the mass-enhancenent factor of four. The reason of the mass-enhancement might be that the doped O-2p hole is itinerating with the spin-polaron cloud due to the antiferromagnetic short-range ordering of the localized Cu-3d holes. Band B can also find its theoretical counterpart about 0.3 eV below the experimental points. Thus, the band structure calculation seems qualitatively in agreement with the angle-resolved photoemission result in the vicinity of the Fermi level. In the higher-binding energy region, the observed bands except for band F are almost dispersionless in sharp contrast to the highly dispersive feature of the calculated bands. As for band F, the direction of energy dispersion is just opposite to that of the calculated Cu3d-O2p bonding states, suggesting a substantial renormalization by the strong electron correlation. The qualitative agreement between the experiment and the band structure calculation near the Fermi level implies that the one-electron approximation may be qualitatively correct at least near the Fermi level. Alternative explanation is that doped O-2p hole states should have a similar band dispersion to that of the band calculation because they should also reflect the symmetry of the crystal. In order to clarify this point, we performed a resonant photoemission measurement.

Figure 3 shows the photon-energy dependence of the spectrum in the vicinity of the Fermi level for the energy regions near the O-2s and Cu-3p core thresholds. The spectra are normalized to the incident-photon flux. The intensity of the electronic states at the Fermi level (band A) exhibits

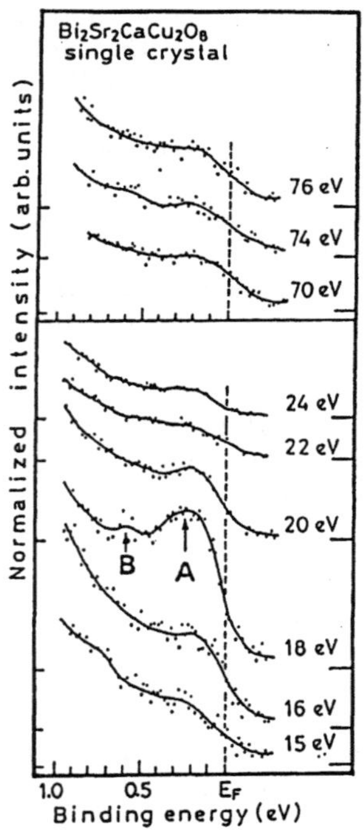

Fig.3: Photoemission spectra of single crystal $Bi_2Sr_2CaCu_2O_8$ in the vicinity of the Fermi level measured with photon energies near the O-2s (lower) and Cu-3p (upper) core thresholds. Photon energy used is indicated on each spectrum. The intensity of photoemission spectra is normalized to incident-photon flux, independently between the O-2s and Cu-3p regions.

a remarkable enhancement at the O-2s core threshold ($h\omega$ = 18 eV) while it gives almost no change when passing the Cu 3d core threshold ($h\omega$ = 74 eV). This is a direct evidence for that the electronic states at the Fermi level in $Bi_2Sr_2CaCu_2O_8$ have a dominant O-2p nature with a far less weight of the Cu-3d states.

3. Evidence of Oxygen p-hole Pairing with s-wave like Symmetry in Y-Ba-Cu-O System [3,4,10].

We observed the NMR spectra of enriched-^{17}O in 90K-"123" compound of Y-Ba-Cu-O. Figure 4 presents the temperature dependence of $1/T_1$ for ^{17}O (solid circle) and ^{63}Cu (open circles) [3,4,10,11]. In normal state, $1/T_1$ for the oxygen sites behave approximately like the Korringa behavior up to 150 K as shown by a solid line (T_1T=constant), while $1/T_1$ above 150 K deviates from a linear temperature dependence [11,12]. In contrast, $1/T_1$ at Cu sites for 90 K-"123" compound shows no Korringa behavior with weak temperature dependence. This striking difference of nuclear spin-lattice relaxation behavior between oxygen and copper sites surpasses our expectation that $1/T_1$ should possesses a common feature due to the strong hybridization between O-2p and Cu-3d states [13,14]. From a microscopic point of view, we can extract the following drastic character of each site. Namely, T_1 at Cu site is remarkably dominated by the spin fluctuation of nearly localized Cu-3d holes. We suppose that the spin fluctuation component generated by two nearest-neighbor Cu spins is canceled out at the oxygen sites due to strong antiferromagnetic correlation among Cu-3d spins. Then the p-hole carriers on the oxygen sites behave like Fermi-liquid states and yield approximately the Korringa behavior below 150 K.

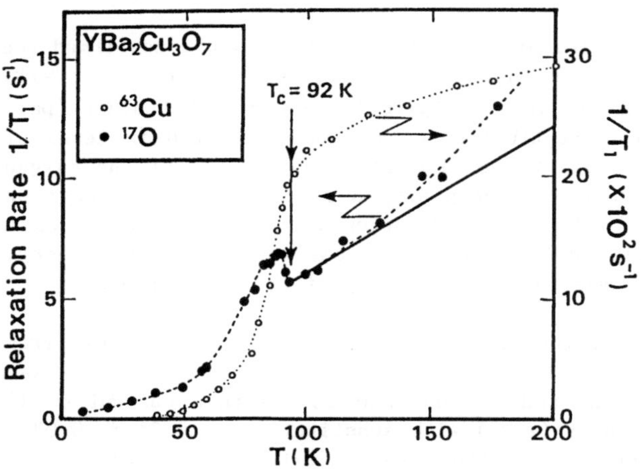

Fig.4: Temperature dependence of the nuclear spin-lattice relaxation rate $1/T_1$ of ^{17}O and ^{63}Cu for $YBa_2Cu_3O_7$ with Tc=92 K. The solid lines indicate the T_1T=const. law for reference.

We focus our interest on the enhancement of $1/T_1$ just below Tc observed at oxygen sites. The relatively large enhancement with $(1/T_1)_s/(1/T_1)_n$ = 1.3 at T/Tc=0.93 should be quantitatively interpreted by taking account of the symmetry of Cooper pair. Although the p-holes introduced on the oxygen interact with the nearly localized Cu-3d holes, the "effective" Coulomb repulsive interaction is expected to be not so large among p-holes due to the low concentration of the p-hole carriers, even if the "raw" Coulomb repulsive interaction of p-hole is as large as 5-6 eV. Then it is likely to apply the p-hole pairing model with s-like symmetry to the high-Tc superconductivity in order to explain quantitatively the distinct enhancement just below Tc characteristic for BCS-like superconductors. In s-wave model, the presence of "coherence factor" resulting from s-symmetry of Cooper pair enhances $1/T_1$ just below Tc together with the increase of the density of state at gap edge.

It is worthwhile that the relaxation rate of the longest component of T_1 of ^{17}O in normal state of 60K-"123" compound decreases at least by one fourth (1/4) as compared with that of 90 K-"123" compound [10]. From the oxygen content dependence of $1/T_1$ of ^{17}O, it has been established from a microscopic point of view that the mobile carriers are predominantly O-2p like rather than Cu-3d like, being consistent with the picture proposed by the photoemission study. Combined with both the NMR and the angle-resolved resonant photoemission studies, we conclude that the theory should be based on a d-p model. The next important subject is to elucidate what sort of p-symmetry (p_x, p_y, p_z) is dominantly favored for the hole doped on the oxygen sites.

4. Evidence of a Measurable but Secondary Contribution to the Superconductivity from Phonons by Oxygen Isotope Effect Study [5,6,20].

We simultaneously made two pairs of Bi-Sr-Ca-Cu-O samples and confirmed the isotope effect on each pair with independent measurements of resistivity and susceptibility at Tohoku University (pair #1) and at NBS-Boulder (pair #2). The resistivity and susceptibility curves were compared in terms of normalized parameters. In order to confirm the substitution of the oxygen isotope, we performed Raman spectroscopy with a resolution of 1.0 cm^{-1} on pair #1. On comparing the ^{16}O- and ^{18}O-enriched samples, we observed a shift from 469 cm^{-1} to 449 cm^{-1} of the best-resolved phonon frequency. Such an isotope shift is consistent with almost perfect substitution.

Figure 5 shows the resistivity of the ^{16}O- and ^{18}O-enriched sample from pair #1 normalized to its value at 273 K. The sample resistance was measured with separate current and voltage contacts, with a current of 10 mA. The observed isotope shift (ΔTc) with substitution is 0.34 $\pm$ 0.03 K for the 110K phase. The shift was determined by extrapolating the linear portions of the curves. Because the linear fits are parallel, ΔTc is the same at the inflection of the curves. Separate resistivity measurements on samples from pair #2, with an ac current of 4 μA at 177 Hz, gave a ΔTc of 0.30 $\pm$ 0.05 K for the 110 K transition. An extrapolation procedure similar

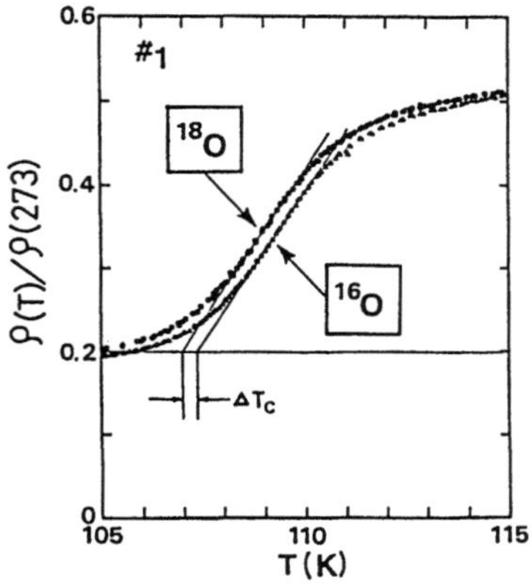

Fig.5: Detailed comparison of the normalized resistivity for the higher-Tc (110 K) phase for ^{16}O- and ^{18}O-enriched samples.

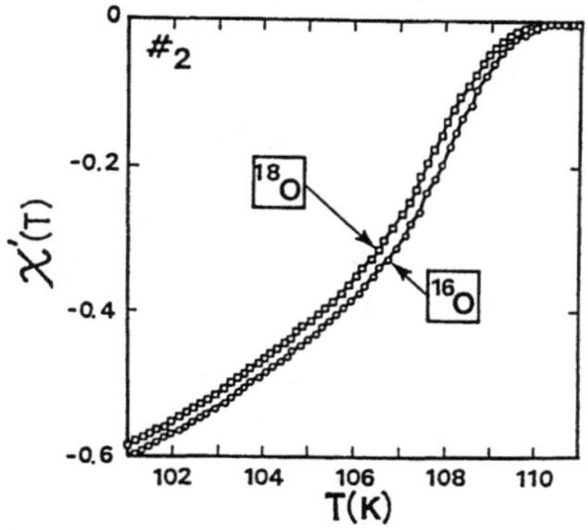

Fig.6: Detailed comparison of the real part χ' of ac susceptibility, normalized to $-\chi'$ at 80 K, for ^{18}O- and ^{16}O-enriched samples.

to that used above gives a ΔTc of about 0.42 ± 0.05 K for the 75K phase. Separate resistivity measurement on samples from pair #2, with an ac current of 100 μA, gave a ΔTc of 0.40 ± 0.05 K for the 75 K phase.

In Fig. 6, we show the real part of the susceptibility (χ') data for pair #2 to study the isotope shift for the 110 K phase. The data are normalized to the value of $-\chi'$ at 80 K. 80 K is well below Tc for the 110 K phase, but above Tc for the 75 K phase. The value of ΔTc is 0.30 ± 0.02 K measured at 108 K, near the inflections of the curves. Separate susceptibility measurements on pair #1 gave a ΔTc of 0.34 ± 0.04 K for the 110 K phase. For the 75 K phases, ΔTc is 0.23 ± 0.05 K measured at 74 K. Separate susceptibility measurements on pair #1 gave a ΔTc of 0.33 ± 0.04 K for the 75 K phase.

In Fig. 7, we plot ΔTc vs. Tc for various high-Tc compounds reported so far [5,6,15-22]. It appears that ΔTc does not change appreciably with Tc. This indicates that an electron-phonon interaction could contribute to the pairing mechanism for the superconductivity in these compounds, but some additional mechanism, like spin fluctuation [23-25], or charge fluctuation [26-28], may play the main role to raise Tc from that of Bi-Pb-Ba-O (Tc=11K) to that of Bi-Sr-Ca-Cu-O (Tc=110K). Because ΔTc does not change appreciably with Tc over all high-Tc compounds, strong electron-phonon coupling alone is insufficient to explain the high value of Tc.

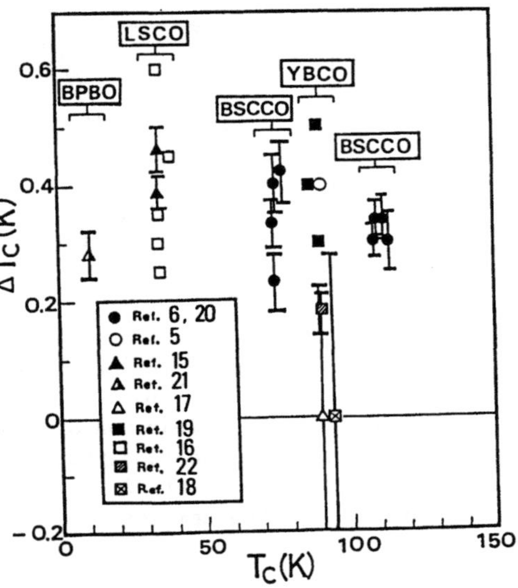

Fig.7: ΔTc vs. Tc for various high-Tc oxide superconductor system: BPBO (Ba-Pb-Bi-O), LSCO (La-Sr-Cu-O), YBCO (Y-Ba-Cu-O) and BSCCO (Bi-Sr-Ca-Cu-O).

5. Electronic Structure of High-Tc Superconductors.

From the angle-resolved resonant photoemission, ^{17}O NMR, and oxygen isotope studies, we propose the unified picture for the electronic structure of the high-Tc superconductors, as follows. Undoped compound, like La_2CuO_4 and $YBa_2Cu_3O_{6.5}$, are essentially charge-transfer semiconductors with large Coulomb repulsion energy (U_{dd}) between the 3d electrons and small charge transfer energy (Δ) from ligand-oxygen to copper [13,14], as shown in Fig. 8a. When they are doped with holes by substituting La with Sr or introducing additional oxygens, they change into metal with the doped O-2p hole state at the Fermi level due to $U_{dd}>\Delta$, as shown in Fig. 8b. Fermi-liquid states with O-2p character provide s-wave Cooper pairing below Tc, as was observed in the BCS-like enhancement of the ^{17}O nuclear spin-lattice relaxation rate just below Tc.

For the undoped compound, like La_2CuO_4 and $YBa_2Cu_3O_{6.5}$, a strongly localized Cu-3d hole enters the antibonding b_{1g} band in which Cu-3d(x^2-y^2) orbital hybridize with the O-2p orbital ($O^{(2)}p\sigma$,x and $O^{(2)}p\sigma$,y as is defined in Fig. 9) in the quasi-two dimensional CuO_2 plane. This b_{1g} band has a dominant Cu-3d character with a less weight of the O-2p character. This is consistent with the neutron diffraction experiment [29]. After extra hole is doped, the system changes into metal with the doped O-2p hole

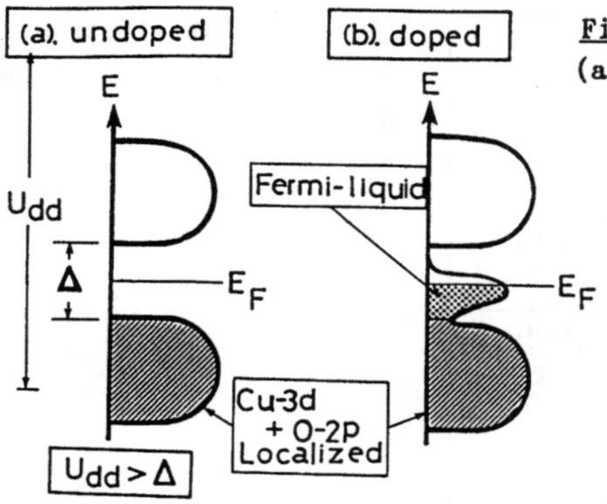

Fig.8: Schematic electronic states of (a) undoped and (b) doped high-Tc oxides.

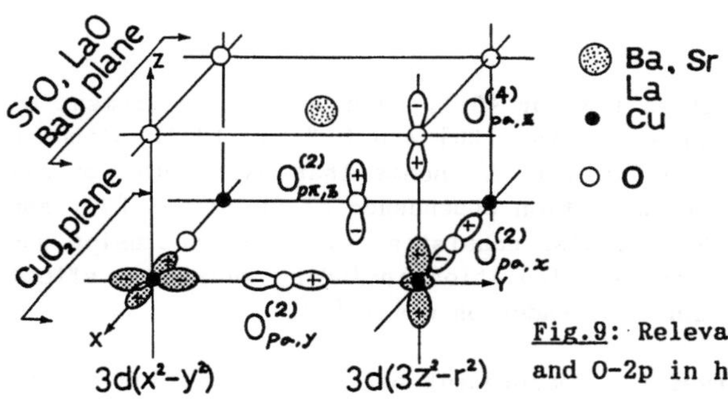

Fig.9: Relevant basis orbitals of Cu-3d and O-2p in high-Tc oxides.

at the Fermi level. In this case, we have two possibilities for the extra hole to enter a_{1g} or b_{1g} band. If the system is completely two-dimensional, extra hole should enter b_{1g} band. b_{1g} band is already occupied by the localized Cu-3d hole, therefore extra hole should enter b_{1g} band with singlet spin configuration of $^1A_1$ spin-multiplet. However, we have considerable p-d hybridization (40% of that in two-dimensional CuO_2 plane, if we assume that hybridization depend on d^{-4}, where d is the atomic distance between Cu and O-site) between the Cu-3d($3z^2-r^2$) orbital in the CuO_2 plane and O-2p orbital ($O^{(4)}p\sigma,z$, as is defined in Fig. 9) in the BaO [Y-Ba-Cu-O] or SrO [Bi-Sr-Ca-Cu-O], or LaO [La-Sr-Cu-O] plane. In this case we have another possibility for the extra hole to enter a_{1g} band with triplet spin configuration of $^3B_1$ spin-multiplet due to Hund's rule. a_{1g} band consists of the basis functions of copper 3d($3z^2-r^2$), and oxygen $O^{(4)}p\sigma,z$, $O^{(2)}p\pi,z$, $O^{(2)}p\sigma,x$, and $O^{(2)}p\sigma,y$, as is shown in Fig. 10. Present angle-resolved resonant photoemission experiment reveals that the electronic states of the doped extra hole at the Fermi level have a dominant O-2p nature with a far less weight of the Cu-3d character. Therefore, b_{1g} or a_{1g} band in which the extra hole enters should have a dominant O-2p character. In order to determine the electronic structure from the

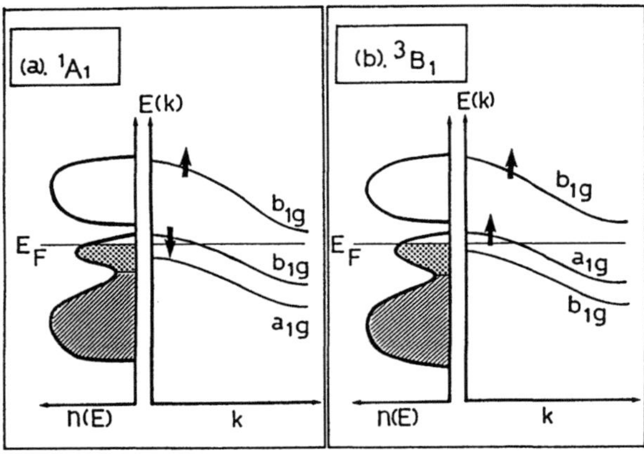

Fig.10: Possible electronic band structures of the doped high-Tc oxides.

above mentioned two possibilities ($^1A_1$ or $^3B_1$), we need the measurement of the polarization dependence of Cu L_3-XAS [30] and O K-XAS [31]. Although available experimental data are limited, it seems that extra hole mainly enters the a_{1g} band from the polarization dependence of Cu L_3-XAS [30] and O K-XAS [31] for $YBa_2Cu_3O_7$. This is also consistent with the conclusion obtained by the electronic structure calculation including many body effect using small cluster [32] and impurity Anderson model [33].

Nücker et al. [34] performed the measurement of the polarization of Cu-2p absorption edge by electron energy loss spectroscopy (EELS), and observed that there is an admixture of about 10% of low-lying Cu-3d($3z^2-r^2$) state with Cu-3d(x^2-y^2) state. On the other hand, almost 100% of the p_x and p_y polarization of the unoccupied O-2p hole states in the plane for $Bi_2Sr_2CaCu_2O_8$ is observed by O-1s absorption edge using EELS. These measurement of the polarization between Cu-2p and O-1s absorption edge is not consistent with each other, however, the admixture between Cu-3d(x^2-y^2) and Cu-3d($3z^2-r^2$) may be understood as follows. Suppose that the doped extra-hole enter b_{1g} band with $^1A_1$ spin multiplet. If the spin of the doped extra-hole at $O^{(2)}p\sigma$,x or $O^{(2)}p\sigma$,y orbital flips from down to up due to the frusration, the charge transfer from $O^{(2)}p\sigma$,x (or $O^{(2)}p\sigma$,y) orbital in b_{1g} band to $O^{(4)}p\sigma$,z orbital in a_{1g} band occurs suddenly because of Pauli exclusion principles. This charge-fluctuation due to the spin-fluctuation may become to be important for the pairing mechanism through the admixture of Cu-3d(x^2-y^2) and Cu-3d($3z^2-r^2$) orbital. In order to obtain a perfect consensus, we need more research of Cu L_3-XAS and O K-XAS.

Combined with both the evidence of the Fermi-liquid states with O-2p character from angle-resolved resonant photoemission experiment and the evidence of O-2p hole pairing with s-wave like symmetry from BCS-like enhancement of $1/T_1$ with Korringa-like behavior, we conclude that the theory should be based on a d-p model. We observed non-zero isotope shift in high-Tc superconductors. This shows a measurable but secondary contribution to the superconductivity from phonons, therefore the mechanism of high-Tc superconductivity, like spin-fluctuation or charge-fluctuation, should include the coupling with the phonons.

Acknowledgments

We thank Professors Y. Endoh, A. Kotani and Dr. K. Okada for valuable discussions. We also thank Professor K. Asayama, Dr. Y. Kitaoka and Mr. K. Ishida for the collaboration of enriched-^{17}O NMR. This work was partially supported by a Grant-in-Aid for Scientific Research on Priority Area from the Ministry of Education, Science, and Culture of Japan.

References

1. T. Takahashi, H. Matsuyama, H. Katayama-Yoshida, Y. Okabe, S. Hosoya, K. Seki, H. Fujimito, M. Sato and H. Inokuchi: Nature 334, 691 (1988).
2. T. Takahashi, H. Matsuyama, H. Katayama-Yoshida, Y. Okabe, S. Hosoya, K. Seki, H. Fujimoto, M. Sato and H. Inokuchi: Submitted to Phys. Rev. B.
3. Y. Kitaoka, K. Ishida, K. Fujiwara, K. Asayama, H. Katayama-Yoshida, Y. Okabe and T. Takahashi: ICCF6, Frankfrut, 1988 to be published in J. Mag. Mag. Mat.
4. K. Ishida, Y. Kitaoka, K. Asayama, H. Katayama-Yoshida, Y. Okabe and T. Takahashi: J. Phys. Soc. Jpn., 57, 2897 (1988).
5. H. Katayama-Yoshida, T. Hirooka, A. J. Mascarenhas, Y. Okabe, T. Takahashi, T. Sasaki, A. Ochiai, T. Suzuki, J. I. Pankove, T. F. Ciszek, and S. K. Deb: Jpn. J. Appl. Phys. 26, L2085 (1987).
6. H. Katayama-Yoshida, T. Hirooka, A. Oyamada, Y. Okabe, T. Takahashi, T. Sasaki, A. Ochiai, T. Suzuki, A. J. Mascarenhas, J. I. Pankove, T. F. Ciszek, S. K. Deb, R. B. Goldfarb and Yangkang Li: Physica C 156, 481 (1988).
7. M. S. Hybertsen and L. F. Mattheiss: Phys. Rev. Lett. 60, 1661 (1988).
8. H. Krakauer and W. E. Pichett: Phys. Rev. Lett. 60, 1665 (1988).
9. S. Massidda, J. Yu and A. J. Freeman: Physica C 52, 251 (1988).
10. Y. Kitaoka, K. Ishida, K. Asayama, H. Katayama-Yoshida, Y. Okabe and T. Takahashi: Submitted to Nature.
11. Y. Kitaoka, S. Hiramatsu, Y. Kohori, K. Ishida, T. Kondo, H. Shibai, K. Asayama, H. Takagi, S. Uchida, H. Iwabuchi and S. Tanaka: Proc. Int. Conf. on High Temperature Superconductors and Materials and Mechanism of Superconductivity, Interlaken, 1988, Physica C 153-155, 83, (1988).
12. T. Imai, T. Shimizu, T. Tsuda, H. Yasuoka, T. Takabatake, Y. Nakazawa, and M. Ishikawa: J. Phys. Soc. Jpn. 57, No. 5 (1988).

13. A. Fujimori, E. Takayama-Muromachi and Y. Uchida: Solid State Commun. 63, 857 (1987).
14. T. Takahashi, F. Maeda, H. Arai, H. Katayama-Yoshida, Y. Okabe, T. Suzuki, S. Hosoya, A. Fujimori, T. Shidara, T. Koide, T. Miyahara, M. Onoda, S. Shamoto and M. Sato: Phys. Rev. B36, 5686 (1987).
15. B. Batlogg, G. Kourouklis, W. Weber, R. J. Cava, A. Jayaraman, A. E. White, K. T. Short, L. W. Rupp and E. A. Rietman: Phys. Rev. Lett. 59, 912 (1987).
16. T. A. Faltens, W. K. Ham, S. W. Keller, K. J. Leary, J. N. Michaels, A. M. Stacy, H. C. zur Loye, T. W. III. Barbee, L. C. Bourne, M. L. Cohen, S. Hoen and A. Zettl: Phys. Rev. Lett. 59, 915 (1987).
17. B. Batlogg, R. J. Cava, A. Jayaraman, R. B. van Dover, G. A. Kourouklis, S. Sunshine, D. W. Murphy, L. W. Rupp, H. S. Chen, A. White, K. T. Short, A. M. Mujsce and E. A. Rietman: Phys. Rev. Lett. 58, 2333 (1987).
18. L. C. Bourne, M. F. Crommie, A. Zettl,H. C. zur Loye, S. W. Keller, K. L. Leary, A. M. Stacy, K. J. Chang, M. L. Cohen and D. E. Morris: Phys. Rev. Lett. 58, 2337 (1987).
19. K. J. Leary, H. C. zur Loye, S. W. Keller, T. A. Faltens, W. K. Ham, J. N. Michaels, and A. M. Stacy: Phys. Rev. Lett. 59, 1236 (1987).
20. H. Katayama-Yoshida, T. Takahashi, H. Matsuyama, Y. Okabe, Y. Kitaoka, K. Ishida, K. Asayama, H. Fujimoto, and H. Inokuchi: To be published in the Proceedings of Progress in High Temperature Superconductivity, Vol. XIV, "Towards a Theoretical Understanding of High Temperature Superconductivity", ICTP, Trieste, Italy, 20 June-29 July, 1988.
21. B. Batlogg, R. J. Cava, and M. Stavola: Phys. Rev. Lett. 60, 754 (1988).
22. D. E. Morris, R. M. Kuroda, A. G. Markelz, J. K. Nickel and J. I. T. Wei: Phys. Rev. B37, 5936 (1988).
23. P. W. Anderson: Science 235, 1196 (1987).
24. M. Imada: J. Phys. Soc. Jpn. 56, 3793 (1987); 57, 3128 (1988).
25. M. Ogata and H. Shiba: J. Phys. Soc. Jpn. 57, 3052 (1988).
26. C. M. Varma, S. Schmitt-Rink and E. Abrahams: Solid State Commun. 62, 681 (1987).
27. M. Tachiki and S. Takahashi: Phys. Rev. B38, 218 (1988).
28. Y. Takada: Submitted to Phys. Rev. B.
29. G. Shirane, Y. Endoh, R. J. Birgeneau, M. A. Kastner, Y. Hidaka, M. Oda, M. Suzuki and T. Murakami: Phys. Rev. Lett., 59, 1613 (1987).
30. A. Bianconi, M. de Santis, A. di Cicco, A. M. Flank, A. Fontaine, P. Legarde, H. Katayama-Yoshida, A. Kotani and A. Marcelli: Physica C 153-155, 1760 (1988).
31. N. Nucker, J. Fink, J. C. Fuggle, P. J. Durham and W. M. Temmerman: Physica C 153-155, 119 (1988).
32. A. Fujimori: Preprint.
33. K. Okada and A. Kotani: Submitted to J. Phys. Soc. Jpn.
34. N. Nucker, H. Romberg, X.X. Xi, J. Fink, B. Gegenheimer and Z.X. Zhao: Submitted to Phys. Rev. B.

春夏冬二升五合.

Extensive Study of the Optical Spectra for High T_c Cuprates and Related Oxides

S. Uchida, S. Tajima, H. Takagi, and Y. Tokura

The University of Tokyo, Tokyo 113, Japan

I. INTRODUCTION

The perovskite-related oxides, $(La_{1-x}Sr_x)_2CuO_4$(LSCO) and $YBa_2Cu_3O_y$(YBCO), have been established to be high-T_c superconductors, which poses a fundamental question: why do such extraordinarily high values of T_c seem to be restricted to Cu oxides? It has also been pointed out that these Cu-oxides show quite peculiar behavior in various physical properties, in particular there is generally a metal to semiconductor transition at a certain composition, $x \simeq 0.025$ for LSCO and $y \sim 6.35$ for YBCO /1/.

In order to elucidate the mechanism of superconductivity in Cu-oxides, it is essential to draw a clear image of the electronic states by investigating the normal state properties as well as the superconducting state. The band calculations for both Cu-oxides reported by many groups fail to explain the fact that La_2CuO_4(LCO) and $YBa_2Cu_3O_6$ are not metallic but semiconducting /2,3/. Similar situation might arise in recently found $Ba_{1-x}K_xBiO_3$ which is also an oxide superconductor with T_c of about 30 K /4/. This material seems to undergo a semiconductor to metal transition at x=0.3 - 0.4. However, as shown later, this is considered to be driven primarily by the strong coupling between conduction electrons and breathing-mode phonons /5/. In its semiconducting phase a spectroscopic evidence is obtained which indicates that the charge-density-wave(CDW) state is formed.

The analysis of the photoemission data has yielded a Coulomb repulsive energy (U_d > 6 eV) on the Cu atoms in the high-T_c cuprates larger than bandwidth ($W \sim 4$ eV) calculated from the band theory. This result provides an evidence that band theory is inadequate and a Mott-Hubbard(M-H) picture may be more appropriate /6/. The fact that the Hall coefficient decreases with increasing dopant content is also better understood by the M-H model rather than the band model /7/.

On the other hand, photoemission and electron-energy-loss experiments /8/ suggest that doping creates holes on oxygen, not on copper, leading to a model in which both copper and oxygen orbitals are incorporated /9/. In such a case, the charge transfer energy Δ between Cu and O is one of the important parameters which control the physical properties or the behavior of the low-lying excitations. The presence of holes on oxygen atoms in the doped oxides indicates that the parent materials, La_2CuO_4 and $YBa_2Cu_3O_6$, might be charge-transfer (CT) insulators for which $\Delta < U_d$ and thus determines the correlation gap. Zaanen et al./10/ have pointed out a trend that Δ becomes smaller and thus determines the character of the insulating state for the late 3d transition metal oxides, CuO and NiO. It is an important issue to demonstrate whether the high-T_c Cu oxides and their relatives also follow this trend. Varma et al./11/ further argue that in the Cu-O superconductors the CT excitations would move down in energy due to the presence of Coulomb repulsion between the nearest neighbor Cu and O and be possible source of binding two holes.

Springer Series in Materials Science, Vol. 11
Mechanisms of High Temperature Superconductivity
Editors: H. Kamimura and A. Oshiyama

In this article, we show the optical reflectivity spectra of high-Tc superconducting Cu-oxides from the far-infrared to the vacuum-ultraviolet region (10 meV - 40 eV) and investigated optical transitions in order to clarify to what extent the band calculation is applicable to these Cu-oxides. For better understanding and for extracting uniqueness of the spectra for Cu-oxides, we have also measured the spectra of non-superconducting (214)-compounds, $(La,Sr)_2MO_4$ (M = Ni,Co and Fe) and Nd_2CuO_4 as superconducting non-Cu oxide $(Ba,K)BiO_3$.

II. EXPERIMENTAL RESULTS

Single crystalline LCO, LSCO(nominal Sr composition:10%), YBCO and Nd_2CuO_4 were grown by the CuO flux method and the other single crystalline (214)-compounds, La_2NiO_4, $(La_{0.75}Sr_{0.25})_2CoO_4$ and $LaSrFeO_4$, by the arc-melt method /12/. For the synthesis of Co and Fe compounds, we need to substitute La with substantial amounts of Sr, since the K_2NiF_4 structure is stable either under reduced oxygen atmosphere or under high Sr doping.

Reflectivity measurements for the ultraviolet energy region above 4 eV have been performed using the synchrotron-orbital-radiation (SOR) as a light source. For the measurements below 6 eV, a BOMEM DA3 Fourier-Transform-Spectroscopy System was employed.

Reflectivity spectra of $(La,Sr)_2CuO_4$

Firstly, we see the gross features of the reflectivity spectra. In Fig. 1 are shown the spectra of undoped and Sr-doped La_2CuO_4 up to 40 eV. The samples are all single crystals and the polarization of the incident light is parallel to the Cu-O 2-dimensional networks (E ⊥ c, we use this convention hereafter) for. As a common feature, there appear three distinct reflectivity edges around 1 eV, 13 eV and 25 eV, each edge indicating the end point of a series of intra- or interband excitations. The lowest energy edge is due to the a collective excitation, the so called plasma edge of free carriers which move along Cu-O 2D planes. The low plasmon energy as compared with conventional metals is a consequence of the low density of carriers in these oxides. The details of the behavior of the plasma reflectivity for varying composition have been described in a separate paper /13/. The second edge indicates the end point of the excitation from the O 2p valence bands to the higher conduction bands which are composed of the orbitals associated with the perovskite A cations for cuprate, La 5d/4f. The third edge observed in the highest energy region originates from the excitations involving all the valence electrons(n_v= 33).

Anisotropy of the spectra

The anisotropy of the reflectivity spectra for LSCO is shown in Fig.2. From the anisotropic crystal structure, the electronic structure is expected to be quasi-two-dimensional and thus, the reflectivity spectrum should be different for E⊥c and for E//c. In fact, as seen in Fig.2, the difference between the two polarizations is significant except in the very high energy region as expected from the crystal structure. For example, in the spectrum of LSCO for E ⊥ c the peak at 9 eV is weaker and the peak at 12.5 eV is stronger than in the spectrum for E⊥c, which indicates that the former peak results from excitation in the plane perpendicular to the c-axis and the latter from excitation parallel to the c-axis. In particular, we are concerned with the absorption in the lowest energy region. We find a very weak peak at about 2 eV in the E⊥c-spectrum of LSCO which is not seen for E// c. Similar anisotropy was found earlier for the spectrum of La_2CuO_4 where a weak 2 eV structure for E⊥c was seen /14/.

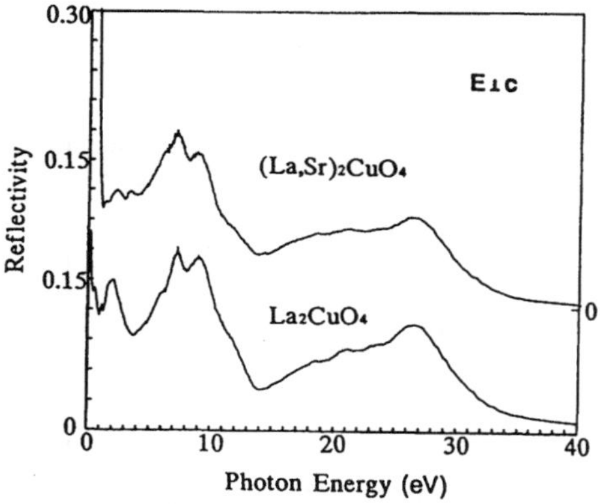

Fig.1 Reflectivity spectra of insulating La_2CuO_4 and metallic Sr doped compound.

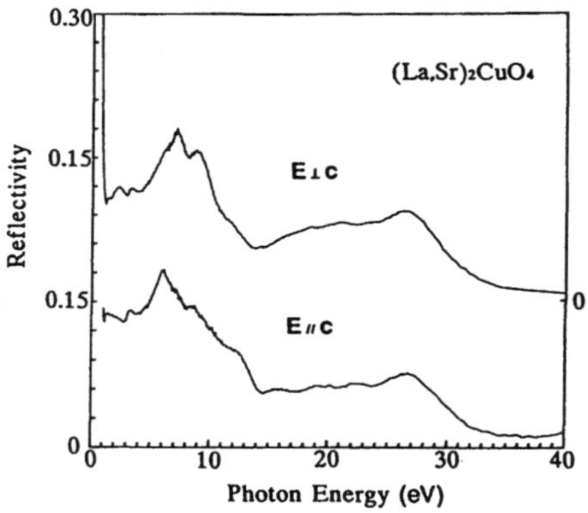

Fig.2 Polarization dependence of the spectrum for LSCO - the top for E⊥c and the bottom for E//c.

Doping effect on the spectrum

As shown in Fig. 1, no significant change can be seen, between spectra of undoped and Sr doped La_2CuO_4 in the energy region higher than 5 eV. On the other hand, in the lower energy region, the spectrum of La_2CuO_4 is characterized by a weak peak at 2 eV, whereas the spectrum of doped material is dominated by a free-carrier plasma reflectivity edge at 0.9 eV and the 2 eV peak becomes very weak. Similar spectral change is also observed for $YBa_2Cu_3O_y$ system when the oxygen content y increases which corresponds to the increase of hole density. In the spectrum if metallic YBCO, a plasma edge is located at 1.5 eV and only very weak structures are seen in the energy range up to 4-5 eV. It seems that the 2 eV absorption in the insulating phase is broadened or weakened by the opening of some decay channel due to doping with holes rather than that its oscillator strength is converted in to the free-carrier-plsmon mode. Recent transmission measurement made by Suzuki at NTT on single crystal films of LSCO has suggested that the intensity of the 2 eV absorption may be transfered to a new absorption observed at 1.5 eV for doped samples /15/.

Spectra of non-Cu oxides and Nd_2CuO_4

To see the contribution of Cu to the spectrum, we have measured the reflectivity spectra of $(La,Sr)_2MO_4$ with various 3d-metal M (M=Cu,Ni,Co and Fe) and Nd_2CuO_4 which has another type of tetragonal crystal structure, so-called T' structure /16/. Figure 3 shows the reflectivity spectra of these four oxides with the K_2NiF_4 structure. We notice that the spectrum above ca. 7 eV is almost invariant both in energy and in spectral shape as M moves from Cu to Fe. This fact provides evidence that the spectrum in this region is determined by optical transitions involving the states associated with the common elements among this series of compounds, e.g. the transitions from occupied O 2p states to empty La 5d and/or La 4f states. The 3d states of the transition metal are apparently not relevant to the strong feature spanning an absorption band between 7 and 12 eV which is convincingly assigned to the O 2p→La 5d/4f transitions.

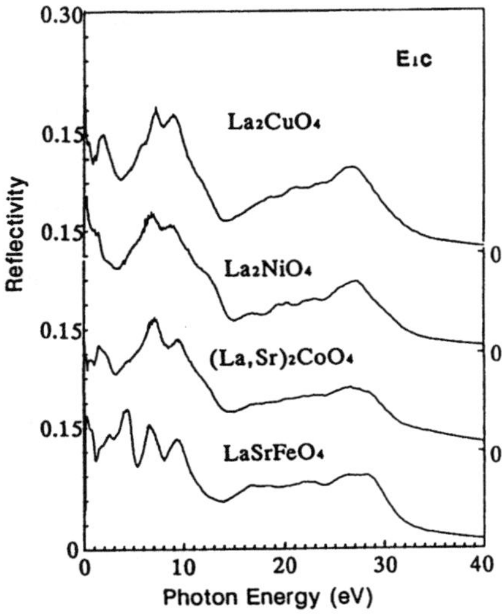

Fig.3 Reflectivity spectra for four (214)-compounds including Cu, Ni, Co and Fe (from the top).

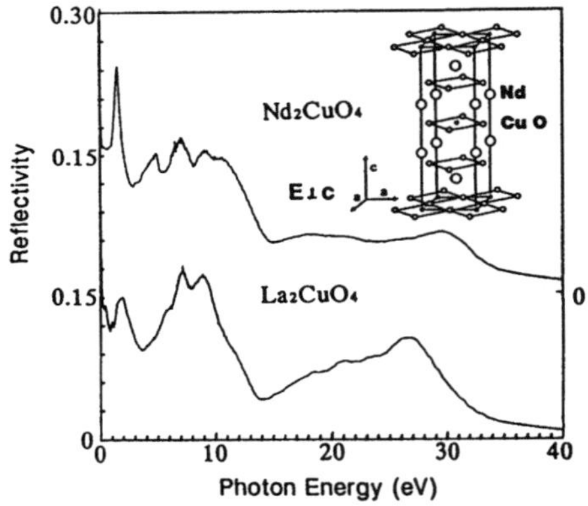

Fig.4 Comparison of the reflectivity spectrum of Nd_2CuO_4 with that of La_2CuO_4. The T'structure of Nd_2CuO_4 is shown in the inset.

On the other hand, the spectrum in the region below 7 eV varies somewhat. In this energy region only a few weak features are seen except for $LaSrFeO_4$ where there is a strong peak centered at 5 eV. No additional strong feature is seen in the lower energy region. The reason might be that $LaSrFeO_4$ is a Mott insulator with $U_d < \Delta$ and we observed interband transmitions across a Mott-Hubbard gap, while other compounds are charge-transfer insulators.

Contrary to the high-T_c Cu-oxides, a pronounced peak is observed at 1.5 eV for Nd_2CuO_4 which is an insulator with similar structure to La_2CuO_4 but perhaps with a different configuration for some of the oxygen sites. In Nd_2CuO_4 the oxygen atoms sandwiched by the Cu-O layers are thought not to occupy the positions just above the Cu atoms so that oxygen octahedra are not formed. The sharpness of the 1.5eV-peak in Nd_2CuO_4 is thought to be due either to the lack of a decay channel as discussed above and to the increase of ionicity. We should note that the 1.5 eV peak is not observed in the configuration E//c /17/ as in the case of the 2 eV peak in LSCO.

III. DISCUSSIONS

Comparison between experiment and band calculation

Figure 5 shows the comparison between the imaginary part of dielectric function of LSCO obtained experimentally and the one calculated from band theory by Mazin et al./18/ for La_2CuO_4. The calculated spectrum is almost unchanged for Sr-doping as far as the virtual crystal approximation is applied. In the energy region between 5 and 10 eV where the absorption peaks originate from the excitation O 2p → La 5d/4f, the agreement between these two spectra is excellent. However, in the lower energy region the experimental result is completely different from the theoretical one. In the theoretical result, an absorption peak at 1.2 eV arises due to the transi-

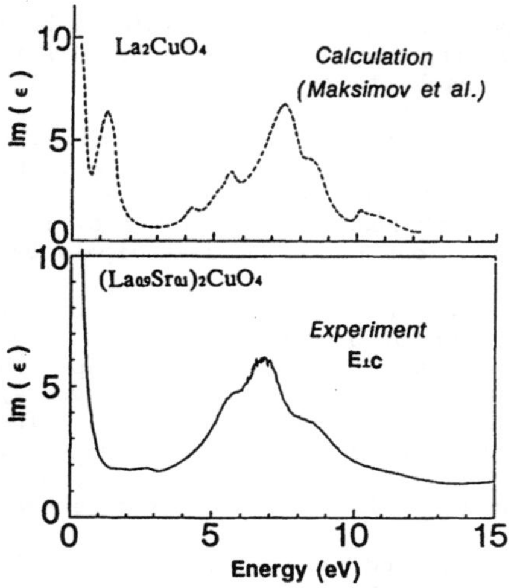

Fig.5 Comparison of the imaginary part of the dielectric function between the band calculation and the present experimental result.

tion between the bands composed of Cu 3d and O 2p orbitals. On the other hand, as we discuss later, corresponding absorption is seen at 2 eV or higher energy in the experimental spectrum.

Another difference is in the plasmon energy. We identify a free-carrier plasma edge in the experimental reflectivity spectrum at almost the same energy as that in the calculated one. However, this is an accidental coincidence, because the presence of the strong 1.2 eV absorption in the calculated spectrum pushes the plasma edge strongly to lower energy. If the absorption is very weak or located at higher energy as in the observed spectrum, then the calculated plasma edge will appear at appreciably higher energy, 2 eV or higher, that is, the observed value of the plasma energy $\omega_p^2 = 4\pi ne^2/m^*$ is smaller by a factor of about 2 than the calculated one. This is suggestive of strong renormalization or strong mass-enhancement of the actual conduction band. A detailed discussion of the plasma reflectivity has been given in our previous paper /13/.

Charge transfer excitation

To date several experiments have been done to determine the value of Δ in the high-T_c cuprates by optical measurements. However, the results are confusing, even controversial. Early experiments on polycrystalline samples have claimed that an absorption peak emerges at around 0.5 eV for both LSCO /19/ and YBCO /20/ upon doping which was ascribed to the CT excitation responsible for an exciton mechanism of superconductivity /11/. However, it is suggested by the experiments on single crystalline samples that the 0.5eV-peak may be an artifact of the Kramers-Kronig analysis on the reflectivity spectrum for a strongly anisotropic polycrystalline medium /21,22/.

The E c optical reflectivity (transmission) measurements on single crystalline samples (oriented thin films) have revealed a few weak absorption maxima in the energy range between 0.5 and 3.0 eV. in both LSCO and YBCO. According to band theory, the conduction bands near the Fermi level are composed of the antibonding band of the Cu 3d hydridzed with O_{2p} orbirtals and higher d-bands of A-site atoms, La 5d for LSCO and Ba 5d and/or Y 4d for YBCO. Therefore, the structure with relatively large oscillator strength observed below 10 eV originates from transitions from the O 2p valence band to these conduction bands. Then, if a strong absorption were observed in the lower energy region, it would be reasonable to assign it to the optically allowed O 2p - Cu 3d charge transfer excitation.

However, the actual situation is not simple. Geserich et al./23/ measured the transmission spectrum of the c-axis oriented YBCO thin film. From the similarity to the spectrum of NiO /24/ they state that the CT excitations start from 4 eV and a couple of weak absorptions below 4 eV correspond to the valence conserving d-d transitions not allowed in dipole but in quadrupole selection rule.

The spectrum of Nd_2CuO_4 seems to be typical of insulating Cu oxide. Judging from the similarity to the spectrum of La_2CuO_4 the spectrum of Nd_2CuO_4 in the region $\hbar\omega > 6$ eV is considered to be dominated by O 2p - Nd 5d/4f transitions, so that such a strong absorption in the lower energy region is likely to be assigned to Cu-O charge transfer excitation. The strong anisotropy of the peak at 2 eV is observed in the spectra of Nd_2CuO_4, indicating that this peak originates from excitation within the Cu-O plane, therefore not from intra-atomic d-d exitation. This is also the case concerning the 2 eV peak in La_2CuO_4.

The apparent decrease in the strength of the corresponding absorption in the high-T_c Cu-oxides, even in the insulating La_2CuO_4, might be due to some as yet unidentified feature inherent to the high-T_c oxides. As a possible cause the peak may be broadened due to the relaxation of the exited holes on the O atoms in the Cu-O plane into O atoms just above (below) Cu which are absent in Nd_2CuO_4.

As Zaanen et al. /10/ studied for transition metal oxides, one may expect that the Δ value can be varied on going from Cu to earlier 3d transition metal elements in the common K_2NiF_4-type structure. Note that these compounds are antiferromagnetic insulators /25/ as Nd_2CuO_4 and undoped La_2CuO_4. One expects therefore a rather strong O 2p $\rightarrow$ M 3d charge transfer absorption in the spectrum, particularly when it is lower in energy than the O - La CT absorption band. However, as seen in Fig. 3, strong absorption is not seen in the low energy region for La_2NiO_4 and $(La,Sr)_2CoO_4$ except for $LaSrFeO_4$. The structure observable in the 2eV region is much weaker than that in Nd_2CuO_4 and even weaker than that in La_2CuO_4. We may not rule out the possibility that this structure would have moved to higher energy and become absorbed in the O 2p $\rightarrow$ La 5d/4f absorption band, but considering that a pronounced peak is observed at 5 eV separated from O-La absorption band for $LaSrFeO_4$, we expected that the CT exitation is in the energy range below 5 eV for Cu oxides and other (214) compounds. No systematic change against 3d-metal substitution is probably because in the perovskite related structure the value of the optically detected energy gap is strongly affected by the mixing of O 2p and M 3d orbitals as well as by the bonding ionicity with respect to A-site cation. On moving in the 3d transition-metal series from right to left, the energy difference between the O 2p and M 3d levels($\varepsilon_d-\varepsilon_p$) increases on one hand and the mixing of these orbitals decreases on the other which reduces the hybridization shift of the energy levels, hence more or less compensating the increase in the energy separation.

It is probable that the charge transfer absorption starts from 2 eV or below in the Cu-oxides other (214)-compounds. A recent photoexcitation experiment on La_2CuO_4 has detected two photoinduced absorptions peaked at 0.5 and 1.4 eV and crossover to photoinduced bleaching above 2.0 eV /26/. This result together with observed luminescence at around 2 eV are strongly suggestive of the presence of an energy gap at this energy.

The charge transfer energy is relevant to the problem of the high T_c oxides in two ways depending on the mechanism of superconductivity. One is the superexchange coupling J between neighboring Cu spins in the two-dimensional Cu-O plane. In the two-band Hubbard model J is roughly given by

$J \sim t_{pd}^4/\Delta^3$. This parameter plays a central role in the magnetic mechanisms of superconductivity. The value of J was deduced from neutron inelastic scattering /27/ and two-magnon Raman scattering on La_2CuO_4 /28/. The value is found to be extremely large ($J \sim 0.15$eV) which seems to favor smaller Δ according to the above formula.

The second is the direct relevance to the charge-transfer mediated pairing mechanism proposed by Varma et al./11/. They emphasize the importance of the nearest neighbor Coulomb repulsion U_{pd} among the parameters involved in the problem. The presence of U_{pd} puts forward the charge fluctuations and makes them low-lying excitations. Varma et al. has suggested that, as a consequence of U_{pd}, an exciton-like CT resonance would be observed in the low energy region (0.1-1.0 eV) of the optical absorption spectrum. Although the spectra for superconducting single crystals of both LSCO and YBCO are primarily Drude-like above 0.1 eV, the presence or otherwise of CT resonance in the low energy region below ω_p is still a debate. The recent careful measurements by Orenstein et al./29/ and Tanaka et al./30/ have indicated an absorption band in the 0.1-0.3 eV region for single crystal $YBa_2Cu_3O_y$ and $EuBa_2Cu_3O_y$.

Optical Spectrum of $(Ba,K)BiO_3$

It is also worthwhile to know the situation in other oxide superconductors which do not contain Cu. $BaPb_{1-x}Bi_xO_3$ and $Ba_{1-x}K_xBiO_3$ are typical ones, in particular the latter shows high-Tc superconductivity of 30 K range. The semiconducting of the end material $BaBiO_3$ is understood as due to charge-density-wave (CDW) formation or charge disproportionation on Bi sites coupled with the breathing-mode phonon /5/.

In Fig. 6 are shown how the optical gap absorption varies on approaching the semiconductor-to-metal transition for both BPBO and BKBO systems. The

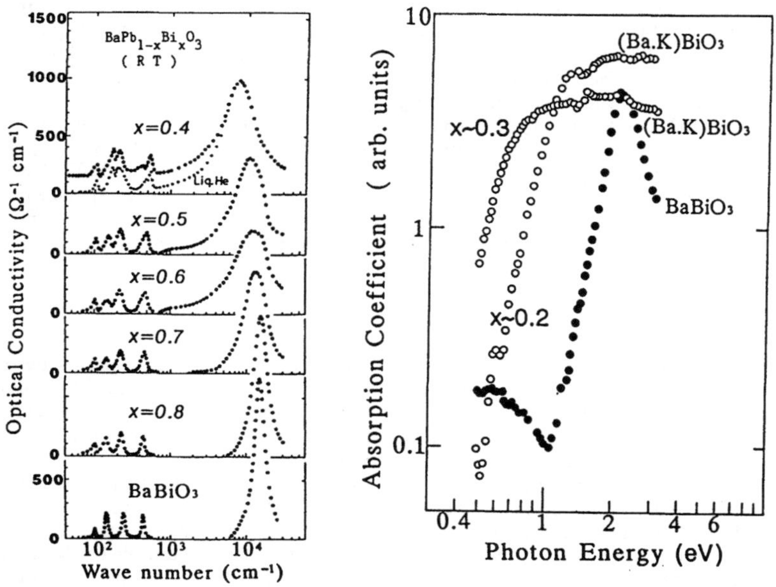

Fig.6 Optical conductivity spectra for $BaPb_{1-x}Bi_xO_3$ (left) and absorption spectra for $Ba_{1-x}K_xBiO_3$ (right).

optical conductivity spectra obtained from the reflectivity spectra via Kramers-Kronig transformation are shown for BPBO.

A sharp dominant peak is observed at 2 eV in the spectrum of $BaBiO_3$ which should correspond to the excitations across a CDW gap. When Bi is partially substituted with Pb, this peak is gradually broadened and simultaneously shifts toward lower energy with a tail extending to much lower energies. The result was interpreted, in terms of local picture, as the formation of the Pb states within the energy gap between the CDW split Bi bands /31/. The semiconductor-metal transition is considered to take place at 65 % Pb substitution when the outermost end of the absorption tail reaches zero energy, that is, when the Pb band edge goes down and touches the lower Bi subband.

In the case of K substitution for Ba, bulk materials, either poly crystals or single crystals, are not available which are suitable for optical measurement because of the difficulty in synthesizing crystals with sufficient density or sufficient size. But it turns out that thin films BKBO can be grown on $SrTiO_3$ substrate at present restricted to the compositional range $x < 0.3$. In this range, the film is still semiconducting but changes its color from gold at $x \sim 0$ to dark purple or blue at larger x, which indicates that substantial amount of K is certainly incorporated.

Figure 6 shows the absorption spectra obtained from the transmission measurement on the films of $BaBiO_3$ and two K-substituted materials. The K composition was estimated from the lattice constant referring to the Argonne group's result /33/. On increasing x the absorption edge shifts towards lower energy, clearly indicating a decrease in the optical energy gap. As compared with BPBO, a distinct feature in BKBO is that the absorption edge remains to show steep rise as in the undoped $BaBiO_3$. The steep edge suggests that the band edges preserve the singularity characteristic of the gap formation triggered by the perfect nesting of the Fermi surface /5/.

This result is, thus, strongly suggestive of the Fermi surface instability, probably CDW, as the primary origin of the semiconductivity of BKBO. However, the result is apparently in contradiction to the rigid band picture where the nesting condition would easily be violated by a slight K doping. In order that CDW persists in the K doped material, it should be local in nature with very short coherence length. An evidence for breathing-mode type distortions in the semiconducting phase has been presented by the recent single crystal X-ray analysis /34/. The X-ray analysis indicates no distortions in the metallic composition $x \simeq 0.4$. From these facts a simple scenario may be written on the metal-semiconductor transition in BKBO that at the critical composition $x \simeq 0.4$ the CDW gap vanishes in correspondence with disappearance of the breathing-mode distortion. In the case of BPBO, CDW remains localized on Bi sites even in the metallic phase /5/.

V. SUMMARY

Optical reflectivity spectra of high-T_c cuprates have been investigated over a wide energy range up to 40eV. The spectra were compared with that for non-superconducting Nd_2CuO_4 as well as for Ni, Co and Fe oxides with the K_2NiF_4 structure and non-Cu perovskite$(Ba,K)BiO_3$ in order to have insight into unique electronic states in the high-T_c oxide materials.

In the high energy region ($\hbar\omega > 5$ eV) it is found that the band picture describes rather well the optical spectra for all these oxides with perovskite-related structures. On the other hand, the observed spectra in the low energy region are considerably different from the band calculation results. The free-carrier plasmon energy is appreciably reduced and the absorption corresponding to Cu-O charge transfer excitations is severely broadened, shifting to higher energy than predicted from the calculations.

These facts indicate that strong renormalization or reorganization is taking place for the energy bands composed of Cu 3d and O 2p orbitals.

La_2NiO_4 and $(La,Sr)_2CoO_4$ are optically quite similar to the superconducting cuprates. We cannot see any systematic change of the optical spectrum as we move from Cu to Co. The low energy absorption, perhaps corresponding to O_{2p}-M_{3d} CT excitation, is obscured, which, if intrinsic, might be a characteristic feature for the CT insulators with perovskite-type structure. In this regard, a strong absorption at 5 eV in $LaSrFeO_4$ might be an evidence for a Mott, not CT, insulator where $U_d<\Delta$. Further experiment is necessary to clarify this point using samples with various Sr content.

In the spectra of Nd_2CuO_4 a sharp exciton-like peak is observed at 1.5 eV probably due to Cu-O charge transfer excitation. We suppose that this distinct feature arises from the more ionic character of these two oxides. The structural difference between Nd_2CuO_4 and La_2CuO_4 might also result in a different valence band structure making the Nd-cuprate non-superconducting. In support of a different valence band structure, we note that doping with Sr is hard to realize metallic conduction in Nd_2CuO_4.

A spectroscopic evidence is obtained for the persistency of the Fermi surface instability, probably a CDW, in the semiconducting region of K doped $BaBiO_3$. The CDW seems to play a key role in the M-S transition and possibly superconductivity of this non-Cu oxied system.

ACKNOWLEDGEMENTS

We wish to thank H.Sato, University of Tokyo, K.Oka and H.Unoki, Electrotechnical Laboratory, Y.Hidaka, M.Suzuki and T.Murakami, Opto-elctronics Laboratories, NTT for the collaboration in sample preparation. The optical measurement in VUV region were made at Synchrotron Radiation Laboratory, ISSP, University of Tokyo under the arrangement by M.Seki and S.Suga. This work was supported by a Grant-in-Aid for Special Distinguished Researhh from the Ministry of Education, Science and Culture of Japan.

REFERENCES

1. For a review on the experimental works, see K. Kitazawa, H. Takagi, K. Kishio, T. Hasegawa, S. Uchida, S. Tajima, S. Tanaka, and K. Fueki, Physica C 153-155, 9(1988).
2. e.g. L.F. Mattheiss, Phys. Rev. Lett. 58, 1028 (1987).
3. e.g. S.Massida,J. Yu, A.J. Freeman and D.D. Koelling, Phys. Lett. 122, 198 (1987).
4. L.F. Mattheiss, E.M. Gyorgy and D.W. Johnson Jr., Phys. Rev 337, 3745 (1988)
5. For a review, see S. Uchida, K. Kitazawa and S.Tanaka, Phase Transitions 8, 95 (1987).
6. A. Fujimori, E. Takayama-Muromachi, Y. Uchida and B. Okai, Phys. Rev. B35, 8814 (1987).
7. N.P. Ong, Z.Z. Wang, J. Clayhold, J.M. Tarascon, L.H. Greene and W.R. McKinnon, Phys. Rev. B35, 8807 (1987).
8. N. Nucker, J. Fink, J.C. Fuggle, P.J. Durham and W.M. Temmerman, Phys. Rev. B37, 5158 (1988).
9. e.g. V.J. Emery, Phys. Rev. Lett. 58, 2794 (1987).
10. J. Zaanen, G.A. Sawatzky and J.W. Allen, Phys. Rev. Lett. 55, 418 (1985).
11. C.M. Varma, S. Schmitt-Rink and E. Abrahams, Solid State Commun.62, 681 (1987) and E. Abrahams, Physica C 153-155, 1622 (1988).

12. Details for crystal growth are described in, T. Matsuura, T. Ishigaki, S. Yamauchi and K. Fueki, Jpn. J. Appl. Phys. 23, 1172 (1984).
13. S. Tajima, T. Nakahashi, S. Uchida and S. Tanaka, Physica C 156, 90 (1988).
14. J. Orenstein, G.A. Thomas, D.H. Rapkine, C.G. Bethea, F. Levine, R. J. Cava, A.S. Cooper, D.W. Johnson Jr., J.P. Remeika and E.A. Rietman, "Novel Superconductivity" eds. S.A. Wolf and V.Z. Kresin (Plenum, 1987) p. 693.
15. M. Suzuki, to be published in Phys. Rev. B.
16. H. Muller-Buschbaum and W. Wollschlager, Z. Anorg. Allg. Chem. 414, 76 (1975).
17. Y. Tokura, unpublished data.
18. I.I. Mazin, E.G. Maksimov, S.N. Rashkeev, S.Yu. Savrasov and Yu.A. Uspenskii, JETP Lett. 47, 113(1988).
19. J. Orenstein, G.A. Thomas, D.H. Rapkine, C.G. Bethea, B.F. Levine, R.J. Cava, E.A. Rietman and D.W. Johnson, Jr., Phys. Rev. B36, 729 (1987).
20. K. Kamaras, C. D. Porter, M. G. Doss, S. L. Herr, D. B. Tanner, D.A. Bonn, J. E. Greedan, A. H. O'Reilly, C. V. Stager and T. Timusk, Phys. Rev. Lett. 59, 919 (1987).
21. S. Tajima, S. Uchida, H. Ishii, H. Takagi, S. Tanaka, U. Kawabe, H. Hasegawa, T. Aita and T. Ishiba, Modern Phys. Lett. B 1, 353 (1988).
22. Z. Schlesinger, R. T. Collins, D. L. Kaiser and F. Holtzberg, phys. Rev. Lett. 59 1958 (1987).
23. H. P. Geserich, G. Scheiber, J. Geerk, H. C. Li, G. Linker, W. Assmus and W. Weber, Europhys. Lett. 6, 277 (1988).
24. R. Newman and R. M. Chrenko, Phys. Rev. 114, 1507 (1959).
25. P. Ganguly and C. N. R. Rao, J. Solid State Chem. 53, 193 (1984).
26. J. M. Ginder, M. G. Roe, Y. Song, R. P. McCall, J. R. Gaine, E. Eh renfreund and A. J. Epstein, Phys. Rev. B 37, 7506 (1988).
27. G. Shirane, Y.Endoh, R. J. Birgeneau, M. A. Kastner, Y. Hidaka, M. Oda, M. Suzuki and T. Murakami, Phys. Rev. Lett. 59, 1329 (1987).
28. K. B. Lyons, P. A. Fleury, J. P. Remeika and T. J. Negran, Phys. Rev. B 37, 2353 (1988).
29. J. Orenstein, G. A. Thomas, D. H. Rapkine, M. Capizzi A. J. Millis, R. N. Bhatt, L. F. Schneemeyer and J. V. Waszczak, Phys, Rev. Lett. 61, 1313 (1988).
30. J. Tanaka. K. Kamiya, M. Shimizu, M. Simada, C. Tanaka, H. Ozeki, K. Adachi, K. Iwahashi, F. Sato, A. Sawada, S. Iwata, H. Sakuma and S. Uchiyama, Physica C 153-155, 1752 (1988).
31. D. Yoshioka and H. Fukuyama, J. Phys. Soc. Jpn. 54,2996 (1985)
32. H. Sato, unpublished
33. D.G. Hinks, B. Dabrowski, J.D. Jorgensen, A.W. Mitchell, D.R.Richards, S. Pei and D. Shi, Nature 333, 836 (1988).
34. L.F. Schneemeyer, J.K. Thomas, T. Siegrist, B. Batlogg, L.W. Rupp, R.L. Opila, R.J. Cava and D.W. Murphy, Nature 335, 421 (1988).

Raman Scattering Spectroscopy in High Temperature Superconductors

S. Sugai

Department of Physics, Faculty of Science, Osaka University, 1-1 Machikaneyama-cho, Toyonaka 560, Japan

Raman scattering spectroscopy from magnetic and electronic excitations as well as phonons provides many kinds of important knowledge for the explication of high T_c superconductivity mechanism. Obtained results are the location of doped holes, the exchange integral between localized spins and its modification by hole doping, the superconducting gap, specific phonon modes with strong electron-phonon interactions, and the intermediate electronic or magnetic polaronic excitations. It is pointed out from the symmetric properties of the anisotropic superconducting gap in $YBa_2Cu_3O_{7-y}$ observed by Cooper et al. and the anomalous phonon mode that the coupling between the $^1A_1$ and the $^3B_1$ hole states in the CuO_5 cluster is important for the superconductivity.

1. Introduction

In the course of the search of high Tc superconductors many materials are discovered, for examples $(La_{1-x}Sr_x)_2CuO_4$, $YBa_2Cu_3O_{7-y}$, $Bi_2Sr_2CaCu_2O_{8+y}$, $Tl_2Ba_2CaCu_2O_{8+y}$, etc. All of these materials have layered structures composed of either CuO_6 octahedra or CuO_5 pyramids, which are supposed to be crucial for the high Tc superconductivity. These materials undergo transitions from metallic states to semiconducting states by a small change of components. Such semiconductivity is essentially due to the localization effect of strongly correlated electrons. It is known that $(La_{1-x}Sr_x)_2CuO_4$, $YBa_2Cu_3O_{7-y}$, and $Bi_2Sr_2Ca_{1-x}Y_xCu_2O_{8+y}$ in the semiconducting phases undergo antiferromagnetic states, so that the magnetic interaction is supposed to be the origin of the superconductivity since the early stage. The conclusive theory is, however, not known yet. Many mechanisms are presented including charge fluctuation, exciton, plasmon, fractional statistics, etc. At present results of many different kinds of experiments are necessary to make a new physics of strongly correlated electron system. This talk presents Raman studies of high Tc superconductors.

2. What Can Be Observed from Raman Scattering?

Raman scattering is the powerful method not only to observe elementary excitations in materials, but also to study the intermediate electronic states and those interactions with elementary excitations. Raman scattering occurs in a third order process where the incident photon creates an electron-hole pair, and then one of them emits a boson such as a phonon, a magnon, or a plasmon, or scatters a fermion such as an electron or a hole, and finally the electron and hole recombine by emitting a photon. The third order process induces resonant scattering, when the incident photon energy ω_i or the scattered photon energy $\omega_s=\omega_i-\omega_0$ is close to the intermediate electronic transition energy Δ. Therefore one can investigate the intermediate electronic states and specify excitation modes coupled strongly with the electronic states.

Springer Series in Materials Science, Vol. 11
Mechanisms of High Temperature Superconductivity
Editors: H. Kamimura and A. Oshiyama
Springer-Verlag Berlin Heidelberg © 1989

Elementary excitations in high T_c superconductors observed by Raman scattering at present are antiferromagnetic magnons,[1-3] phonons,[4-32] and single particle excitations of holes.[33-37] The magnon is observed as 2-magnon scattering in which two magnons at +k and -k are emitted simultaneously. From this scattering the exchange integral between localized spins at Cu sites is obtained. High temperature superconductors show resonant Raman scattering in the visible region. Phonons with strong electron-phonon interactions show multi-phonon scattering in the resonant condition. [3,6] It is supposed that holes form magnetic polarons clothed with localized spins of Cu ions. The single particle excitations from these holes show a superconducting gap below T_c reflecting the quasi-particle density of states. With varying the hole density, the interaction between a localized spin at Cu site and an itinerant hole spin at O site as well as the interaction between a phonon and a hole can be investigated.

3. Electronic States of CuO_5 and CuO_6 Clusters

High T_c superconducting oxides undergo transitions from metallic to semiconducting states by small variations of the components. These transitions are caused by the localization of Cu 3d electrons at each atomic site from the strong correlation, especially from the large intraatomic Coulomb interaction among Cu 3d electrons. The semiconducting phases of high Tc superconductors are charge transfer insulators in which the Coulomb interaction U between d electrons is larger than the energy difference between the metal d orbital and the ligand p orbital.[38,39] A cluster model is applied to calculate the strongly correlated electron system in place of the band theory. The method treats a Cu atom and the nearest neighbor 5 or 6 O atoms. Here I introduce a part of the calculation by FUJIMORI[40] as far as it is concerned in the analysis of the present Raman spectroscopy.

The ground state in the CuO_5 pyramidal cluster with one doped hole is the $^1A_1$ or $^3B_1$ state. In the $^1A_1$ state the doped hole enters into a molecular orbital with x^2-y^2 symmetry composed of only in-plane O $p\sigma_{x,y}$ orbitals and forms a singlet with the Cu d^9 spin. In the $^3B_1$ state the hole enters into an orbital with $3z^2-r^2$ symmetry composed of the apex O $p\sigma_z$ orbital and in-plane O $p\sigma_{x,y}$ orbitals and forms a triplet with the Cu spin. Which state is the lowest energy is dependent on the energy difference between the p orbitals at the apex and the in-plane oxygens, $\Delta\varepsilon_p=\varepsilon(p\sigma_z)-\varepsilon(p\sigma_{x,y})$, and the relative atomic distance of the out-of-plane Cu-O bond to the in-plane one, r. The $^3B_1$ state is stabilized for large $\Delta\varepsilon_p$ and small r. With the increase of the hole concentration in $YBa_2Cu_3O_{7-y}$, r decreases [41] and consequently the triplet state is stabilized. It is supposed that the lattice vibration which modulates r couples strongly with the electronic transition between the $^1A_1$ and the $^3B_1$ states, which is accompanied by the charge transfer between O atoms and the spin flip.

It is an important problem for the investigation of superconductivity mechanism whether the doped hole enters into the in-plane O site or the apex O site. Recent experiments of polarized Cu L_3-edge x-ray absorption by BIANCONI et al.[42] suggested that the doped holes are in the orbitals of $3z^2-r^2$ symmetry, that is doped holes in $YBa_2Cu_3O_{7-y}$ are mainly located at apexes. The general consensus is, however, not obtained yet. As shown later the polarized Raman spectroscopy suggests that the doped holes enter in-plane O sites in $(La_{1-x}Sr_x)_2CuO_4$ and apexes in $YBa_2Cu_3O_{7-y}$. The two states $^3B_1$ and $^1A_1$ are close in energy and their coupling is supposed to be a key for the high Tc superconductivity mechanism.

When the hole concentration increases, the low energy background scattering below 1000 cm^{-1} increases in Raman spectra. This scattering is due to the single particle excitation of holes, or magnetic polarons clothed with localized spins at Cu sites. Below Tc the scattering intensity decreases below the superconducting gap 2Δ and piles up above 2Δ reflecting the density of states of quasiparticles.[33-37] Recently COOPER et al.[37] observed that the superconducting gap of B_{1g} symmetry is larger than that of A_{1g} symmetry in $YBa_2Cu_3O_{7-y}$. This anisotropic gap may be explained as follows. The superconducting gap opens in the $^3B_1$ band in $YBa_2Cu_3O_{7-y}$, because the holes are in this band. The A_{1g} spectra correspond to break the $^3B_1$ hole pair and create two $^3B_1$ quasiparticles. The B_{1g} spectra correspond to break the $^3B_1$ pair and create the $^3B_1$ and $^1A_1$ quasiparticles, or to break the $^3B_1$-$^1A_1$ pair and create two $^1A_1$ quasiparticles. The anisotropic gap suggests the strong coupling of the $^3B_1$ and $^1A_1$ states in the superconducting state. The coupling of these two states are also related to the decrease of the B_{1g} phonon energy below Tc in $YBa_2Cu_3O_{7-y}$.[11,34]

4. 2-Magnon Raman Scattering

Semiconducting phases of $(La_{1-x}Sr_x)_2CuO_4$, $YBa_2Cu_3O_{7-y}$, and $Bi_2Sr_2Ca_{1-x}Y_x$-Cu_2O_{8+y} undergo antiferromagnetic states at low temperatures. With the increase of the hole concentration the antiferromagnetic transition temperatures (T_N's) decrease and the superconducting states appear. Neutron scattering experiments revealed that the antiferromagnetic spin order in $(La_{1-x}Sr_x)_2CuO_4$ is preserved even above T_N on the 2-dimensional CuO_2 layer as a fluctuating state with a long correlation length.[43-45]

The antiferromagnetic magnon has an acoustic phonon like dispersion, $E_k=(SJZ+g\mu H_A)^2-(2JZ\gamma_k)^2$.[46] Here spin S=1/2, J is the exchange integral, Z is the number of the nearest neighbor atoms, H_A is the anisotropy field which is nearly equal to 0, and $\gamma_k=(\cos k_xa+\cos k_ya)/2$. For 1-magnon scattering only $E_k\sim 0$ mode is observed, because Raman scattering examines only $k\sim 0$ mode from the momentum conservation among the incident light , the scattered light, and the elementary excitation. While for the 2-magnon scattering simultaneous creation of -K and +k magnons gives spectra reflecting the magnon density of states in which the zone boundary modes are dominant. The peak energy of the 2-magnon scattering is lower than the energy of two magnons created independently, because the interaction between the magnons created at the nearest neighbor sites is strong. The decrease of the energy is larger, as the spin becomes smaller. The peak energy is estimated at about 2.7 J in the present case of S=1/2, but no reliable theory is presented, because the quantum effect is very large in the present case.

Figure 1 shows polarized Raman spectra of La_2CuO_4 with the excitation of λ_i=5145 Å at 30 K. The 3200 cm^{-1} peak in the (x, x) spectra is assigned to the 2-magnon scattering. The symbol (x, x) denotes that the incident and the scattered lights are polarized along the x axis. No observation of the 2-magnon peak in the (y, x) and the (z, z) spectra is consistent with the selection rule of the 2-magnon scattering from the antiferromagnetic spin order on the 2-dimensional quasi-square lattice. No rapid decrease of the 2-magnon peak intensity is observed even above T_N (240 K), which indicates that the observed scattering is attributed to the spin dynamics in the 2-dimensional antiferromagnetic domain as a fluctuating state consistently with the results of neutron scattering experiments.[43-45] The peaks below 800 cm^{-1} are assigned to mainly 1-phonon scattering and the peaks between 800 and 1500 cm^{-1} to 2-phonon scattering.

Figure 2 shows the Sr concentration dependence of the (x, $\frac{x}{y}$) Raman spectra in $(La_{1-x}Sr_x)_2CuO_4$. With the increase of Sr concentration the hole concentration increases monotonically, and T_N decreases rapidly. The samples with x=0.035 and 0.058 undergo superconducting states at about 10 K. As indicated by bars the energies of 2-magnon peaks decrease with the increase of x. This is explained as follows. When an O atom is introduced between two Cu atoms, the antiferromagnetic spin order is destroyed, wheth-

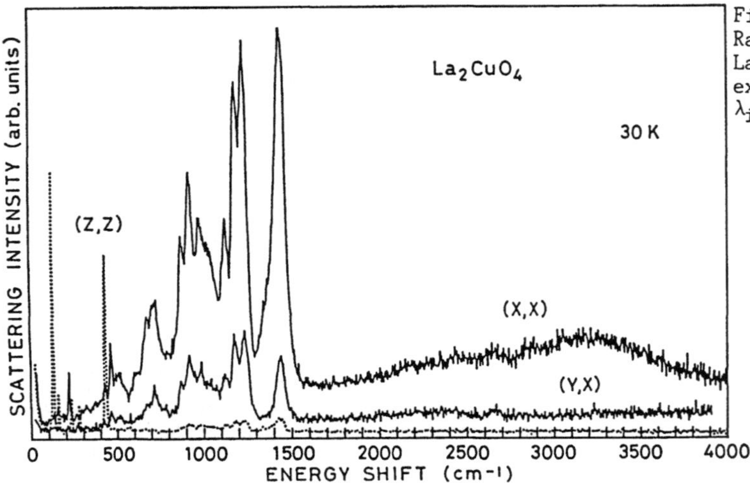

Fig.1 Polarized Raman spectra of La_2CuO_4 at 30 K excited with λ_i=5145 Å.

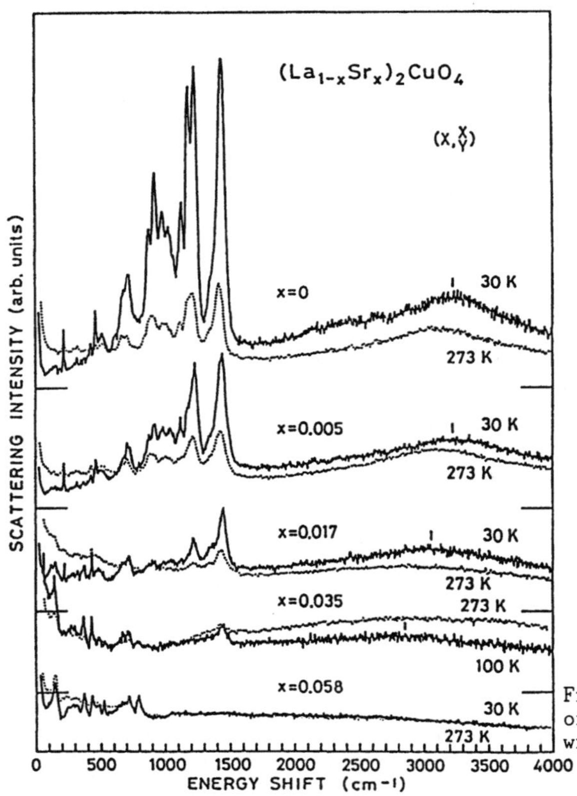

Fig.2 Sr concentration dependence of the $(x,{}^{x}_{y})$ Raman spectra excited with λ_i=5145 Å.

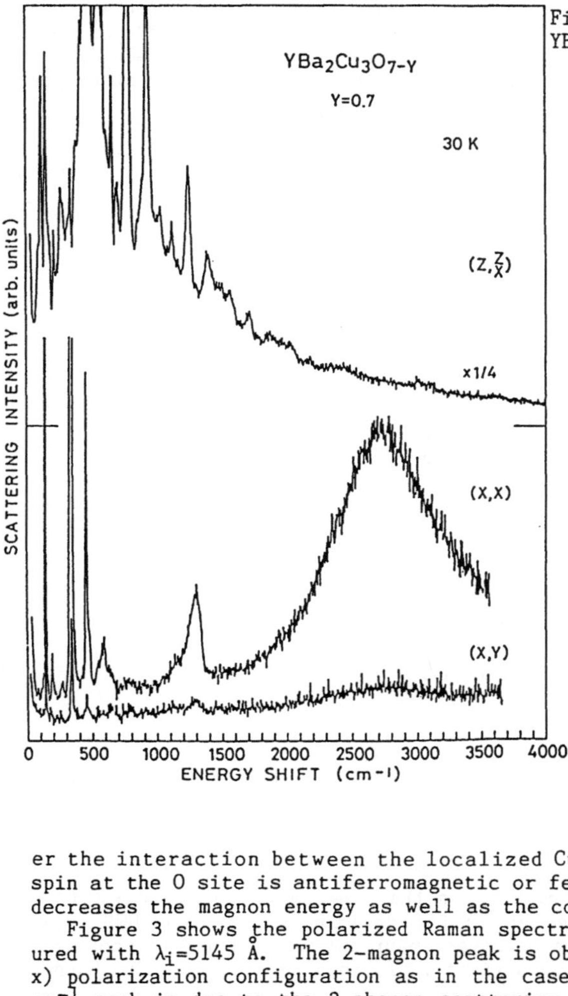

Fig.3 Polarized Raman spectra of $YBa_2Cu_3O_{6.3}$ excited with λ_i=5145 Å.

er the interaction between the localized Cu spin and the itinerant hole spin at the O site is antiferromagnetic or ferromagnetic.[47] This process decreases the magnon energy as well as the correlation length.

Figure 3 shows the polarized Raman spectra of $YBa_2Cu_3O_{6.3}$ at 30 K measured with λ_i=5145 Å. The 2-magnon peak is observed at 2740 cm^{-1} in the (x, x) polarization configuration as in the case of $(La_{1-x}Sr_x)_2CuO_4$. The 1298 cm^{-1} peak is due to the 2-phonon scattering. Similar 2-magnon scattering is observed at 3080 cm^{-1} in $Bi_2Sr_2Ca_{0.5}Y_{0.5}Cu_2O_{8+y}$.[48]

Whether the magnon exists or not in the high hole concentration region enough to induce the superconductivity is very important to know whether the spin correlation is the origin of the superconductivity or not. Only the elementary excitation with sufficiently high energy compared with T_c can cause the attractive force between holes. Figure 4 shows the Raman spectra of $(La_{1-x}Sr_x)_2CuO_4$, $YBa_2Cu_3O_{7-y}$, $Bi_2Sr_2CaCu_2O_{8+y}$ with the highest hole concentration in our experiments. The measurement was made at 30 K with λ_i=5145 Å light. No obvious 2-magnon peak is observed. However large scattering intensity ranging from low energy to over 4000 cm^{-1} is noticed. Usually no phonon peak above 1000 cm^{-1} nor single particle excitation of holes or electrons above 100 cm^{-1} is observed in oxides. Then the scattering intensity is assigned to the overdamped 2-magnons caused by the interaction with holes. The effective exchange integral $J^*=\omega_{2-magnon}/2.7$ is 1200 cm^{-1} in La_2CuO_4,[2,3] 1140 in $Bi_2Sr_2Ca_{1-x}Y_xCu_2O_{8+y}$,[48] and 1010 in

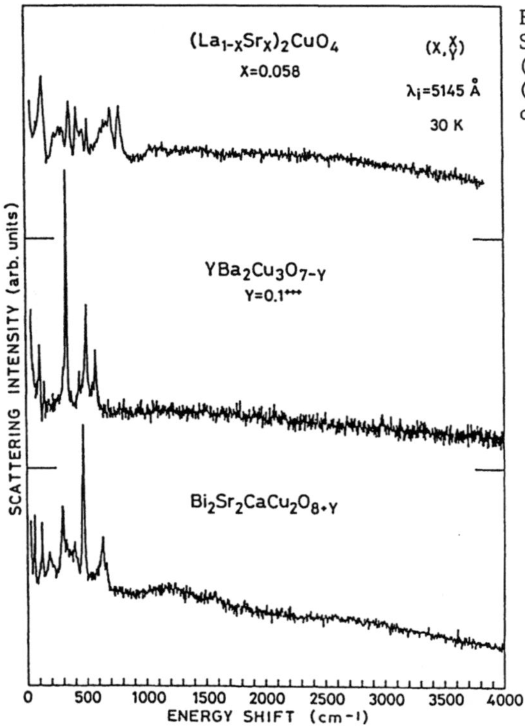

Fig.4 Raman spectra of $(La_{1-x}Sr_x)_2CuO_4$ (Tc=10 K), $YBa_2Cu_3O_{7-y}$ (Tc=92 K), and $Bi_2Sr_2CaCu_2O_{8+y}$ (Tc=80 K) with sufficient oxygen concentration.

$YBa_2Cu_3O_{6.3}$.[1,49,50] The J^* decreases as the increase of the Cu-Cu distance.

5. Phonon

There is a consensus that the high T_c in the superconducting oxides cannot be explained by the electron-phonon interaction only in the BCS type. However the measurement of phonons gives important information on the physics of high temperature superconductivity. In this section specific phonon modes are discussed.

It is abnormal that the 2-phonon scattering intensity in La_2CuO_4 is larger than 1-phonon intensity in the (x, x) spectra of Fig. 1. This indicates that the scattering is in the resonant condition and the electron-phonon interactions of these modes are very large. The energies of the 2-phonon peaks are decomposed into sums of four modes, p, r, u, and v as shown in Fig. 5.[6] Figure 6 shows the atomic displacement of these modes. The modes p and r are quadratic vibrational modes (B_{3g}) of O atoms on the CuO_2 layer and the u and v modes are breathing modes (B_{2g}). The 2-phonon scattering intensity decreases with the increase of the hole concentration as shown in Fig. 5. The 2-phonon mode which observed up to the higher hole concentration is the breathing v mode. The change of the scattering intensity with the hole concentration is the same as the 2-magnon scattering. This indicates that the magnon and the breathing phonon have some correlation and they are enhanced by the resonant effect via the same intermediate electronic states.

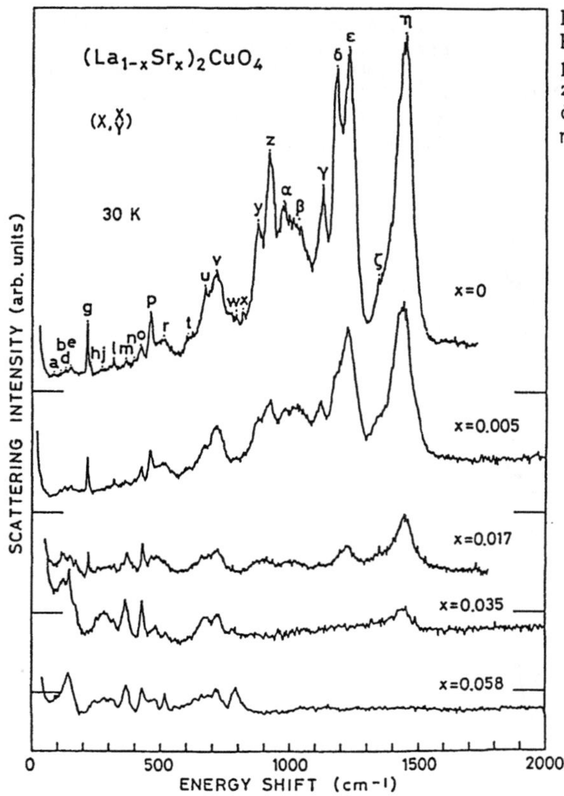

Fig.5 1- and 2-phonon part of the Raman spectra in La_2CuO_4. The 2-phonon peaks are decomposed as z(2p), α(p+r), β(2r), γ(p+u), δ(r+u, p+v), ε(r+v), ζ(2u), and η(2v).

The 2-phonon scattering from the breathing mode is also observed in $YBa_2Cu_3O_{7-y}$. The 1298 cm^{-1} peak in the (x, x) spectra of Fig. 3 is the 2-phonon scattering. The intensity decreases correlatively with the 2-magnon peak on the increase of the hole concentration. In $Bi_2Sr_2Ca_{0.5}Y_{0.5}Cu_2O_{8+y}$ the 2-phonon peak is observed at 1354 cm^{-1} and has the same dependence on the hole concentration.[48]

In the (z, z) polarization configuration multi-phonon scattering is not observed in $(La_{1-x}Sr_x)_2CuO_4$, while it is observed in $YBa_2Cu_3O_{7-y}$. Figure 7 shows the multi-phonon scattering of the 468 cm^{-1} mode in $YBa_2Cu_3O_{6.3}$.[50] This mode is the vibrational mode of O atoms at the apexes of CuO_5 pyramids as shown in Fig. 8 (c). With the increase of the hole concentration new modes appear at 444 and 501 cm^{-1} besides 471 cm^{-1} (468 cm^{-1} at y=0.7), and the original 468 cm^{-1} mode disappears in the sample with sufficient O concentration. Among the new modes the 444 cm^{-1} mode is assigned to the same mode as the 454 cm^{-1} mode which is observed in the (x, x) spectra at y=0.7. The activity in the (z, z) polarization is attained by the relaxation of the selection rule by the hole doping. Multi-phonon scattering is observed in the 471 cm^{-1} mode, but not in the 501 cm^{-1} mode. The 501 cm^{-1} mode is assigned to the same vibrational mode as the 468 cm^{-1} mode, but the only difference is the location of a hole at the apex. The location of hole decreases the interatomic distance between the O atom at the apex and the in-plane Cu atom[41] and in consequence increases the interatomic force constant.

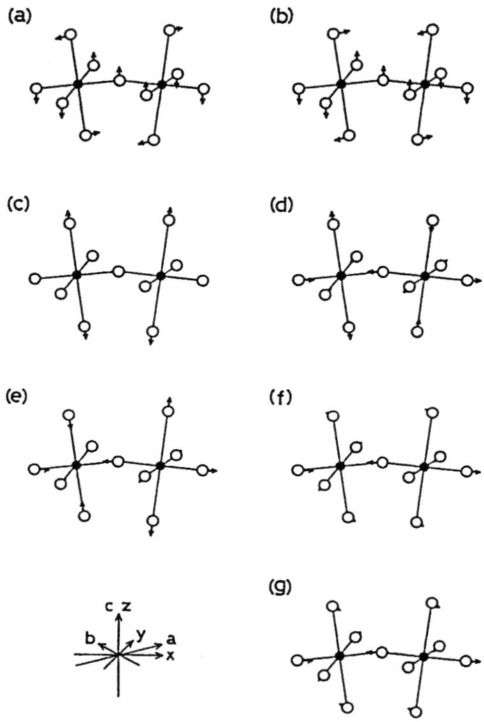

Fig.6 Atomic displacements of the specific normal modes in $(La_{1-x}Sr_x)_2CuO_4$. (a) A_g(c) 126 cm^{-1}, (b) A_g (j) 273, (c) A_g(o) 426, (d) B_{2g}(u) 673, (e) B_{2g}(v) 716, (f) B_{3g}(p) 462, and (g) B_{3g}(r) 511.

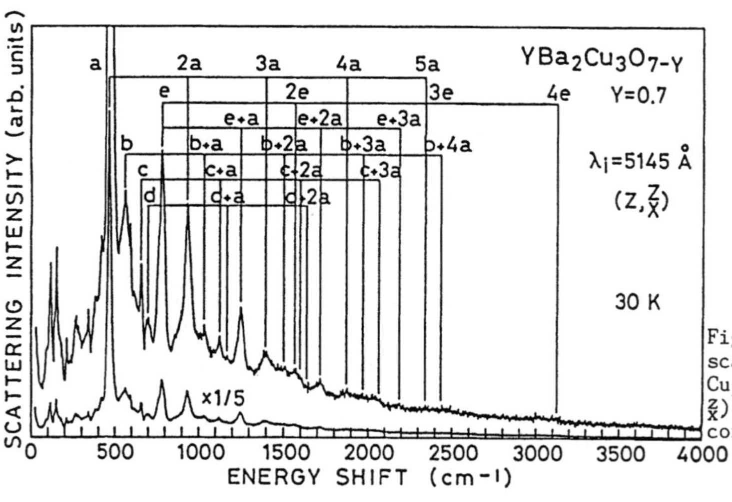

Fig.7 Multi-phonon scattering of $YBa_2Cu_3O_{6.3}$ in the (z, $\frac{z}{x}$) polarization configuration.

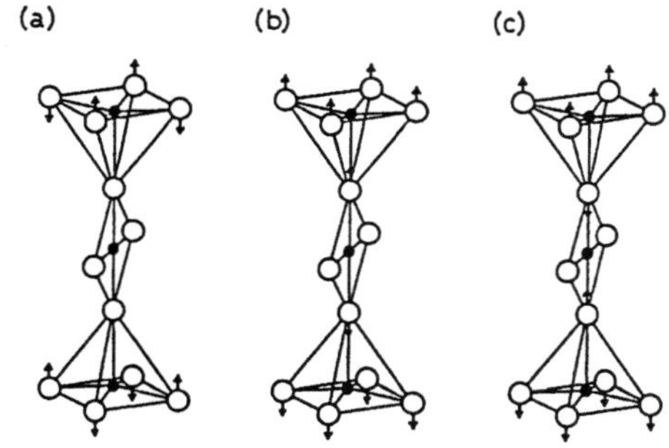

Fig.8 Normal modes of oxygen vibrations in $YBa_2Cu_3O_{7-y}$. (a) B_{1g} 344 cm^{-1}, (b) A_{1g} 454, and (c) A_{1g} 468.

The location of holes can be obtained from the anomaly of the polarized Raman spectra. In $(La_{1-x}Sr_x)_2CuO_4$ the (z, z) spectra are simple. The 2-phonon scattering is weak and the number of peaks is the same as expected from the group theory. While in the (x, x) spectra the 2-phonon scattering is abnormally strong and unusual number of peaks appear in the 1-phonon scattering. The case of $YBa_2Cu_3O_{7-y}$ is the contrary. As shown in Figs. 3 and 7 the 1-phonon scattering in the (x, x) spectra is simple, while strong multi-phonon scattering is observed in the (z, z) spectra. These can be explained as follows. If holes are located on the O $2p\sigma_{x,\ y}$ orbitals on the CuO_2 layers, the single particle excitation of holes is allowed in the (x, x) polarization configuration. While if holes are on the O $p\sigma_z$ orbitals at apexes, it is allowed in (z, z). As a consequence of the hole-phonon interaction many phonon modes which are forbidden in themselves become active in the polarization configuration where the hole scattering is allowed. Therefore the following results are induced. Holes in $(La_{1-x}Sr_x)_2CuO_4$ are located at O atoms on the CuO_2 layers and those in $YBa_2Cu_3O_{7-y}$ at apexes of the CuO_5 pyramids. In $Bi_2Sr_2CaCu_2O_{8+y}$ the (x, x) spectra include much larger number of peaks than expected from group theory, and therefore the location of holes is tentatively assigned to the O atoms on the CuO_2 layers, although the (z, z) spectra are not obtained yet.

6. Intermediate Electronic States

The Raman scattering from $(La_{1-x}Sr_x)_2CuO_4$, $YBa_2Cu_3O_{7-y}$, and $Bi_2Sr_2Ca_{1-x}Y_xCu_2O_{8+y}$ are under the resonant condition for the incident light in the visible region. It is obvious from the existence of the strong multi-phonon scattering and the strong dependence of the spectra on the incident wavelength. Figure 9 shows the Raman spectra of La_2CuO_4 excited with three wavelengths. For the incident light of 5145 Å 2-phonon peaks at 800-1500 cm^{-1} are strongly enhanced. The enhanced region shifts to higher energy, when the incident wavelength is shortened. For λ_i=4579 Å the 2-magnon scattering at 3000 cm^{-1} is strongly enhanced. The shift of the resonant region suggests that the present resonance is in the out-going resonant condition, $\omega_i=\Delta+\omega_0$. The intermediate electronic transition energy Δ is estimated at 18000 cm^{-1} (2.2 eV).

With the increase of hole concentration the 2-magnon and 2-phonon peaks disappear simultaneously and the spectra change into structureless in the energy region above 1000 cm^{-1}. Such spectra, however, have strong dependence on the incident wavelength. Figure 10 shows the incident wavelength dependence on the spectra of $YBa_2Cu_3O_{6.9}$. The region of the energy shift below 1300 cm^{-1} is Raman scattering in which the energy shifts of peaks are the same for the different incident wavelengths, while the region of higher energy shift is different for different incident wavelengths. The absolute

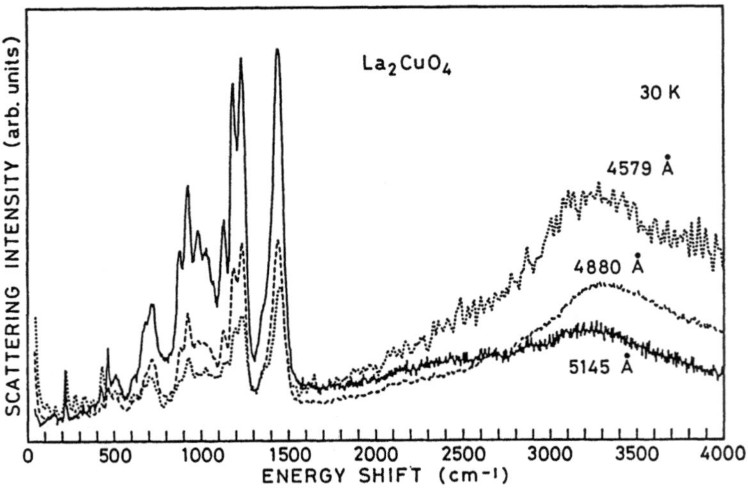

Fig.9 Resonant Raman effects in La_2CuO_4 for three incident light wavelengths.

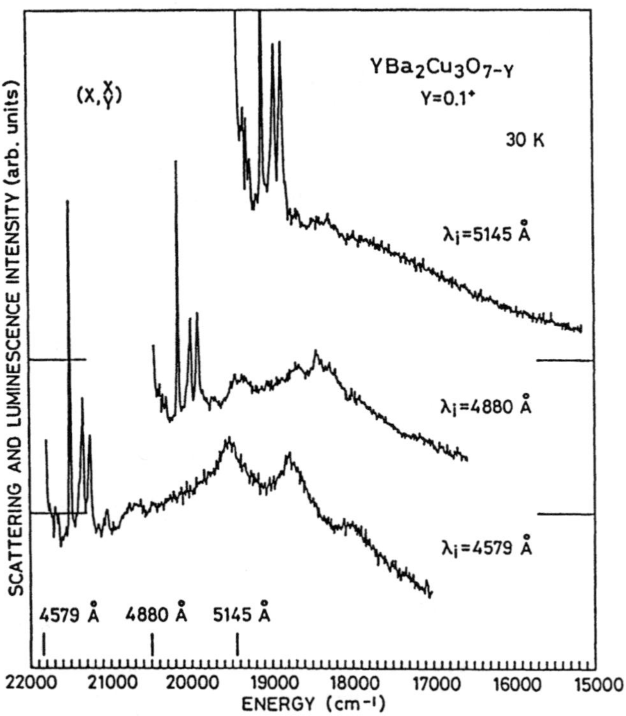

Fig.10 Incident light wavelength dependence of Raman ($\Delta\varepsilon \lesssim 1300$ cm^{-1}) and luminescence spectra in $YBa_2Cu_3O_{6.9}$ (Tc=92 K).

wavelengths of the peaks at 18030 cm^{-1} (2.24 eV), 18770 (2.33), and 19520 (2.42) observed with λ_i=4579 Å change a little, when the incident wavelength is elongated, although small red shift and broadening is observed. This indicates that the origin of these peaks is hot-luminescence. The similar spectra are also observed in $(La_{0.942}Sr_{0.058})_2CuO_4$. No structure is observed in the light reflection spectra in this energy region, which is in contrast with the case of $BaPb_{1-x}Bi_xO_3$ in which the intermediate transition energy of the resonant scattering is directly observed in the light reflection spectra.[51] This means that some energy relaxation occurs in the high T_c superconductors. The origin of the hot-luminescence is supposed to be a magnetic polaron exciton, because (1) the center wavelength of the luminescence for λ_i=4579 Å which gives the largest intensity is the same as the wavelength of the 2-magnon scattering in the semiconducting phase on excited with the same wavelength, and (2) the difference from $BaPb_{1-x}Bi_xO_3$ is the existence or nonexistence of the antiferromagnetic state. The hot-luminescence is so weak that it is hidden by the 2-magnon scattering peak in the semiconducting antiferromagnetic phase.

It is an important problem why the resonant Raman scattering from the 2-magnon and the 2-phonon processes disappears in the metallic phase with large hole concentration in spite of the presence of the electronic levels which give the hot-luminescence with almost the same energy as the intermediate electronic transition in the resonant Raman scattering in the semiconducting phase. The presence and absence of the multiphonon scattering according to the absence and presence of holes is clearly seen in one spectrum for the vibrational mode of O atoms at the apexes in $YBa_2Cu_3O_{7-y}$ as presented above. In the (z, z) spectra the 471 cm^{-1} mode shows multi-phonon scattering, while the 501 cm^{-1} mode does not. The intermediate electronic states are tentatively assigned to the states corresponding to the charge transfer excitation from the O 2p orbital to the Cu 3d orbital. In the semiconducting phase where a localized hole is located at the Cu 3d orbital and no itinerant hole exists, the transition $d^9 \rightarrow d^{10}\underline{L}$ is sharp and gives strong resonant scattering, while in the metallic phase where an itinerant hole is located at O 2p orbital the transition $d^9\underline{L} \rightarrow d^{10}\underline{L}^2$ is broad and does not give strong resonance. The observed hot-luminescence is tentatively assigned to the photo emission from the magnetic polaron exciton of $d^{10}\underline{L}^2$. The study of the magnetic polaron is very important to investigate the new physics of the strongly correlated electron system composed of localized S=1/2 spins and itinerant S=1/2 hole spins as in high T_c superconductors.

7. Conclusions

Raman scattering experiments present many important data for the explication of the high Tc superconductivity mechanism. The obtained results at present are antiferromagnetic exchange integral J from the 2-magnon scattering, assignment of the location of doped holes, the superconducting gap, the characteristic phonon modes which have strong electron-phonon interaction, intermediate electronic or magnetic polaronic states from the resonant Raman scattering and the hot-luminescence. At present the consensus is not reached on the superconductivity mechanism in which what is essential among the magnetic interaction between spins, the charge transfer, excitons, or other excitations. Raman scattering experiments suggest the correlation between the magnon and the breathing phonon mode. The symmetric properties that the different gap energy is observed between the A_{1g} and the B_{1g} symmetries in $YBa_2Cu_3O_{7-y}$[37] and that the phonon mode influenced most strongly by the gap generation is the B_{1g} mode suggest that the interaction between the lowest two levels $^1A_1$ and $^3B_1$ in the CuO_5 cluster is important for the superconductivity. The transition between $^1A_1$ and $^3B_1$

states implies that the spin flip, the charge transfer between O sites at the CuO_2 layer and the apex, and phonons contribute to the appearance of the superconductivity cooperatively.

Acknowledgments

The author thanks M. Sato, S. Shamoto, S. Uchida, H. Takagi, H. Takei for the supply of good single crystals. This work is supported by a Grant-in-Aid for Scientific Research on Priority Areas "Mechanism on Superconductivity" from the Ministry of Education, Science and Culture, Japan, the Iwatani Naoji Foundation's Research Grant, and the Murata Science Foundation.

References

1. K.B.Lyons, P.A.Fleury, L.F.Schneemeyer, and J.V.Waszczak: Phys. Rev. Lett. 60, 732 (1988)
2. K.B.Lyons, P.A.Fleury, J.P.Remeika, A.S.Cooper, and T.J.Negran: Phys. Rev. B37, 2353 (1988)
3. S.Sugai, S.Shamoto, and M.Sato: Phys. Rev. B38, 6436 (1988)
4. S.Sugai, M.Sato, and S.Hosoya: Jpn. J. Appl. Phys. 26, L495 (1987)
5. N.Koshizuka, H.Unoki, K.Oka, K.Hayashi, T.Okuda, and Y.Kimura: Jpn. J. Appl. Phys. 27, L1300 (1988)
6. S.Sugai: submitted to Phys. Rev.
7. R.J.Hemley, and H.K.Mao: Phys. Rev. Lett. 58, 2340 (1987)
8. M.Stavola, C.M.Krol, W.Weber, S.A.Sunshine, A.Jayaraman, G.A.Kourouklis, R.J.Cava, and E.A.Rietman: Phys. Rev. B36, 850 (1987)
9. A.Yamanaka, F.Minami, K.Watanabe, K.Inoue, S.Takekawa, and N.Iyi: Jpn. J. Appl. Phys. 26, L1404 (1987)
10. Y.Morioka, M.Kikuchi and Y.Syono: Jpn. J. Appl. Phys. 26, L1499 (1987)
11. R.M.Macfarlane, Hal Rosen, and H.Seki: Solid State Commun. 63, 831 (1987)
12. S.Nakashima, M.Hangyo, K.Mizoguchi, A.Fujii, A.Mitsuishi, and T.Yotsuya: Jpn. Appl. Phys. 26, L1794 (1987)
13. G.A.Kourouklis, A.Jayaraman, B.Batlogg, R.J.Cava, M.Stavola, D.M.Krol, E.A.Rietman, and L.F.Schneemeyer: Phys. Rev. B36, 8320 (1987)
14. D.M.Krol, M.Stavola, W.Weber, L.F.Schneemeyer, J.V.Waszczak, S.M.Zahurak, and S.G.Kosinski: Phys. Rev. B36, 8325 (1987)
15. C.Thomsen, R.Liu, M.Bauer, A.Wittlin, L.Genzel, M.Cardona, E.Schonherr, W.Bauhofer, and W.Konig: Solid State Commun. 65, 55 (1988)
16. J.Chrzanowski, S.Gygax, J.G.Irwin, and W.N.Hardy: Solid State Commun. 65, 139 (1988)
17. C.Thomsen, R.Liu, A.Wittlin, L.Genzel, M.Cardona, W.Konig, M.V.Cabanas, and E.Garcia: Solid State Commun. 65, 219 (1988)
18. M.Hangyo, S.Nakashima, K.Mizoguchi A.Fujii, A.Mitsuishi, and T.Yotsuya: Solid State Commun. 65, 835 (1988)
19. C.Thomsen, M.Cardona, W.Kress, R.Liu, L.Genzel, M.Bauer, E.Schonherr, and U.Schroder: Solid State Commun. 65, 1139 (1988)
20. Gerald Burns, F.H.Dacol, F.Holtzberg, and D.L.Kaiser: Solid State Commun. 66, 217 (1988)
21. D.Kirillov, J.P.Collman, J.T.McDevitt, G.T.Yee, M.J.Holcomb, and I.Bozovic: Phys. Rev. B37, 3660 (1988)
22. R.Bhadra, T.O.Brun, M.A.Beno, B.Dabrowski, D.G.Hinks, J.Z.Liu, J.D.Jorgensen, L.J.Nowicki, A.P.Paulikas, Ivan K.Schuller, C.U.Segre, L.Soderholm, B.Veal, H.H.Wang, J.M.Williams, K.Zhang, and M.Grimsditch: Phys. Rev. B37, 5142 (1988)
23. R.Liu, C.Thomsen, W.Kress, M.Cardona, B.Gegenheimer, F.W. de Wette, J.Prade, A.D.Kulkarni, and U.Schroder: Phys. Rev. B37, 7971 (1988)

24. R.Nishitani, N.Yoshida, Y.Sasaki, and Y.Nishina: Jpn. J. Appl. Phys. 27, L1284 (1988)
25. R.M.Macfarlane, H.J.Rosen, E.M.Engler, R.D.Jacowitz, and V.Y.Lee: Phys. Rev. B38, 284 (1988)
26. H.J.Rosen, R.M.Macfarlane, E.M.Engler, V.Y.Lee, and R.D.Jacowitz: Phys. Rev. B38, 2460 (1988)
27. K.F.McCarty and J.C.Hamilton, R.N.Shelton, D.S.Ginley: Phys. Rev. B38, 2914 (1988)
28. M.Krantz, H.J.Rosen, R.M.Macfarlane, and V.Y.Lee: Phys. Rev. B38, 4992 (1988)
29. S.Sugai, H.Takagi, S.Uchida, and S.Tanaka: Jpn. J. Appl. Phys. 27, L1290 (1988)
30. M.Cardona, C.Thomsen, R.Liu, H.G. von Schnering, M.Hartweg, Y.F.Yan, and Z.X.Zhao: Solid State Commun. 66, 1225 (1988)
31. G.Burns, G.V.Chandrashekhar, F.H.Dacol, M.W.Shafer, P.Strobel: Solid State Commun. 67, 603 (1988)
32. M.Stavola, D.M.Krol, L.F.Schneemeyer, S.A.Sunshine, R.M.Fleming, J.V.Waszczak, and S.G.Kosinski: Phys. Rev. B38, 5110 (1988)
33. K.B.Lyons, S.H.Liou, M.Hong, H.S.Chen, J.Kwo, and T.J.Negran: Phys. Rev. B36, 5592 (1987)
34. S.L.Cooper, M.V.Klein, B.G.Pazol, J.P.Rice, and D.M.Ginsberg: Phys. Rev. B37, 5920 (1988)
35. R.Hackl, W.Glaser, P.Muller, D.Einzel, and K.Andres: Phys. Rev. B38, 7133 (1988)
36. A.Yamanaka, T.Kimura, F.Minami, K.Inoue, and S.Takekawa: Jpn. J. Appl. Phys. 27, L1902 (1988)
37. S.L.Cooper, F.Slakey, M.V.Klein, J.P.Rice, E.D.Bukowski, and D.M.Ginsberg: submitted to Phys. Rev.
38. A.Fujimori, E.Takayama-Muromachi, Y,Uchida, and B,Okai: Phys. Rev. B35, 8814 (1987)
39. A.Fujimori, E.Takayama-Muromachi, Y,Uchida: Solid State Commun. 63, 857 (1987)
40. A.Fujimori: submitted to Phys. Rev.
41. H.Asano, K.Takita, T.Ishigaki, H.Akinaga, H.Katoh, K.Masuda, F.Izumi, and N.Watanabe: Jpn. J. Appl. Phys. 26, L1341 (1987)
42. A.Bianconi, M.De Santis, A.Di Cicco, A.M.Flank, A.Fontaine, P.Lagarde, H.Katayama-Yoshida, A.Kotani, and A.Marcelli: Phys. Rev. B38, 7196 (1988)
43. G.Shirane, Y.Endoh, R.J.Birgeneau, M.A.Kastner, Y.Hidaka, M.Oda, M.Suzuki, and T.Murakami: Phys. Rev. Lett. 59, 1613 (1987)
44. Y.Endoh, K.Yamada, R.J.Birgeneau, D.R.Gabbe, H.P.Jenssen, M.A.Kastner, C.J.Peters, P.J.Picone, T.R.Thurston, J.M.Tranquada, G.Shirane, Y.Hidaka, M.Oda, Y.Enomoto, M.Suzuki, and T.Murakami: Phys. Rev. B37, 7443 (1988)
45. R.J.Birgeneau, D.R.Gabbe, H.P.Jenssen, M.A.Kastner, P.J.Picone, T.R.Thurston, G.Shirane, Y.Endoh, M.Sato, K.Yamada, Y.Hidaka, M.Oda, Y.Enomoto, M.Suzuki, and T.Murakami: Phys. Rev. B38, 6614 (1988)
46. J.B.Parkinson: J. Phys. C (Solid State Phys.) ser. 2, 2, 2012 (1969)
47. A.Aharony, R.J.Birgeneau, A.Coniglio, M.A.Kastner, and H.E.Stanley: Phys. Rev. Lett. 60, 1330 (1988)
48. S.Sugai, S.Shamoto, and M.Sato: in preparation
49. S.Sugai: in Proceedings of the 4th Japanese-Soviet Symposium on Ferroelectricity, Tsukuba, Japan, 20-26 Aug. 1988
50. S.Sugai, S.Shamoto, M.Sato, H.Takagi, S.Uchida, and S.Tanaka: in preparation
51. S.Tajima, S.Uchida, A.Masaki, H.Takagi, K.Kitazawa, S.Tanaka, and S.Sugai: Phys. Rev. B35, 696 (1987)

Tunneling and the Energy Gap in the High-Temperature Superconductors

M. Lee, A. Kapitulnik, and M.R. Beasley

Department of Applied Physics, Stanford University, Stanford, CA94305, USA

Abstract: We argue that, when taken as a whole, the available tunneling data provide strong evidence for an energy gap in 123 YBaCuO and 2212 BiSrCaCuO. A value of $2\Delta/k\ T_c \simeq$ 6-7 is deduced using data for which both the gap and the gap opening transition temperature are available. The data are compared with recent Raman and infrared reflectance data.

Historically, tunneling studies of superconductors have been one of the most powerful probes of the nature of superconductivity and its underlying mechanisms. Not surprisingly, there has been great interest in tunneling into the new, high-temperature superconducting oxides, and numerous attempts have been reported. Normal tunneling is also of interest because even the existence of a Fermi liquid in the normal state of these materials has been called into question. Various gap-like structures and anomalous normal tunneling have been observed, but no consistent picture has so far emerged. In particular, no widely accepted value of the energy gap or the important ratio $2\Delta/k\ T_c$ have been established. Making tunnel junctions on superconductors with short coherence lengths is nontrivial, and in any particular case it is difficult to assess the influence of materials problems, either in the barrier or in the superconductor itself. In addition, for superconductive tunneling, it is not clear *a priori* to what degree the simple predictions of BCS theory provides a useful guide in interpreting the data.

Despite these problems, as we argue in this paper, the situation is much clearer when the available tunneling data are taken as a whole. We find that there is strong evidence for an energy gap of 17 to 23 and 24 to 30 [meV] in 123 YBaCuO and 2212 BiSrCaCuO, respectively. Evidence for such gaps exist in sandwich tunnel junctions, point contact junctions and break junctions. Moreover, it is possible to understand the differences between the tunneling curves obtained in these various junction types, provided the average versus local nature of their tunneling is taken into account.

Let us begin by examining the tunneling characteristics obtained with sandwich-type junctions. Figure 1 shows the differential tunneling conductance dI/dV versus bias voltage for one of our junctions formed on a highly oriented a-axis YBaCuO thin

Springer Series in Materials Science, Vol. 11
Mechanisms of High Temperature Superconductivity
Editors: H. Kamimura and A. Oshiyama
Springer-Verlag Berlin Heidelberg © 1989

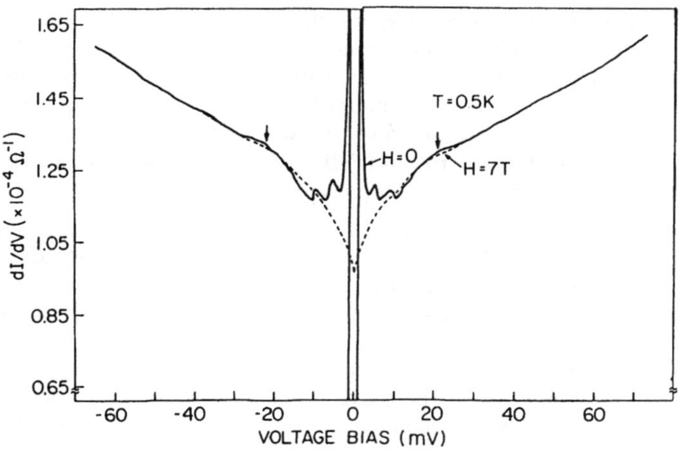

Fig. 1. Differential conductance of a sandwich-type tunnel junction on 60 [K], 123 YBaCuO formed with the native barrier and a *Pb* counter electrode

film using the native barrier and a *Pb* counter electrode [1]. Clearly evident are a depressed conductance for bias voltages less than 20 [meV], along with clear *Pb* gap and *Pb* phonon structure near zero bias. There also is a linearly increasing differential conductance at high bias first observed by our group in 214 (LaSr)CuO [2]. There is a weak peak above this linear backgound conductance at about 20 [meV]. The dashed curve shows the tunneling conductance when superconductivity of the *Pb* has been quenched by the application of magnetic field. All these features are reproducible in our experience. Similar behavior has been observed by DYNES [3] with sandwich junctions formed on single crystal YBaCuO, and more recently by GEERK, XI and LINKER [4] with thin-films samples. It seems to be representative of tunneling into the natural surface of 123 YBaCuO.

A definite peak in the tunneling conductance at 20 [meV] was observed earlier by IGUCHI et al. [5], using sandwich tunnel junctions on polished ceramic samples. Iguchi and coworkers also saw related structure at approximately 40 and 60 [meV]. Evidence for structure at 40 [meV] is also present in our data, but it is very weak and not evident in the figure. It is stronger in other samples. As reported by Iguchi, and also found by us, this gap-like depression of the tunneling density of states goes to zero at a temperature of 60-70 [K], not 90 [K] where the resistive transition is observed. At a minimum, this observation demonstrates the need to measure T_c by gap closing when analyzing tunneling data on these new superconductors. It also suggests that the region near the surface being probed by the tunneling is likely the oxygen-deficient 60 [K] phase of 123 YBaCuO. Further support for this contention can be found in the work of ARKO [6] who found in photoemission studies that even freshly cleaved surfaces of 123 YBaCuO appear to be oxygen deficient, unless they are cleaved and maintained continuously at low temperatures. Nonstoichiometry of the metallic constituents near the surface has also been reported by many groups. The picture that emerges is of an oxygen gradient near the surface of 123 YBaCuO across which the electronic properties go from insulating to metallic to a 60 [K] superconductor to a 90 [K] superconductor as a function of depth. The existence of an oxygen-deficient region near the surface might also explain the strong presence of SIN-like tunneling by the Pb counter electrode. Similarly, the existence of a metal-insulator transition, with the concomitant possibility of strong electron-electron interactions, may account for the cusp seen at zero bias when the superconductivity of the Pb has been quenched. In any event, clearly some kind of proximity junction is present at the surface of YBaCuO. Nonetheless, a gap-like depression in the conductance for biases below 20 [meV] is universally seen.

Figure 2 shows the tunneling conductance as a function of temperature obtained using a sandwich junction on a freshly cleaved (at room temperature) 2212 BiSrCaCuO single crystal with a native barrier and a Nb counter electrode [7]. The junction was formed with the tunneling direction nominally along the c-axis, although the $a-b$ planes may be exposed at steps on the crystal. Again, there is a clear depression of the tunneling conductance at low bias, but here there is a strong zero bias conductance anomaly rather than gap structure due to the conventional superconducting counter electrode. This conductance anomaly was found to be insensitive to magnetic field. Also evident in the data is a small peak in the tunneling conductance at 26 [meV]. The temperature dependence of this peak is shown in Fig. 3, and it is seen to go to zero at $\approx$83 [K] very close to the observed resistive transition ($\approx$85 [K]) of this BiSrCaCuO crystal. MAEDA et al. [8], have reported tunneling data on 2212 BiSrCaCuO polished ceramics using an amorphous Si barrier and an Ag counter electrode. They also see a peak in the tunneling conductance at 30 [meV], but there is no zero-bias anomaly evident in their data.

It seems clear that there is gap-like structure in the tunneling I-V curves of these high-temperature superconductors quite independently of how the junctions are made. This similarity is even more striking given the very different surface quality of these two materials. We have already mentioned the problems at the surface of YBaCuO.

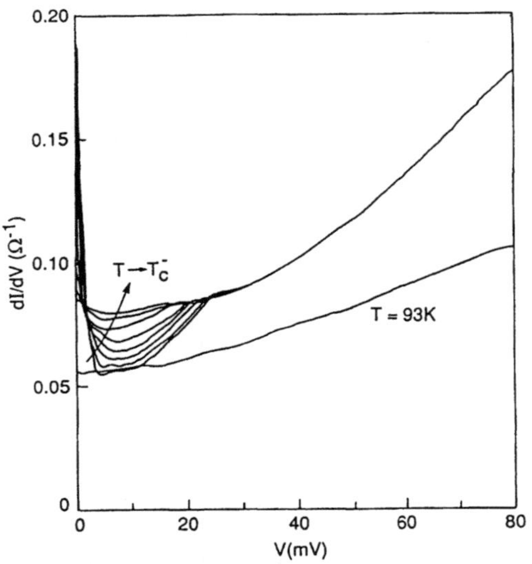

Fig. 2. Differential conductance of a sandwich-type tunnel junction on 212 BiSrCaCuO formed with the native barrier and a Nb counter electrode

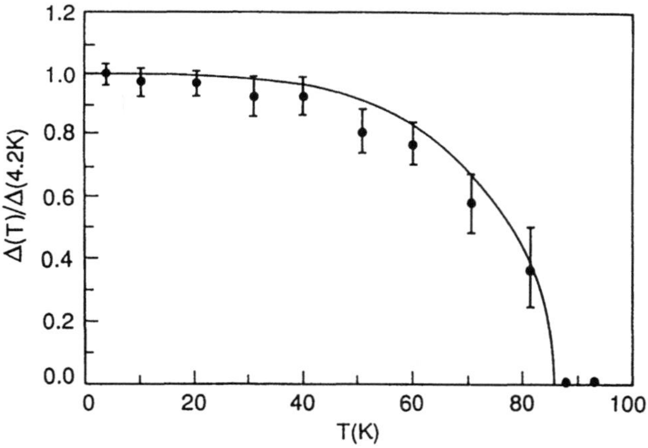

Fig. 3. Temperature dependence of the peak in dI/dV from the data in Fig. 2

By contrast the surface of 2212 BiSrCaCuO single crystals seem highly ideal. The crystals cleave cleanly and readily between the weakly van der Wals bonded Bi oxide planes. The surfaces of these crystals yield outstanding LEED patterns that are stable for long periods of time in an ultra-high vacuum [9]. Also, as reported by KIRK et al. [10], stable atomic resolution scanning tunneling microscope images are possible under UHV conditions on the surface of the same freshly-cleaved BiSrCaCuO as used in the sandwich tunneling studies, whereas atomic resolution microscopy has never been possible on YBaCuO. Scanning tunneling microscopy with lower resolution is even possible in an ambient atmosphere [11]. (Note that scanning tunneling microscopy was not possible on BiSrCaCuO at low temperatures because the surface became insulating.)

Given the ubiquity of these gap-like features observed with sandwich-type junctions, it is interesting to reconsider the earlier point contact data reported in the literature. As noted by researchers attempting point contact tunneling, a wide variety of I-V curves are obtained with this technique. Gap-like features at various voltages, multiple structures at high bias, linear background conductances with and without gap-like structure and shorts are all found. Reproducibility is poor, even from point to point on a given sample, not only for clearly inhomogeneous samples such as bulk ceramics and polycrystalline thin films, but even on single crystals. Nonetheless, distinct superconducting gap-like features are commonly seen. Some of these seem to bear a relationship to the data seen with sandwich junctions. For example, Fig. 4 shows data reported by KIRTLEY et al. [12], with point contact junctions on thin films and single crystals of 123 YBaCuO. The peaks in the conductance occur at 18-20 [meV], close to the value seen in Fig. 1.

If we accept this correspondence, we may conclude that, under some circumstances, it is possible to get clean gap-like structures with low conductance below the gap in the oxide superconductors. Whether it is the local nature, the particular location, the particular tunneling direction, or some other aspect of the point contact tunneling that is important, has not been established, but an existence proof for clean gap structure does seem to exist. As also seen in the figure, additional structure is generally present

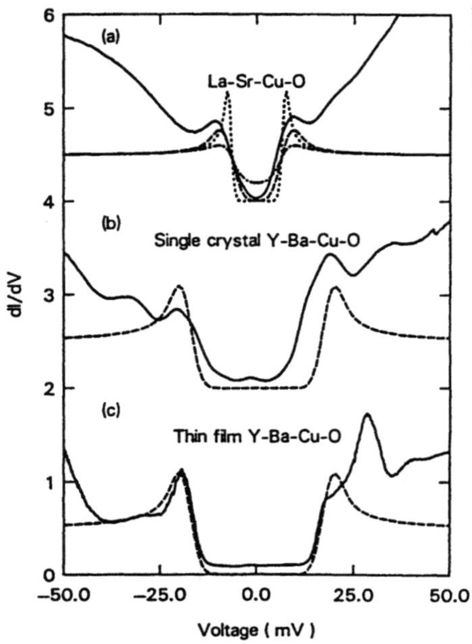

Fig. 4. Point contact tunneling conductance on LaSrCuO and YBaCuO. (From Ref. 12)

in the I-V curve at high bias with point contact tunneling. Evidently this high-bias structure does not exist, or is averaged out, in the sandwich junctions. Such averaging may also contribute to the substantial conduction below the gap seen in sandwich tunnel junctions. Recall, however, the persistent weak structure at 40 [meV] in our sandwich-type junctions.

Additional evidence for clean gap structure can be gleaned from point contact tunneling on less homogeneous samples. Figure 5 shows point contact tunneling results obtained by KIRK et al. on ceramic YBaCuO [13], and a polycrystalline BiSrCaCuO film [11]. Again, the data represent commonly, but not universally, observed behavior. In any event, in both cases the differential conductance is very small at low bias and then rises abruptly, peaking at voltages of approximately 40 and 60 [meV] for 123 YBaCuO and 2212 BiSrCaCuO, respectively. These values correspond closely to twice the values found from the data discussed above, and lend themselves naturally to the interpretation that they represent SIS grain boundary tunnel junctions in these inhomogeneous samples. The role of the point contact in these cases thus seems to be to select a particular grain and/or grain orientation. Similar data was obtained by MORELAND [14] using break junctions formed entirely at liquid helium temperatures using ceramic YBaCuO. His dI/dV data exhibit low conductance at low bias followed by a sharp peak at 40 [meV]. The obvious interpretation again appears to be SIS tunneling, either at the break or at an internal grain boundary. In summary, we see that, while a comprehensive understanding of all the observations with such junctions is not possible, selected data can be understood and lend further support to our contention that well-defined superconductive gaps exist in the oxide superconductors. Moreover, while charging effects may arise, it seems unlikely they would lead systematically to structure at the observed energies in such a wide variety of junction types.

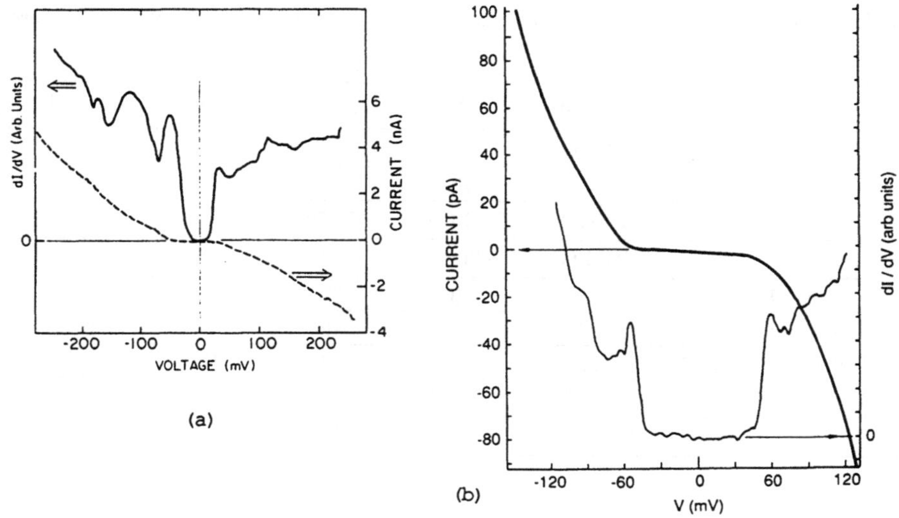

Fig. 5. Point contact tunneling on (a) 123 YBaCuO, and (b) 2212 BiSrCaCuO. The structure is interpreted as grain boundary tunneling

With this improved understanding of the tunneling data on 123 YBaCuO and 2212 BiSrCaCuO, let us address the question of the ratio $2\Delta/k\ T_c$, restricting ourselves to the data for which the gap closing T_c is also available. A collection of such data from the literature is shown in Table 1. In calculating $2\Delta/k\ T_c$, to be definite we have taken the peak in the differential conductance as a measure of the gap. For a conventional BCS superconductor with thermal smearing this choice would tend to overestimate the gap. Similar uncertainties exist if strong inelastic lifetime broadening is present. In the case of thermal broadening, the accepted procedure is to choose the point below the peak at which the observed conductance crosses the extrapolated normal state conductance. On the other hand, for tunneling data that averages over highly anisotropic gaps, the peak should come at the most heavily weighted gap. Since these materials are clearly anisotropic (if only due to their orthorhombic structure), and the data are taken at low temperatures, we have taken the peak as a measure of the gap. In any event, despite these uncertainties, the evidence is strong for a $2\Delta/k\ T_c$ value that is larger than the largest value observed in conventional superconductors (i.e, H_g with a value of 4.6.)

Raman scattering and infrared reflectivity studies provide alternative means of detecting the existence of a superconductive energy gap. Figure 6 shows the low-energy normalized (superconducting/normal) Raman spectrum of a cleaved single crystal of 2212 BiSrCaCuO similar to the one used to obtain the data in Fig. 2 [19]. As with the tunneling data, there is uncertainty about how precisely to interpret such Raman data. Nonetheless, the well-defined peak in the spectrum at 450 [cm^{-1}] ($\simeq$56 [meV]) corresponds closely to the 2Δ values inferred from the tunneling data. A similar conclusion can be drawn on the basis of the Raman data on 123 YBaCuO by COOPER et al. [20].

Table 1. Ratio of $2\Delta/k$ T_c from various experiments in which both Δ and T_c were directly measured

Group	Material	Δ	T_c(gap)	$2\Delta/kT_c$	Method
Edgar et al.[15]	YBCO	20	~90	4-5	P
Sera et al.[16]	YBCO	20	~90	4-5	P
Iguchi et al.[5]	YBCO	20,40,60	60-70	7.1	S
Lee et al.[1]	YBCO	~20,40	60	7.5	S
Tsai et al.[17]	YBCO	20	77	6	B
Tsai et al.[17]	YBCO	10	40	5.9	B
Bulaevskii et al.[18]	EuBCO	36	90	10	P
Iguchi et al.[5]	ErBCO	20,40,100	60-70	7.1	S
Lee et al.[7]	BSCCO	24	83	6.7	S

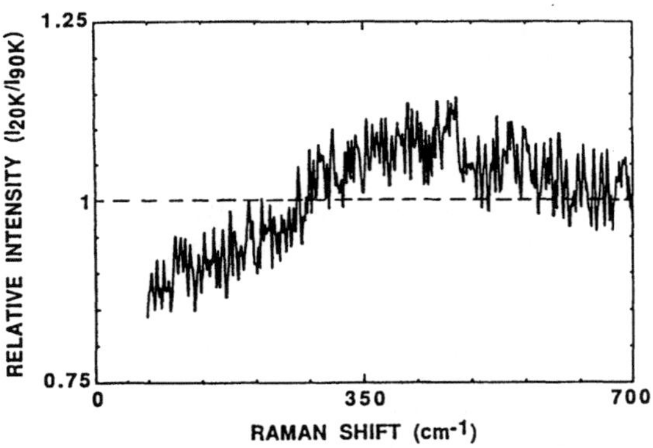

Fig. 6. Normalized (superconducting/normal) Raman spectrum of 2212 BiSrCaCuO at low energies

Recent infrared reflectivity data and values for the energy gap have been reported by THOMAS et al. [21]. Based on when the reflectivity goes to unity, these authors deduce an energy gap 2Δ at about 100 [cm^{-1}] ($\simeq$ 12 [meV]) for 123 YBaCuO with a T_c = 50 [K]. This is distinctly smaller than the values quoted above. We note, however, that there is a definite shoulder in the reflectivity data of these authors at 500 [cm^{-1}] ($\simeq$ 60 [meV]). Thus, it is not clear that the data are in substantive disagreement.

Finally, let us turn to the high-bias tunneling conductance. As noted by essentially all researchers, the high-bias tunneling conductance generally has a strong voltage dependence. A linear increase in dI/dV with V is commonly seen. An example of such behavior is seen in Fig. 1. In order to gain a clearer understanding of this phenomenon, we have examined it as a function of temperature. In Fig. 7, we show the temperature

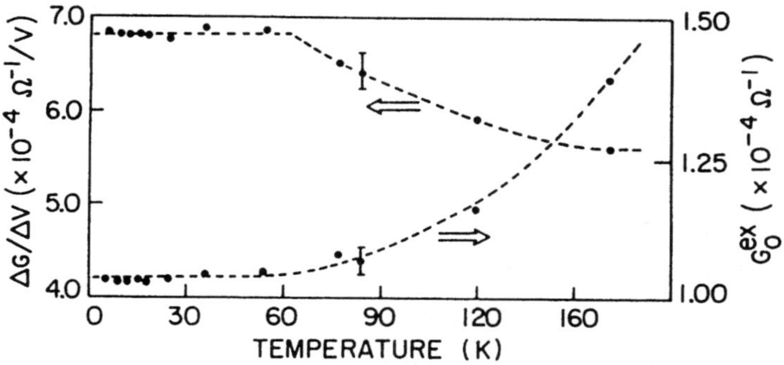

Fig. 7. Temperature dependence of the slope and intercept of the high bias tunneling conductance of 123 YBaCuO

dependence of the slope $\Delta G/\Delta V$ and the intercept G_0 of this linear differential conductance. Remarkably, both the slope and the intercept are temperature independent below T_c ($\simeq$60 [K]) and become temperature dependent above T_c.

The linearly increasing conductance observed in these materials is unusual. It has been suggested that it reflects the inapplicability of the Fermi liquid concept in these materials, and is a consequence of the presence of the holons and spinons of RVB theory [22,23]. Even within the framework of Fermi liquid theory no explanation has been put forward. Hopefully, the observed temperature dependence shown in Fig. 7 will help resolve the origin of this effect.

In conclusion we have shown that the available tunneling data are beginning to form a self-consistent picture. The data taken as a whole suggest that there is a gap in the new oxide superconductors and that $2\Delta/k\ T_c \simeq 6\text{-}7$.

We thank M. Kirk, D. Mitzi and M. Naito for their critical participation in the Stanford work described here. This work was supported by the U.S. Office of Naval Research under contracts nos. N00014-83-K-0391 and N00014-89-K-0327. One of us (M. Lee) would like to acknowledge an NSF Graduate Fellowship.

References

1. M. Lee, M. Naito, A. Kapitulnik, and M. R. Beasley: Bul. Am. Phys. Soc. 33, 594 (1988); submitted for publication.
2. M.D. Kirk, D.P.E. Smith, D.B. Mitzi, J.Z. Sun, D.J. Webb, K. Char, M.R. Hahn, M. Naito, B. Oh, M.R. Beasley, T.H. Geballe, R.H. Hammond, A. Kapitulnik, and C.F. Quate: Phys. Rev. B 35, 8850 (1987).
3. R.C. Dynes, as reported in reference 22.
4. J. Geerk, X.X. Xi, and G. Linker: submitted to Z. Physik.
5. Iguchi et al.: Physica B 148, 322 (1987).
6. A.J. Arko et al.: as presented at the Rare Earth Research Conference, Sept. (1988).

7. Mark Lee, D.B. Mitzi, A. Kapitulnik, and M.R. Beasley: submitted for publication.
8. A. Maeda et al.: Jap. J. Appl. Phys. 27 (4), L661 (1988).
9. Z.X. Shen et al.: submitted for publication.
10. M.D. Kirk et al.: submitted for publication.
11. M.D. Kirk et al.: Appl. Phys. Lett. 52, 2071 (1988).
12. J.R. Kirtley et al.: Phys. Rev. B 35, 8846 (1987).
13. M.D. Kirk et al.: Phys. Rev. B 35, 8850 (1987).
14. J. Moreland et al.: Phys. Rev. B 35, 8856 (1987).
15. A. Edgar et al.: J. Phys. C 20(36), L1009 (1987).
16. M. Sera et al.: Solid State Commun. 65, 997 (1988).
17. J.S. Tsai et al.: preprint.
18. L.N. Bulaevskii et al.: preprint.
19. K. Kirillov, I. Bozovic, T.H. Geballe, A. Kapitulnik, and D.B. Mitzi: submitted for publication.
20. S.L. Cooper, M.V. Klein, H.G. Pazol, J.P. Rice, and D.M. Ginsberg: Phys. Rev. B 37, 5920 (1988).
21. G.A. Thomas et al.: Phys. Rev. Lett. 11, 1313 (1988).
22. P.W. Anderson and Z. Zou: Phys. Rev. Lett. 60, 132 (1988).
23. R.B. Laughlin: to appear in Science.

Energy Gap Measurement Made on Cryogenically Cleaved Y-Ba-Cu-O and Bi-Sr-Ca-Cu-O Surfaces

J.S. Tsai[1], *I. Takeuchi*[1], *J. Fujita*[2], *T. Yoshitake*[2], *S. Miura*[2], *S. Tanaka*[1], *T. Terashima*[3], *Y. Bando*[3], *K. Iijima*[4], *and K. Yamamoto*[4]

[1]Microelectronics Research Laboratories, NEC Corporation, Kawasaki 213, Japan
[2]Fundamental Research Laboratories, NEC Corporation, Kawasaki 213, Japan
[3]Institute for Chemical Research, Kyoto University, Uji 611, Japan
[4]Research Institute for Production Development, Kyoto 606, Japan

The surfaces of the cryogenically cleaved Y-Ba-Cu-O film and Bi-Sr-Ca-Cu-O films are studied. Energy gaps are measured at these cleaved surfaces by the tunneling technique. (001), (103) and (110) oriented Y-Ba-Cu-O epitaxial films are broken in a cryogenic environment along the appropriate directions together with the $SrTiO_3$ substrate. Pb electrode is brought close to the *in situ* clean broken film edge. The normalized energy gap $2\Delta(0)/k_BT_c$ measured in the direction along and perpendicular to the Cu-O plane are found to be 6.0 ± 0.2 and 3.6 ± 0.2 respectively. These values are independent of the variation in the values of T_c within the examined range of 40K~90K. The gap difference structure at Δ_{YBCO}-Δ_{Pb} is observed which helps identifying the value of energy gap of the oxide superconductor unambiguously. (001) oriented epitaxial Bi-Sr-Ca-Cu-O films are also studied. The gap voltage along ab-plane is near 20 mV, but the gap opening temperature of Bi-Sr-Ca-Cu-O junction is not identified in the experiment.

1. INTRODUCTION

The high-T_c oxide superconductors such as $YBa_2Cu_3O_{7-\delta}$ have a marked anisotropic crystal structure. The more recently found Bi-Sr-Ca-Cu-O compounds and Tl-Ba-Ca-Cu-O compounds have even stronger anisotropy. The existence of the layered Cu-O planes in those materials might suggest the presence of notable anisotropic electrical properties in these crystals. From the magnetic measurements, anisotropy in the critical field, coherence length and penetration depth were discovered[1]. Anisotropy in the critical current was also found in thin film samples[2]. EBISAWA *et al.*[3] have predicted an anisotropic energy gap, based on the discrepancy in the observed gap data obtained from tunneling experiments and infrared experiments. Several attempts were made[4] but no evidence of gap anisotropy was discovered. In the previous tunneling studies, varieties of the gap values $2\Delta(0)/k_BT_c$ were reported, ranging typically from 4 to 6[5]. Some of the reported ambiguity may be attributed to inhomogeneity and anisotropy of the samples. The multiple gap like structures observed by many authors might have further complicated the issue. We have developed a novel broken film edge junction technique, with which a clean and preferentially oriented tunneling interface was realized. With such Y-Ba-Cu-O junction, $\Delta(T)$ was monitored so that T_c where $\Delta(T)=0$ could be extrapolated. The reduced energy gap $2\Delta(0)/k_BT_c$ obtained in this way showed systematic and reproducible anisotropy in the two chosen crystal orientations, namely, along c-axis and ab-plane[6]. Bi-Sr-Ca-Cu-O cleaved surface was also studied with the same technique.

Springer Series in Materials Science, Vol. 11
Mechanisms of High Temperature Superconductivity
Editors: H. Kamimura and A. Oshiyama
Springer-Verlag Berlin Heidelberg © 1989

2. EXPERIMENTAL

Superconductive $YBa_2Cu_3O_{7-\delta}$ films (thickness 100~300nm) were prepared by ion beam sputtering[7] and activated reactive evaporation[8]. They were epitaxially grown on $SrTiO_3$ substrates and X-ray studies and RHEED analysis proved these films to be of single-phase having single orientations. In order to obtain the clean surfaces for tunneling, films were broken at liquid He temperature. The Pb electrodes were then brought to contact with the exposed film edges to form tunnel junctions. Use of pre-cut grooves in substrates made tearing of the films along desired directions in straight lines possible. SEM microphotographs of severed film edges revealed that the breaks actually took place in the middle of the grains rather than at the grain boundaries (Fig. 6). (001) films were used to prepare tunneling surfaces parallel to the Cu-O plane (ab-plane), and breaking (110) films in appropriate directions provided surfaces either parallel or perpendicular to the Cu-O plane. The direction of the breaks were along the crystal plane. So breaking (110) film along ab-plane would obtain a smooth surface with occasional atomic-size steps. For the film edge made along c-axis, the contribution of (001) plane should be quite limited for the same reason. The direction of the break made in (001) films was randomly chosen, but such film edges should not contain much of the (001) plane either. The resistance of the tunnel junctions was controlled manually. It was demonstrated from the contact resistance measurement on top surfaces of the films that virtually all samples were covered with a non-superconductive layer. This indicates that the direction-controlled tunnelings at low-impedance broken film edges are quite legitimate. $dI/dV-V$ measurements were conducted to obtain the gap voltage. Temperature dependence of the energy gaps were traced and they were extrapolated to provide the values of T_c which represent the dominant phase along the junction. By doing so, one can obtain the T_c of the tunneling surface itself.

Ion beam sputtered[9] and rf sputtered[10] epitaxial Bi-Sr-Ca-Cu-O films were also used to make the broken film edge junctions. So far, only the (001) films were used, so that the cleaved surface was supposed to have the (100)/(010) orientation. The contact resistance between the electrode and the natural surface of the Bi-Sr-Ca-Cu-O film was rather low compared to that of the Y-Ba-Cu-O film. The Y-Ba-Cu-O film typically had a surface contact resistance of $10^6\sim10^8$ $\Omega/\mu m^2$, but the natural surface contact resistance of the Bi-Sr-Ca-Cu-O film was typically around $10^3\sim10^5$ $\Omega/\mu m^2$. To prevent the possible leakage through the low impedance top surface of Bi-Sr-Ca-Cu-O, a insulating layer (3000 Å thick SiO) was evaporated on top of the oxide superconductor film.

3. RESULTS AND DISCUSSIONS

3.1 Y-Ba-Cu-O

Figure 1 (a) (b) show $dI/dV-V$ curves of a junction made with a (110) Y-Ba-Cu-O film broken along (001) plane. The curve (a) taken at 4.2K reveals small peaks near 5mV and another structure near 8mV which corresponds to the gap difference peaks $(\Delta_{YBCO}-\Delta_{Pb})/e$ and the gap sum peaks $(\Delta_{YBCO}+\Delta_{Pb})/e$ respectively. At temperatures above the T_c of Pb(T=10K), the small peak in dI/dV vanished, and the position of the main gap structure shifted by the amount of $\Delta Pb/e$ (Fig. 1(b)). The observation of such gap difference and sum structures strongly indicates that one is observing no other than the energy gap of $YBa_2Cu_3O_{7-\delta}$. Since the energy gap of Pb is well established, the energy gap of $YBa_2Cu_3O_{7-\delta}$ can be determined conclusively from these curves. Energy gap in the Cu-O plane was obtained using (001) films and (110) films broken along the c-axis. Fig. 1(c) shows dI(V)/dV curve of a junction formed by breaking the same film used in Fig. 1(a) and (b) along (110) plane. A larger gap value was observed compared to the Fig. 1(b).

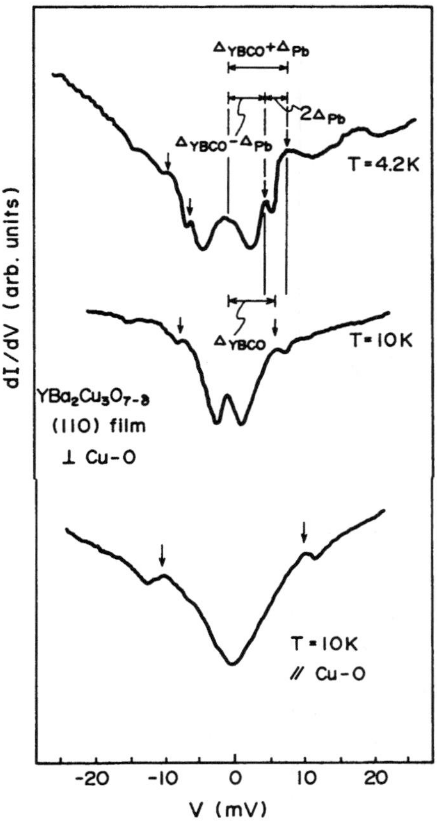

Fig. 1 dI(V)/dV of junctions made with (110) film broken along c-plane (a), (b); and broken along c-axis (c). (a) and (b) are dI(V)/dV curves taken below and above the T_c of Pb. Gap difference and sum structures can be seen in (a).

To obtain the normalized value of the gap $2\Delta(0)/k_BT_c$, a valid determination of T_c is required. Temperature dependence of the gap $\Delta(T)$ was traced in order to find $T_c = T(\Delta = 0)$, for T_c obtained in this manner would most likely represent the T_c of the dominant phase at the junction surface. Fig. 2 shows such $\Delta(T)$. Gap structure near T_c was quite difficult to make out, thus, T_c value was extrapolated from $\Delta(T)$ with an appropriate error. $2\Delta(0)/k_BT_c$ values for tunneling parallel to the Cu-O plane and perpendicular to the Cu-O plane were scattered about 6 and 3.6 respectively signifying the anisotropy. This is illustrated in Figure 3 where $\Delta(0)$ were plotted against the T_c for the Y-Ba-Cu-O broken film-edge tunnel junction. In Fig. 3, both of the parameters were extrapolated from $\Delta(T)$ curves like Fig. 2 for each point. Even with large error values in critical temperatures, they fit quite accurately on two straight lines (top for parallel and bottom for perpendicular to the Cu-O plane) passing through the origin. Two distinct values of gap energy were observed along two different crystal orientations at the surfaces that were carefully prepared. It seems that the observed gap voltage systematically correlates with the direction of the tunneling. The smaller normalized gap value along c axis was also reported recently in a sandwich type tunnel junction[11]. Perhaps such data indicate that in our experiment, the electrons were truly tunneling through the surface barrier directionally along the chosen crystal axis. As for the physical meaning of the apparent anisotropy observed in our experiment, one can only speculate some of the possibilities. As

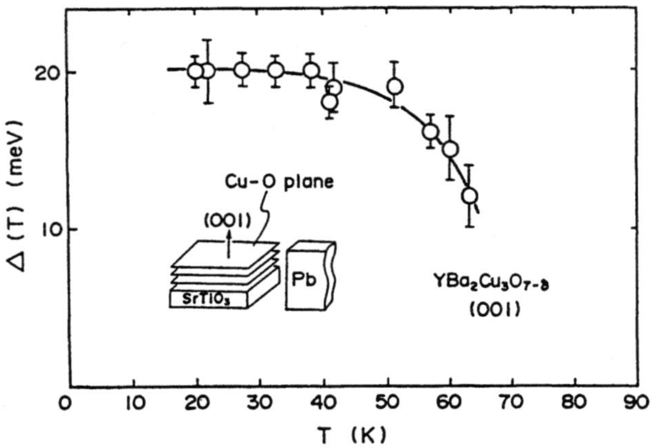

Fig. 2 Temperature dependence of the gap voltage. This Δ(T) curve was for a junction made with (001) film.

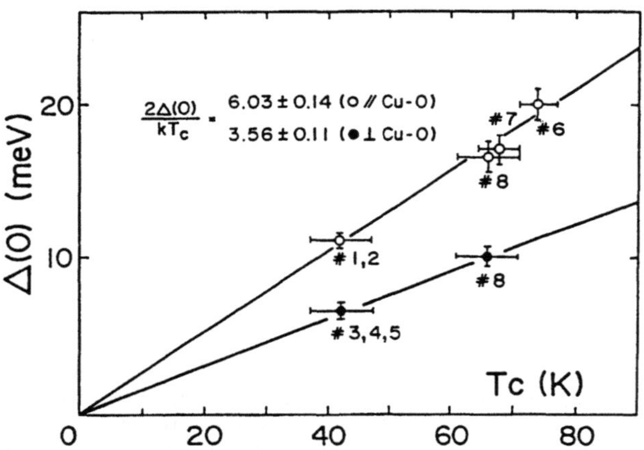

Fig. 3 Δ(0) vs T_c for 8 different samples. Closed circles are for the tunneling done perpendicular to the Cu-O plane; open circles are for the tunneling parallel to the Cu-O plane.

previously observed in a SN sample by BINNING and HOENIG[12], we might be observing the true energy gap anisotropy of the Y-Ba-Cu-O system. Anisotropic gap suppression near S/I interface due to short and anisotropic coherence length[13] might also be responsible for the observed effect.

Impedance of the tunnel junction could be controlled externally, but the observed gap voltage was independent of the impedance values ranging between $10{\sim}10^6\Omega$. Fig. 4 shows the impedance dependence of observed gap voltage for two different junctions. Further reduction in the impedance of the junction would lead to an observation of Josephson current. In some samples, the Josephson current between Pb/Y-Ba-Cu-O junction remained finite even above the T_c of the Pb electrode. Fig. 5 shows the observed $I_c(T)$. AC Josephson effect was also observed

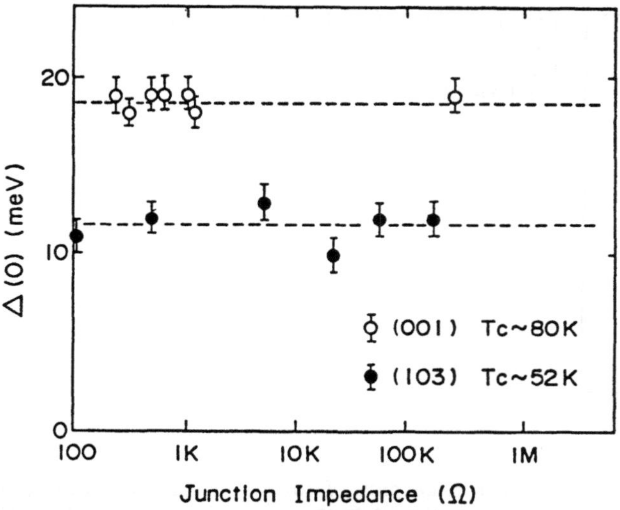

Fig. 4 Impedance dependence of the gap voltages for two samples.

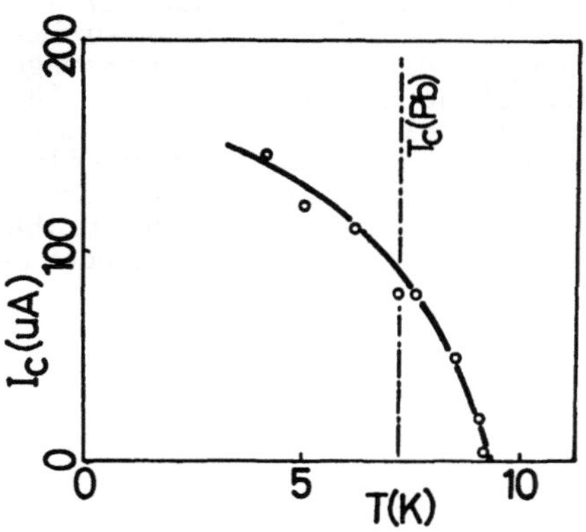

Fig. 5 Temperature dependence of the Josephson current I_c of the junction. T_c enhancement was observed.

up to about 9.2K. The Josephson frequency we found in such junction was observed up to about 9.2K. The Josephson frequency we found in such junction was the usual 2eV/h. One interpretation of such phenomenon is the proximity-induced Josephson effect discussed by HAN *et al.*[14].

3.2 Bi-Sr-Ca-Cu-O

The I-V and σ-V characteristics of Bi-Sr-Ca-Cu-O broken film-edge junction did not show gap like structures of the oxide superconductor in most of the trials. But occasionally it showed the gap like structure around 20 mV as shown in Fig. 6. Compare to the Y-Ba-Cu-O junctions the irreproducibility of the gap structure in the Bi-Sr-Ca-Cu-O junctions was quite evident. Nonetheless, the gap structure of Pb was quite noticeable even for these junctions not showing the gap of Bi-Sr-Ca-

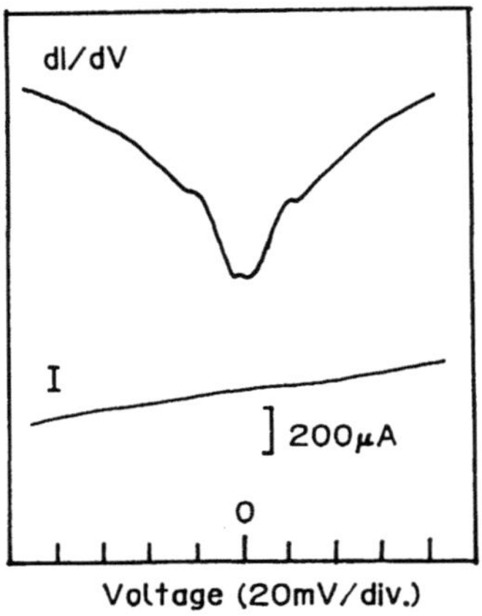

Fig. 6 σ-V and I-V characteristics of Bi-Sr-Ca-Cu-O broken film edge junction. Gap structure near 20 mV can be seen.

Cu-O compound. This implies that a large part of the Bi-Sr-Ca-Cu-O junction surface was covered with a normal conductive material, so that the tunneling signal of the superconductive Bi-Sr-Ca-Cu-O surface was overshadowed by the predominant Pb/I/N SIN tunneling signal and a nontunneling leakage signal. The origin of such a large normal conductive surface in the cleaved junction was attributed to the top natural surface of the Bi-Sr-Ca-Cu-O film. Such top surface is probably mostly normal. Even with the SiO protective top layer, the cleavage probably would expose a considerable portion of the top normal surface. The contact resistance through the top surface of Bi-Sr-Ca-Cu-O film was about $10^3 \sim 10^5$ Ω/μm². Assuming the area of such a normal surface in the broken film-edge junction to be about 0.1×100 μm², that would produce a shunt resistance of about $10^2 \sim 10^4$ Ω due to the Pb/I/N, SIN tunneling. This value is close to the observed typical junction impedance, indicating that such a junction model is quite conceivable.

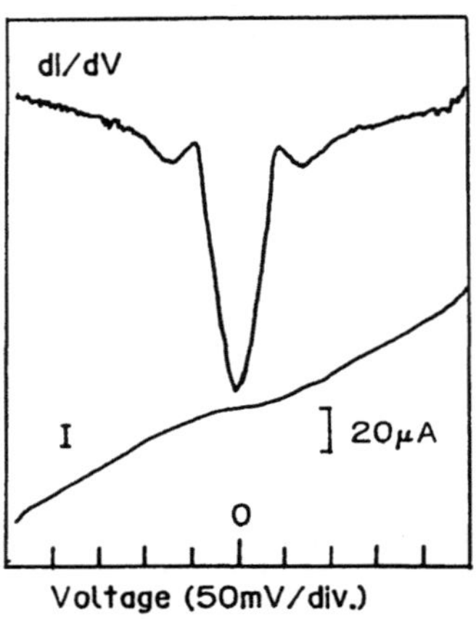

Fig. 7 σ-V and I-V characteristics of Bi-Sr-Ca-Cu-O broken film-edge junction showing S-I-S like gap structure near 40 meV.

The normalized energy gap can not be established because the gap opening temperature T_c could not be traced out in our Bi-Sr-Ca-Cu-O junctions due to the instability of the structures. The resistively found T_c (end point) was about 70 K (with transition width of ~10 K) giving the normalized energy gap of 6.6. This value agrees with the value obtained for the Y-Ba-Cu-O in the same crystal direction (in the ab plane). A larger energy gap was found sometime near 40 mV as shown in Fig. 7. It could probably be attributed to the internal (intergranular) Bi-Sr-Ca-Cu-O/I/Bi-Sr-Ca-Cu-O junction[16].

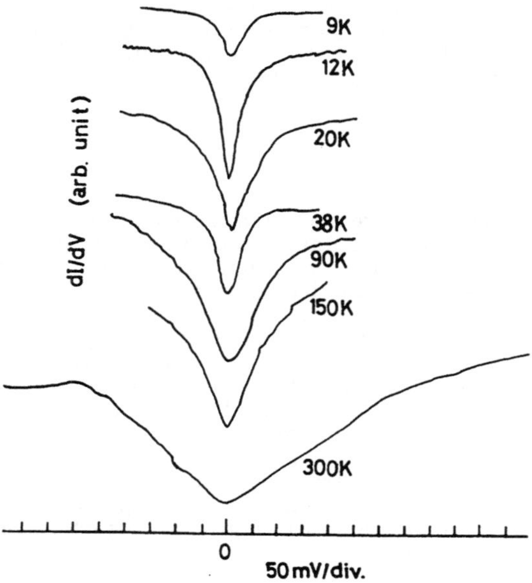

Fig. 8 Background σ-V characteristics of Bi-Sr-Ca-Cu-O junction.

Even when the σ-V characteristics of the Bi-Sr-Ca-Cu-O junction did not show obvious gap like peaks at high voltage (they always show clear Pb gap at low voltage), they displayed very nonlinear characteristics with a large temperature dependence as shown in Fig. 8. Similar characteristics were also reported by other groups [15]. As one can see from Fig. 8, at low temperature, the linear-V conductance saturates to a fixed value above a certain characteristic voltage V_c. As the temperature increases, V_c also increases, and the conductance does not quite saturate to a

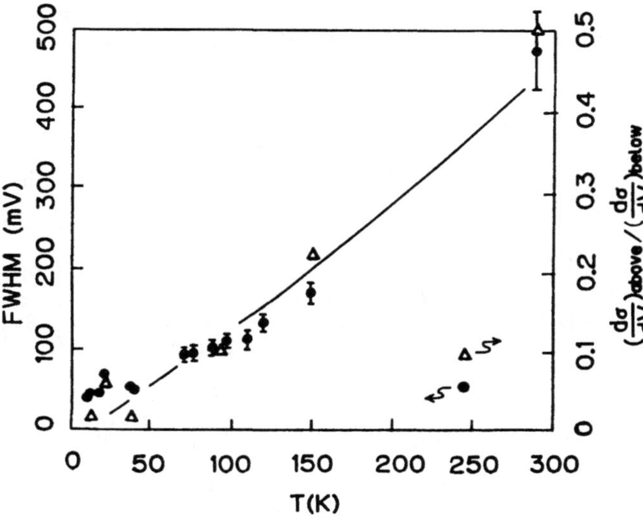

Fig. 9 Temperature dependence of full width half maximum and dσ/dV above Vc taken from Fig. 8.

constant value. To characterize such a tendency, full width at half maximum (FWHM) voltage of the central depression ($\approx V_c$) and the slope of σ-V curve above Vc are plotted against temperature in Fig. 9. A continuous change of the FWHM as well as the dσ/dV can be seen from Fig. 9 even up to room temperature. The nature of such background σ-V characteristics is still not clear but as previously discussed, it probably originated from the Pb/I/N tunneling and/or the non-tunneling leakage.

4. CONCLUSION

In conclusion we have measured the energy gap in Y-Ba-Ca-O and Bi-Sr-Ca-Cu-O compounds by using a unique thin film breaking technique to prepare clean and preferentially oriented superconducting surfaces. *In situ* tunneling measurements were carried out on such surfaces with Pb counterelectrode. From the Y-Ba-Cu-O junctions the normalized energy gap $2\Delta(0)/k_BT_c$ of 6.0 ± 0.2 and 3.6 ± 0.2 were observed consistently and reproducibly in the directions parallel and perpendicular to the Cu-O plane, respectively. These values are independent of T_c, even when T_c was as low as 40K in some samples where the one dimensional oxygen chain ceases to exist. The anisotropy ratio of the gaps in two crystal orientations was 1.6 ± 0.1 by these tunneling studies. The Bi-Sr-Ca-Cu-O broken film edge junctions also showed the energy gap near 20 mV along the ab-plane. The gap opening temperature of this junction was not identified. The normalized gap value calculated from the $T_c(R=0)$ agrees well with the value found in Y-Ba-Cu-O. Very nonlinear σ-V background characteristics were observed in the Bi-Sr-Ca-Cu-O junctions.

ACKNOWLEDGEMENTS

We are in debt to H. Igarashi for the most helpful discussions. We are also grateful for the support of H. Abe and M. Yonezawa, throughout the course of this research.

REFERENCES

1. T.K. Worthington, W.J. Gallagher and T.R. Dinger Phys. Rev. Lett. 59, 1160 (1987)
2. Y. Enomoto, T. Murakami, M. Suzuki and K. Moriwaki, Jpn. J. Appl. Phys. 26, L1248 (1987)
3. H. Ebisawa, Y. Isawa and S. Maekawa, Jpn. J. Appl. Phys. 26, L992, (1987)
4. J.R. Kirtley, W.J. Gallagher, Z. Schlesinger, R.L. Sandstrom, T.R. Dinger and D.A. Chance, 1987, preprint. Phys. Rev. B35, 8846 (1987)
   Z. Schlesinger, R.T. Collins, D.L. Kaiser, and F. Holtzberg, Phys. Rev. Lett. 59, 1958 (1987)
5. A. Barone, Physica C, 153-155, 1712 (1988)
6. J.S. Tsai, I. Takeuchi, J. Fujita, T. Yoshitake, S. Miura, S. Tanaka, T. Terashima, Y. Bando, K. Iijima, K. Yamamoto, Physica C 153-155, 1385 (1988)
7. J. Fujita, T. Yoshitake, A. Kamijyo, H. Igarashi and T. Satoh, to be published, Extended Abstracts MRS Spring Meeting (1988)
8. T. Terashima, K. Iijima, K. Yamamoto, Y. Bando and H. Mazaki, Jpn. J. Appl. Phys. 27, L91, (1988)
9. J. Fujita, T. Tatsumi, T. Yoshitake, H. Igarashi, To be published in Proceedings of Conference on the Science and Technology of Thin Film Superconductor Nov. 14-18, Colorado Springs
10. S. Miura, unpublished

11. J. Takada, H. Mazaki, T. Terashima, K. Iijima, K. Yamamoto, K. Hirata and Y. Bando, Extended Abstracts of the 20th Conference on Solid State Devices and Materials, Tokyo, 455 (1988)
12. G. Binnig and H.E. Hoenig, Z. Physik. B32, 23 (1978)
13. G. Deutscher and K.A. Müller, Phys. Rev. Lett. 59, 1745 (1987)
14. S. Han, K.W. Ng, E.L. Wolf, A. Millis, J.L. Smith and Z. Fisk, Phys. Rev. Lett. 57, 238 (1986)
15. S. Zhao, H. Tao, Y. Chen, Y. Yan and Q. Yang, Solid State Communi. 67, 1179 (1988)
16. M. Lee, A. Kapitulinik, M.R. Beasley, This proceedings.

Transport and Magnetic Properties of $(La_{1-x}Sr)_2CuO_4$

H. Takagi[1], *Y. Tokura*[2], *and S. Uchida*[1]

[1]Engineering Research Institute, University of Tokyo, Bunkyo-ku, Tokyo 113, Japan

[2]Department of Physics, University of Tokyo, Hongo, Bunkyo-ku, Tokyo 113, Japan

Transport and Magnetic Properties of $(La_{1-x}Sr_x)_2CuO_4$ are systematically investigated over a wide composition range up to x=0.175 including nonsuperconducting metal phase in the highly doped region. Remarkable change associated with the superconductor to nonsuperconductor transition is observed both in the Hall coefficient and the magnetic susceptibility, suggesting the modification of both charge and spin states.

I. INTRODUCTION

While the mechanism of the high-Tc superconductivity in Cu oxides is not understood yet, much effort has been paid for identifying the physical properties which are essential for the occurrence of high-Tc superconductivity. Among the high-Tc cuprates, $(La_{1-x}Sr_x)_2CuO_4$ system is the simplest and can be a prototype of these materials. La_2CuO_4 for which the average valence of [Cu-O] complex is just zero is regarded as the mother compound of the system. The intriguing feature of the mother compound is that it is the antiferromagnetic insulators as a consequence of the strong on site coulomb interaction on Cu site [1]. The high-Tc superconductivity appears in the vicinity of this antiferromagnetism. The increase of the average valence of the Cu-O complex with doping leads to the transition from the insulator to the superconductor. In this system, Tc increases monotonously up to x=0.075. Above this composition, however, the physical properties reported were quite ambiguous. This had been ascribed to the presence of oxygen vacancies which will be introduced more easily in the high Sr concentration range. Recently, Torrance et al. have carefully synthesized a series of specimens without oxygen vacancy and have established the relationship between the average valence of Cu-O complex and Tc [2]. It had been pointed out in the earlier stage of the study that the higher average valence lead to the higher-Tc. They have, however, clearly demonstrated a saturation of Tc against the average valence and the disappearance of the superconductivity at higher valence state, giving a new test for the proposed mechanism.

In order to elucidate the novel mechanism, the detailed cooperative study on both the charge and spin states in the superconductor and nonsu-

Springer Series in Materials Science, Vol. 11
Mechanisms of High Temperature Superconductivity
Editors: H. Kamimura and A. Oshiyama
Springer-Verlag Berlin Heidelberg © 1989

perconductor will be useful. At the present stage, $(La_{1-x}Sr_x)_2CuO_4$ system is the only one system for which we can trace the details of the change in the electronic states over a wide range of the average valence of [Cu-O] complex which covers from the antiferromagnetic state to the nonsuperconducting phase through the superconducting phase. Furthermore, its simple K_2NiF_4-type structure consisting of single CuO_2 layer and the precisely controllable average valence should allow much simpler interpretation for the physics behind than the other Cu oxide systems. In these contexts, we have investigated the effect of the Sr doping on the transport and magnetic properties of $(La_{1-x}Sr_x)_2CuO_4$ in detail with emphasis on the charasteristic change in the physical properties associated with the composition induced transition from superconductor to nonsuperconductor.

II SAMPLE PREPARATION AND EXPERIMENTAL PROCEDURE

In order to investigate the details of the doping effect, it is essentially important to prepare the samples with good homogeneity and precisely controlled stoichiometry. We have employed the spray-dry technique so as to satisfy the above requirement. Furthermore, a particular attention was paid for incorporating oxygen in the highly doped region. Details of the sample preparation will be described elsewhere [3].

These series of samples were confirmed to be a single phase of K_2NiF_4 type structure by the powder X-ray diffraction. The oxygen content was determined by the iodometric titration technique described by Nazzal et al. [4] Oxygen content is kept almost 4 within our resolution at least up to x=0.175 by applying high O_2 pressure or long-period annealing. Because of the difficulty in controlling the stoichiometry particularly the oxygen content, it has been pointed out the chemically determined valence of the [Cu- O] complex describes the system as a better variable than the Sr content. In this paper, however, we employ the nominal Sr content as a variable, since we did not detect any substantial difference between the valence determined chemically and that expected from the nominal Sr content. We have measured the electrical resistivity, the Hall coefficient and the magnetic susceptibility on these series of samples.

III. RESULTS AND DISCUSSIONS

First, we will show the resistivity data in order to see the gross feature of the system. Reflecting the well controlled stoichiometry, the resistivity changes very systematically with doping as shown in Fig.1. As well known, La_2CuO_4 shows semiconducting behavior over the measured temperature range except a trace of very minor superconductivity at low temperatures. With only one percent or less Sr doping, the resistivity drastically decreases to the order of 10^{-3}ohm-cm and T-linear resistivity appears at high temperature, indicating mobile carriers are supplied by very small amount of doping. This feature is quite distinct from isostructural La_2NiO_4 or La_2CoO_4 [5]. In these materials Sr seems to form localized electronic states within an insulating gap. At the low doping level below x=0.03, the supplied carriers tend to localize at low temperatures which is evidenced by the variable range hopping type resistivity increase. As seen from the systematic decrease of the onset temperature of the low temperature resistivity increase, the localization is suppressed by doping, which can be ascribed to the screening of the random potential by the introduction of carriers . At around x=0.03, the superconductivity appears. Tc increases rapidly with doping, having a maximum with Tc=39 K at around x=0.075.

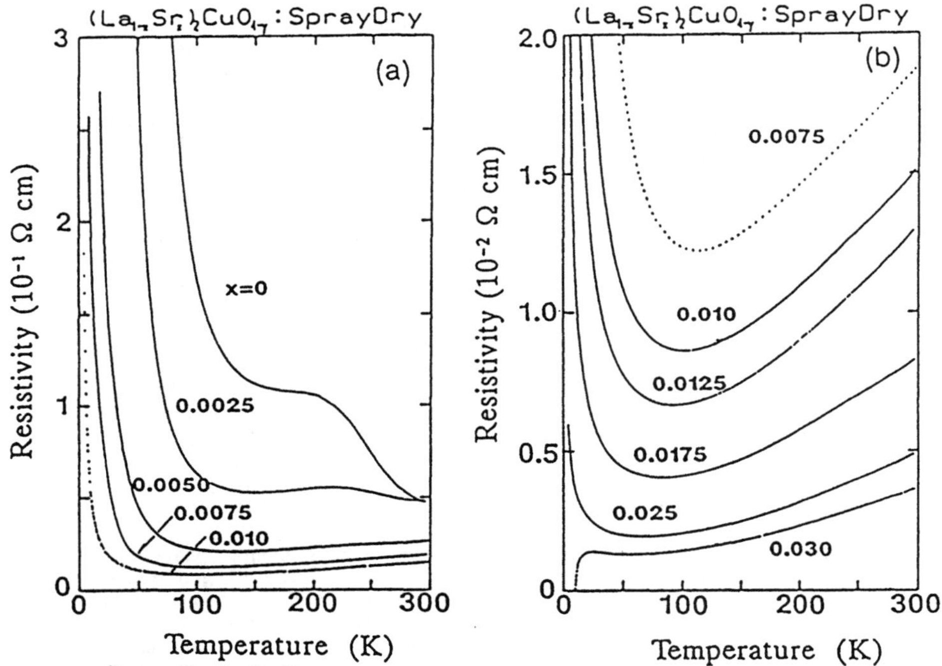

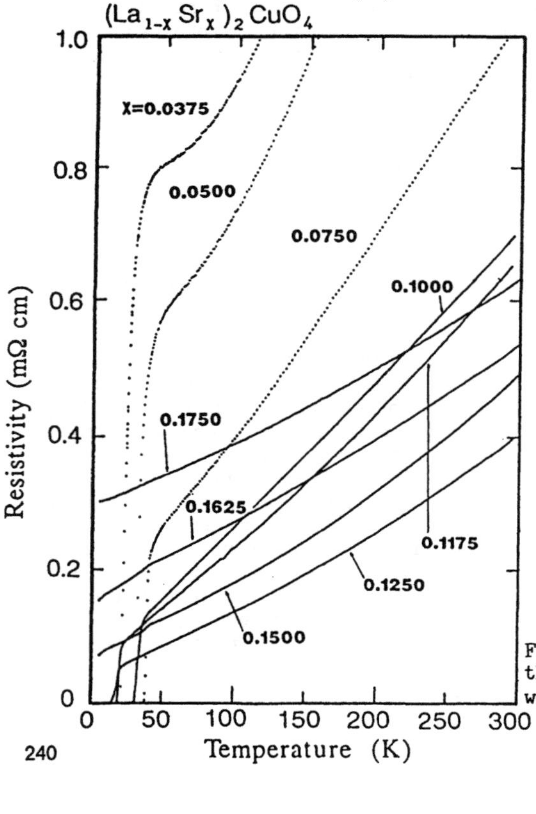

Fig.1 Temperature dependence of the resistivity for $(La_{1-x}Sr_x)_2CuO_4$ with various Sr content.

An intriguing event occurs above x=0.075 [2]. In contrast to the case for x < 0.075, Tc decreases rapidly, although the magnitude of the resistivity is still decreasing. Above x=0.13, we cannot see the superconducting transition any more down to 1.8 K. The system becomes a nonsuperconducting metal. No remarkable change associated with the disappearance of the superconductivity cannot be seen in the temperature dependence of the resistivity. With further increase in doping, the upturn of the resistivity at low temperatures become significant above x=0.15. This might be ascribed to the scattering due to the disorder as a consequence of the very small amount oxygen vacancy which we cannot detect within our resolution.

The transition temperature Tc determined from the Meissner measurement is summarized as a function of the Sr content in Fig. 2. The superconductivity is seen from x=0.03 to x=0.125. Since the magnitude of the Meissner signal is remarkably small at around x=0.03 and x=0.125, the presence of inhomogeneity seems to make the boundary broadened. Therefore, the real superconducting phase is likely to be confined to the narrower composition range than that experimentally observed. Recently, Moodenbaugh et al. have found an anomaly composition dependence of Tc in $(La_{1-x}Ba_x)_2CuO_4$. Tc shows an anomalous dip at around x=0.06 [6]. In the present Sr doped system, however, we do not see such pronounced anomaly at around x=0.06 as in Ba doped system, although a small trace of the degradation of the superconductivity is observed both in the resistivity and the Meissner signal. Therefore the anomaly seems to be more specific to the Ba doped system, possibly related to the larger difference of ionic radius between La^{3+} and Ba^{2+}. Hereafter, we denote the composition ranges 0 < x < 0.01, 0.01 < x < 0.03, 0.03 < x < 0.125 and x > 0.125 as the antiferromagnetic phase, the intermediate region, the superconducting phase and the nonsuperconducting metal phase, respectively.

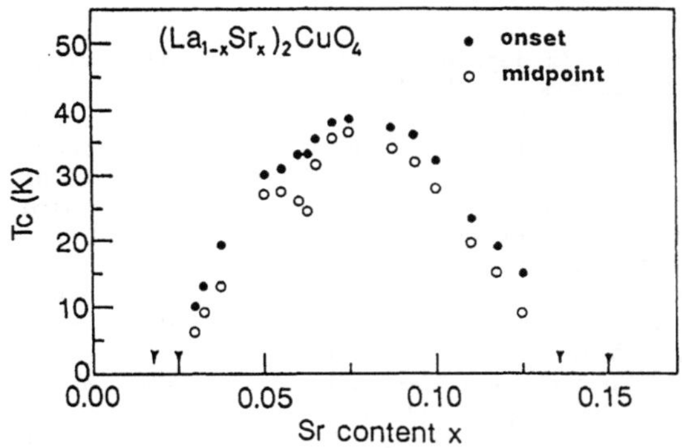

Fig.2 Sr content x dependence the superconducting transition temperature Tc. Tc was determined by the meissner measurements.

Now we come to show the Hall coefficient and the magnetic susceptibility in order to see how the electronic states are modified by doping. The useful information about the electronic states can be obtained from the Hall coefficient . In Fig. 3, the Hall coefficient at 80 K and 300 K are shown against the Sr content. As already observed in the earlier stage of the study [7,8], the positive Hall coefficient decreases almost in proportion to the inverse of the Sr content 1/x in the low composition range. This behavior has been explained in terms of the Mott-Hubbard picture in which a Sr doping provides one mobile hole in the

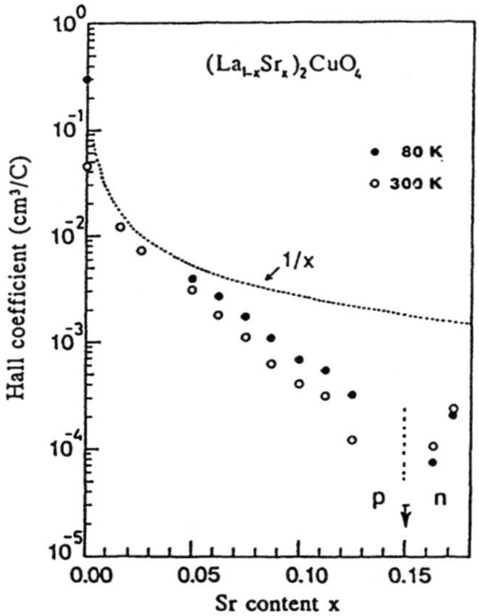

Fig.3 Sr content dependence of the Hall coefficient at 80 K and 300 K.

lower filled band in the rigid band manner. With further doping, the Hall coefficient decreases more rapidly than expected from the Mott-Hubbard picture above x=0.05. In the nonsuperconducting metal phase the Hall coefficient becomes as small as of the order of $10^{-4}cm^3/C$ or less and changes its sign from positive to negative. This rapid decrease was first reported by Ong et al. [8]. In their report, however, the deviation from 1/x is significant at around x=0.1 and the decrease is much steeper than the present result.

It has been argued whether the 1/x dependence of the Hall coefficient can be understood within the framework of the Fermi liquid theory. Although attempts have been made, any theory based on the Fermi liquid description has not succeeded in explaining the anomalous 1/x dependence satisfactory. This has been casting a doubt about the applicability of the Fermi liquid description [9]. If we presume that the Mott-Hubbard picture is appropriate in the low x region, the rapid decrease of the Hall coefficient indicates the breakdown of the simple Mott-Hubbard picture in the highly doped region. The naive interpretation for this might be the two carrier description. We believe, however, the rapid decrease should be sought to the modification of the electronic states. Even if we assume the two carrier description that the holes are introduced to the another band in the highly doped region, it should provide the positive contribution within the rigid band picture. Therefore, it can not explain the rapid decrease or the change of the sign. As far as the highly doped region is concerned, the very small or negative Hall coefficient seems to be not so different from what is expected from Fermi liquid theory. Thus, one of the plausible explanations might be that the rapid decrease of the Hall coefficient indicates a crossover from the Mott-Hubbard picture to the Fermi liquid like picture. The high-Tc superconductivity seems to be realized just in the crossover region.

We should note that an alternative explanation is possible for the rapid decrease of the Hall coefficient. In this paper, we have assumed that the Hall coefficient of the present polycrystals reflects that within the plane with the current parallel to the plane and the magnetic field perpendicular to the plane because of the large anisotropy of the resistivity between the parallel and the perpendicular to the plane. In the low x region, the presence of the anisotropy has been clearly demonstrated by the anisotropic resistivity measurement and the upper critical field measurement on single crystal [10,11]. As for the highly doped region, however, no information about the anisotropy is available yet because of the difficulty in synthesizing the single crystal with high Sr content. If the conduction become remarkably isotropic with doping in the highly doped region, it is possible that the negative contribution bring about the rapid decrease of the Hall coefficient. Recently, Suzuki et al. reported the Hall coefficient of single crystalline thin film [12]. In their report, the decrease of the Hall coefficient in the highly doped region is not so rapid as the present result, although the gross feature is qualitatively similar to the present result. In order to clarify this point, single crystals with precisely controlled stoichiometry is required.

The other point to be noted is the temperature dependence of the Hall coefficient shown in Fig. 4. In the antiferromagnetic phase, the Hall coefficient exhibits strong semiconducting temperature dependence with the activation energy of the order of 10meV estimated at around 300 K. In the intermediate region, it becomes temperature independent like a typical metal, consistent with the appearance of the T-linear resistivity. In contrast to the intermediate region, however, the Hall coefficient again shows negative temperature dependence in spite of the metallic resistivity in the superconducting phase. For x=0.075, the Hall coefficient increases by a factor of more than 2 from 300 K to 50 K. The negative temperature dependence of the Hall coefficient seems to correlate with the appearance of the superconductivity. The observation of the remarkable negative temperature dependence in the superconducting phase is common among the high-Tc cuprate system. Particularly, it is most pronounced in $Ba_2YCu_3O_{7-y}$, for which the Hall coefficient obeys well with the empirical relationship $R_H=A/T$ [13]. At the present stage, the

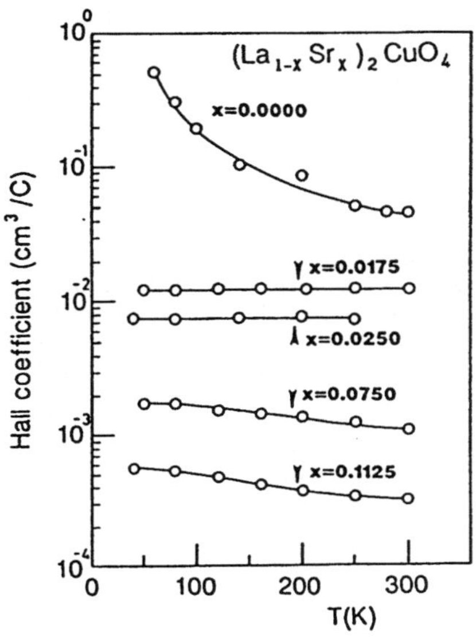

Fig.4 Temperature dependence of the Hall coefficient for various Sr content x.

origin of the peculiar temperature dependence is not clear. Several mechanisms had been proposed as the origin of this remarkable temperature dependence. These are the skew scattering as a consequence of the scattering by the localized moment, two carrier model in which two types of carries with different temperature dependence of the mobility contribute to the conduction. As for the skew scattering, we can not see any evidence of the magnetic scattering in the resistivity as in the heavy fermion system. In the case of $Ba_2YCu_3O_{7-y}$ system, it has been proposed that the presence of inequivalent two Cu site result in the two carrier contribution. Apparently, it is not the primary origin, since the present system which does not contain Cu-O chain shows remarkable temperature dependence.

We shall proceed to the effect of doping on the magnetic susceptibility. All the samples investigated shows a weak paramagnetism of the order of 10^{-7}emu/g as shown in Fig. 5. La_2CuO_4 is an antiferromagnet with T_N 240 K, as demonstrated by the presence of the peak in the temperature dependence of the susceptibility. The first remarkable change with Sr doping is the disappearance of the peak, indicating the suppression of the antiferromagnetic order. The peak temperature rapidly shifts to the lower temperatures and disappears at around x=0.01, which agrees well with the previous study by Fujita et al [14]. In the intermediate region above x=0.01, the susceptibility first shows the negative temperature dependence, changing sign with doping. On the other hand, at the appearance of the superconductivity at around x=0.03, no remarkable change can be seen in the magnetic susceptibility. The susceptibility shows the positive temperature dependence in the superconducting phase. This is consistent with the results of the neutron scattering measurements which suggests that the magnetism does not change appreciably across the insulator to superconductor transition [15].

For higher doping, a notable change appears, which seems to be correlated with the disappearance of the superconductivity. In the superconducting phase, the positive temperature dependence can be seen over the measured temperature range up to 400 K below x=0.075. For x>0.075, there appears a broad peak in the temperature dependence. The broad peak shifts to the lower temperature with increasing doping and disappears at around x=0.12 where Tc rapidly decreases and the superconductivity fades out. Judging from the decrease of the peak temperature, it is likely that the samples with x<0.075 should have also a peak above 400 K. Actually, the presence of the broad peak was observed for low x by Johnston et al [16].

In the nonsuperconducting metal phase, the susceptibility shows Curie-like increase with lowering temperature. Since the oxygen vacancy can be easily introduced in the highly doped region, we have to take account of the defects created by the oxygen vacancy as the origin of the Curie-like behavior. We can rule out the possibility of the oxygen vacancy in the following way. In Fig. 6 is shown the temperature dependence of the magnetic susceptibility of the same sample with x=0.125 after the various heat treatments. As seen from the figure, after the heat treatment under the reduced atmosphere which reduces the oxygen content in the sample, the Curie-like behavior disappears and the temperature coefficient becomes positive like the samples with lower x. Obviously, the oxygen vacancy is not the origin of the Curie-like behavior. The oxygen vacancy seems to act as a donor and to reduce x effectively.

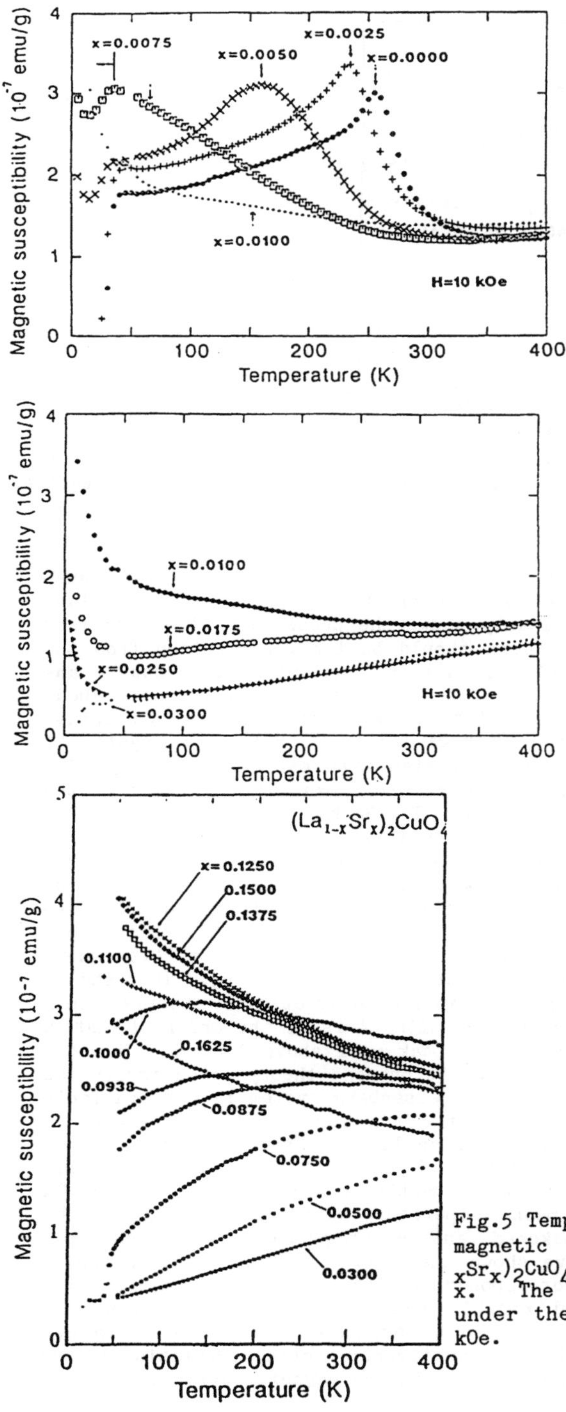

Fig.5 Temperature dependence of the magnetic susceptibility for $(La_{1-x}Sr_x)_2CuO_4$ with various Sr content x. The measurement was performed under the constant field of H=10 kOe.

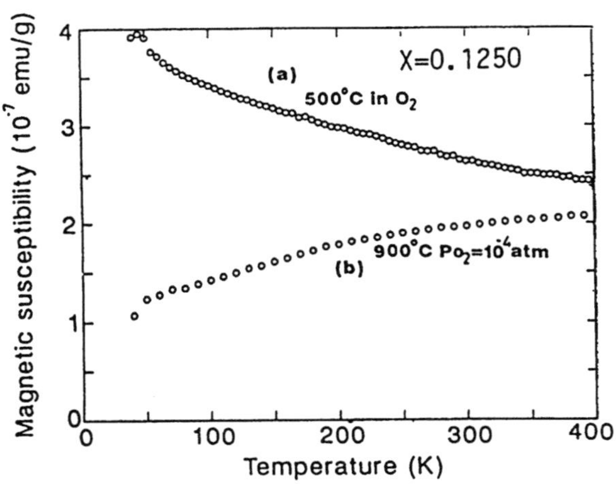

Fig.6 Temperature dependence of the megnetic susceptibility for the sample with x=0.125. (a) as prepared, (b) annealed under the oxygen partial pressure of 10^{-4} atm at 800 C°.

We ascribe the positive temperature dependence of the magnetic susceptibility primarily to the Cu spin contribution. The extensive inelastic neutron experiments have clearly shown the presence of the localized moments on Cu sites, so that there should be the contribution from the localized moments. These localized Cu spins instantaneously couple antiferromagnetically over the scale of the coherence length ξ . How about the contribution to the susceptibility from these localized Cu spins? The mother compound La_2CuO_4 can be regarded as 2D Heisenberg system of S=1/2. For the 2D S=1/2 Heisenberg system with exchange coupling J<0, the calculation based on the high temperature expansion predicts that the susceptibility shows a peak at around $T=1.8J/k_B$ like 1D system [17]. This has been actually observed experimentally in some Cu quadratic layer systems. In the case of La_2CuO_4, the susceptibility in our measured temperature range below 400 K is now understood to be dominated by the contribution from the very small moment perpendicular to plane originating from the orthorhombic distortion [18]. In the tetragonal phase above 500 K, Johnston et al. have demonstrated that the susceptibility shows positive temperature coefficient at least up to 1000 K [16]. This should come from the contribution from 2D antiferromagnetically coupled spin system, since the in-plane exchange coupling J is of the order of 1000 K thus the susceptibility should have a peak well above 1000 K. Now the problem is the effect of the doping. At present theory which treats the susceptibility of the doped 2D S=1/2 Heisenberg system is not available yet. Nevertheless, it is likely that the behavior is not so different at least when dopant concentration is not so large. From these reasons, we believe that the positive temperature dependence observed over a wide composition range originates from the antiferromagnetically coupled Cu spins. The decrease of the peak temperature with higher doping can be interpreted in terms of weakened antiferromagnetic correlation upon doping. Birgeneau et al. have shown that the correlation length varies as $\xi=3.9/\sqrt{2x}$ (Å) which corresponds to the average distance between doped holes [15]. At around x=0.125 where the broad peak disappears, is as small as 7.8 Å which is about the twice of Cu-Cu distance. Therefore, it is reasonable that the antiferromagnetic correlation is too weak to overcome the Curie-like increase in the magnetic susceptibility.

We imagine that the modification of the electronic states suggested from the Hall coefficient might play a role in this drastic change of the magnetic susceptibility which is suggestive of the weakened antiferromagnetic correlation. Certainly, there is a paramagnetic contribution which might be ascribed to the Pauli paramagnetism in the observed magnetic susceptibility in the nonsuperconducting phase after subtracting the Curie contribution.

From the view point of the spin mechanism of the superconductivity, the observed correlation between the superconductivity and the magnetism is favourable. In these theories, the coexistence of the mobile carriers and the antiferromagnetic correlation is essential for the high-Tc superconductivity. For example, in the magnetic frustration model proposed by Birgeneau et al. [19], Tc is determined by two competing effects, the increase of the density of states and the destruction of the antiferromagnetic order. Thus in the highly doped region, Tc decreases with doping because of the dominant contribution of the latter effect.

V SUMMARY

We have shown how the physical properties change with doping. Particularly, we have found the change in both charge and spin states in the highly doped region. The characteristic features are summarized for various composition regions. Theoretical models for the high-Tc superconductivity have to explain these changes upon doping.

(i) $0 < x < 0.01$
The system is an insulator with the semiconducting temperature dependence of the resistivity and the Hall coefficient. The long range antiferromagnetic order is formed which is rapidly suppressed with doping.

(ii) $0.01 < x < 0.03$
Metallic temperature dependence of the resistivity appears with the temperature independent Hall coefficient like a typical metal. At the low temperatures, however, localization plays a dominant role in the conduction and thus makes the system insulating. After the disappearance of the long range antiferromagnetic order, the susceptibility has a positive temperature coefficient which we ascribe to the presence of the strong antiferromagnetic correlation between Cu spins.

(iii) $0.03 < x < 0.12$
The superconductivity is observed. The positive Hall coefficient decreases more rapidly than that expected from the Mott-Hubbard picture, indicating the breakdown of the Mott-Hubbard picture. On the appearance of the superconductivity, a remarkable negative temperature coefficient of the Hall coefficient is observed, which is common among high-Tc oxides. The antiferromagnetic correlation is still strong enough to preserve the positive temperature coefficient of the magnetic susceptibility, although it was weakened rapidly with doping.

(iv) $0.12 < x$
In this region the compound is metallic but not superconducting. The Hall coefficient decreases to the order of 10^{-4} cm^3/C, changing its sign from positive to negative. In the framework of the Fermi liquid description, the negative sign can be understood rather easily than the $1/x$ dependence in the low x region. Possibly because the antiferromagnetic correlation length is as short as of the order of the shortest Cu-

Cu distance, the magnetic susceptibility does not exhibit the negative temperature coefficient. Instead, the temperature dependence is characterized by the Curie-like behavior.

IV. ACKNOWLEDGMENT

The authors would like to thank M.Uota, T.Ido, S.Ishibashi for their help in the experiments. This work was supported by a grant-in-Aid of Scientific Research of the Special Project Research on the High Temperature oxide Superconductors from the Ministry of Education, Science and Culture.

References

1. K.Kitazawa, H.Takagi, K.Kishio, T.Hasegawa, S.Uchida, S.Tajima, S.Tanaka and K.Fueki, for a comprehensive review on the experimental works, see also other articles in this volume.
2. J.B. Torrance, Y. Tokura, A.I. Nazzal, A. Bezinge, T.C. Huang and S.S.P. Parkin: Phys. Rev. Lett. 61, 1127 (1988).
3. H. Takagi et al. to be submitted.
4. A.I. Nazzal et al.: physica 135-156C, 1367 (1988).
5. J. Gopalakrishnan, G. Colsmann and B. Reuter: J. Solid State Chem. 22, 145 (1977).
6. A.R. Moodenbaugh, Y. Xu, M. Suenaga, T.J. Folkerts and R.N. Shelton: Phys. Rev. B38, 4596 (1988).
7. S. Uchida, H. Takagi, H. Ishii, H. Eisaki, T. Yabe, S. Tajima and S. Tanaka: Jpn. J. Appl. Phys. 26, 440 (1987).
8. N.P. Ong, Z.Z. Wang, J. Clayhold, J.M. Tarascon, L.H. Green and W.R, Mackinon: Phys. Rev. B35, 8807 (198 7).
9. H. Fukuyama and Y. Hasegawa: Physica 148B, 204 (1987).
10. S.W. Tozer, A.W. Kleinsasser, T. Penny, D. Kaiser and F. Holtzberg: Phys. Rev. Lett. 59, 1768 (1987).
11. Y. Iye, T. Tamegai, H. Takeya and H. Takei: Jpn. J. Appl. Phys. 26, L1057 (1987)
12. M. Suzuki: Phys. Rev. B, Submitted.
13. Y. Iye, T. Tamegai, T. Sakakibara, T. Goto, N. Miura, H. Takeya and H. Takagi: Physica C153-155, 26 (1988)
14. T. Fujita, Y. Aoki, Y. Maeno, J. Sakurai, H. Fukuba and H. Fujii: Jpn.J.Appl.Phys. 26, L368 (1987).
15. R.J. Birgeneau, D.R. Gabbe, H.P. Jenssen, M.A. Kastner, P.J. Picone, T.R. Thurston, G. Shirane, Y. Endoh, M. Satoh, K. Yamada, Y. Hidaka, M. Oda, Y. Enomoto, M. Suzuki and T. Murakami: Phys Rev. B37, 7443 (1988).
16. D.C. Johnston, S.K. Sinha, A.J. Jacobson and J.M. Newsan: Physica C153-155, 572 (1988).
17. H.A. Algra, L.J. de Jongh and R.L. Carlin: Physica 93B, 24 (1978).
18. R.J. Birgeneau, M.A. Kastner, A. Aharony, G. Shirane and Y. Endoh: Physica C153-155, 515 (1988).
19. R.J. Birgeneau, M.A. Kastner and A. Aharony: Preprint.

Anisotropic Transport in Y-Ba-Cu-O and Bi-Sr-Ca-Cu-O

A. Zettl, A. Behrooz, G. Briceno, W.N. Creager, M.F. Crommie, S. Hoen, and P. Pinsukanjana

Department of Physics, University of California at Berkeley, and Materials and Chemical Sciences Division, Lawrence Berkeley Laboratory, Berkeley, California, CA 94720, USA

The anisotropic normal state transport properties of the superconducting oxides Y-Ba-Cu-O and Bi-Sr-Ca-Cu-O are investigated by dc resistivity, thermoelectric power, high frequency conductivity, and uniaxial stress effects in single crystals. We also explore the superconducting state by measurements of T_c under c-axis stress, and oxygen isotope substitution. Energy gap structure is investigated by break junction single-crystal tunneling.

1. INTRODUCTION

The new classes of oxide superconductors based on Cu-O_2 sheets have unusual superconducting and normal state properties. The sheet structure gives rise to quasi-two-dimensional electronic structure with large anisotropy in the normal state. The low dimensionality has been exploited in numerous models of high-T_c superconductivity. The unusually high transition temperatures, together with the observed reduced isotope effect, suggest a new electron pairing mechanism. A good understanding of the superconducting properties of a material necessitates a good understanding of the normal state properties.

We here explore anisotropy in the normal states, and to a lesser degree in the superconducting states, of the oxide superconductors Y-Ba-Cu-O and Bi-Sr-Ca-Cu-O. We find that the electronic conduction in the normal state is not well described by conventional mechanisms, in particular in the direction perpendicular to the Cu-O_2 planes. Our findings place restrictions on the type of transport possible in the normal state, and indirectly on the superconductivity mechanism.

2. ANISOTROPIC TRANSPORT IN $YBa_2Cu_3O_{7-\delta}$

2.1 Resistivity and Thermoelectric Power

As first demonstrated by TOZER et al[1], the resistivity tensor in $YBa_2Cu_3O_7$ single crystals suggests a substantial temperature-dependent anisotropy. In fact, for many crystals, the c-axis resistivity appears "semiconductor-like" while the a-b plane resistivity appears metallic. For some crystals, on the other hand, the "upturn" in the c-axis resistivity starts only very close to T_c[2]. The difference may be a combination of impurities (including oxygen vacancy) and degree of twinning

Springer Series in Materials Science, Vol. 11
Mechanisms of High Temperature Superconductivity
Editors: H. Kamimura and A. Oshiyama

in the a-b plane. For any crystal, the upturn is easily enhanced by depleting the oxygen content, i.e. increasing δ.

For a typical high purity crystal with full oxygen content, the c-axis resistivity at room temperature is of order 10-20 mΩcm. In the tight-binding approximation, the electronic mean free path is given by[3]

$$\Lambda = v_F\tau = h\sigma/4ne^2a \qquad (1)$$

where n is the carrier density, σ is the conductivity in the direction in question, and a is the lattice constant in the direction in question. The carrier concentration in $YBa_2Cu_3O_7$ is of order $10^{22}/cm^3$. Eq. (1) then gives for the c-axis mean free path $\Lambda \approx 10^{-5}$Å, clearly an unphysical value. This immediately suggests that band transport is inappropriate to $YBa_2Cu_3O_7$, at least along the c-axis. In sections below we test various predictions of band theory for the c-axis conduction, and consistently find discrepancies.

Fig. 1 shows[2] the normalized c-axis resistivity for $YBa_2Cu_3O_7$ corresponding to three different oxygen concentrations, δ ≈ 0, 0.5, and 0.7. We respectively label these samples pristine oxygenated (PO), oxygen deficient (OD), and very oxygen deficient (VOD). The VOD state is not superconducting at any temperature. It is tempting to describe the temperature dependence as "semiconductor like". Fig. 2a shows the data of Fig. 1 plotted as log(σ) vs 1/T. Only at high temperatures does the conductance appear thermally activated, with corresponding activation energies E_g= 2.4meV, 41meV, and 450meV for PO, OD, and VOD samples, respectively. At low temperatures, the data curve away from exponential behavior in a manner similar to low-dimensional disordered metals, where carriers become localized with decreasing temperature. In a regime of strong localization, one expects for a three dimensional system a temperature dependence

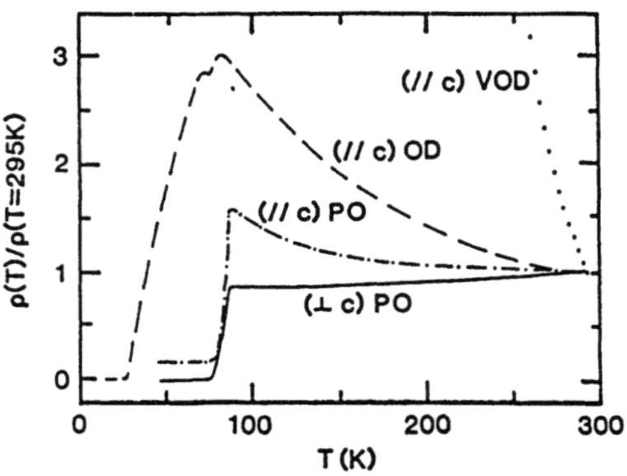

Figure 1. Normalized c-axis resistivity of $YBa_2Cu_3O_y$ for different oxygen contents. The a-b plane resistivity is also shown for a fully oxygenated sample.

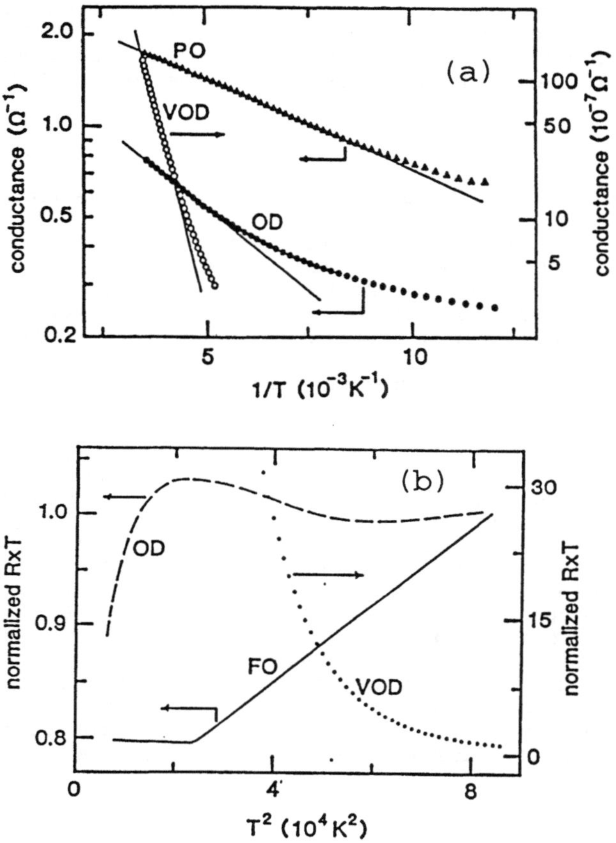

Figure 2. a) Fits to activated "semiconductor-like" conductivity for c-axis conductance in $YBa_2Cu_3O_y$. Data for three different oxygen contents are shown. The data are activated only at high temperatures, where the activation energy increases with decreasing oxygen content. b) Fits to the Anderson-Zou hole soliton c-axis tunneling formula, Eq. (4). The $YBa_2Cu_3O_y$ data fit the formula only for full oxygen content and only at high temperatures.

$$\sigma = \sigma_o \exp[-(T_o/T)^{1/4}], \tag{2}$$

while in the regime of weak localization one has

$$\sigma = \sigma_o + 2\pi e^2 T^{3/2}/h\pi^3\gamma. \tag{3}$$

The data of Fig. 1 fits neither Eq. (2) or (3), nor their analogs for two or one dimensional systems. We also note that we have observed no unusual magnetoresistance effects in the c-axis conduction, again giving evidence against standard localization behavior.

ANDERSON and ZOU[4] have suggested that if the normal state of $YBa_2Cu_3O_7$ is described by a resonating valence bond (RVB) state, the c-axis conduction is dominated by tunneling between planes of hole solitons with an expected temperature 1/T dependence, i.e.

$$\rho_c = A/T + BT \tag{4}$$

where the term linear in T accounts for experimental "contamination" from
a-b plane conduction. To test Eq. (4), one plots $\rho_c T$ vs T^2, as was originally done by HAGEN et al[5]. Fig. 2b shows our c-axis resistivity plotted in this way to test Eq. (4). It is apparent that a reasonable (linear) fit occurs only for high oxygen content, and even then only over a restricted temperature range.

A general empirical expression has been suggested[6] for the c-axis resistivity in $YBa_2Cu_3O_7$,

$$\rho_c \sim T^{\alpha} \exp[\varepsilon_g/k_B T], \tag{5}$$

where α is a constant between 0.5 and 1.0 and ε_g represents a reduced or effective gap for activated charge transport. Eq. (5) appears to fit well the c-axis conductivity for different $YBa_2Cu_2O_7$ crystals with $\varepsilon_g \approx 25$meV. This is shown in Fig. 3, where $\ln[\rho_c/T^{\alpha}]$ is plotted versus 1/T, using data from two different research groups. One physical interpretation[6] of Eq. (5) is that the exponential term arises from activated behavior similar to the conductivity in amorphous semiconductors, while the T^{α} term comes from the temperature dependence of the mobility, and hence the scattering time τ. Possible sources of the temperature dependence of the mobility are the phonon occupation number and the average carrier velocity.

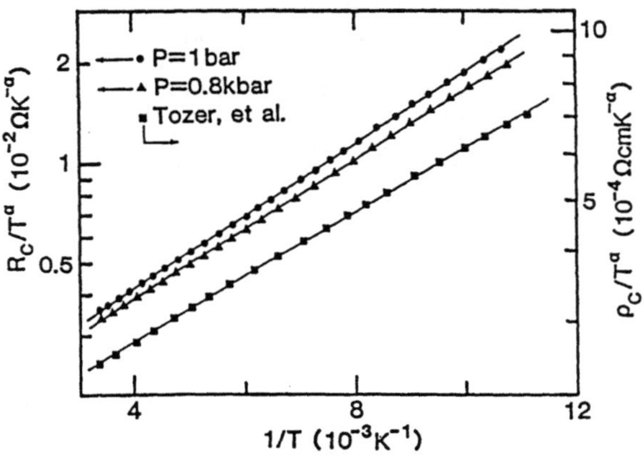

Figure 3. Fits of c-axis conductivity in $YBa_2Cu_3O_7$ to Eq. (5), with the specimen at ambient pressure and under c-axis pressure. Fits to data of ref. 1 are also shown.

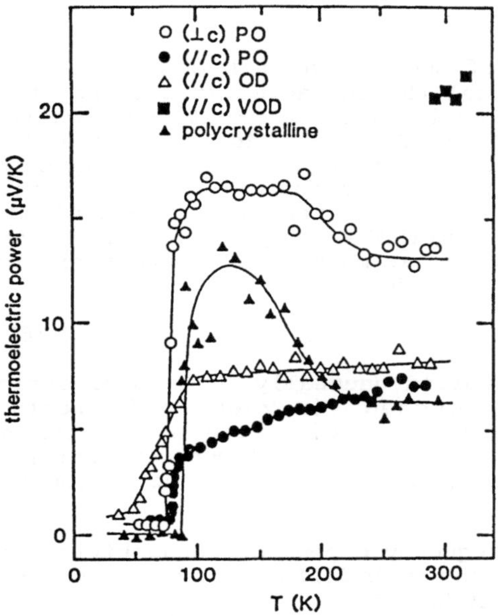

Figure 4. Thermoelectric power for $YBa_2Cu_3O_y$ for different directions in the crystal and for different oxygen contents. The polycrystalline result for full oxygen content is also shown.

Another transport coefficient complementary to the resistivity is the thermoelectric power (TEP). For a metal one expects a TEP linear in temperature, neglecting phonon drag effects. For a semiconductor with gap E_g, the TEP is proportional to E_g/k_BT. In the superconducting state, the superconducting electrons to first order short out any thermally induced EMF, hence the TEP is zero. Fig. 4 shows[6] the TEP for different crystal directions and oxygen contents of $YBa_2Cu_3O_{7-\delta}$. Also shown is the TEP for a polycrystalline sample. The a-b plane TEP is not linear in T (and is similar to the polycrystalline result), in contrast to what might be expected from the a-b plane metallic resistivity. The c-axis TEP is linear in T, and hence not of the semiconductor form.

ALLEN et al[7] have investigated the phonon-induced resistivity $\rho_{\alpha\beta}$, Hall coefficient $R^H{}_{\alpha\beta\gamma}$, and TEP $S_{\alpha\beta}$ for $YBa_2Cu_3O_y$ based on band structure calculations using a variational solution of the Boltzman transport equations. Some of the data in Fig. 4 are consistent with these predictions, but discrepencies exist. For example, the measured TEP is positive (holelike) both in the a-b plane and along the c-axis, while ALLEN et al predict that S_{xx} and S_{yy} will be negative, and the sign of S_{zz} is dependent on the choice of $\tau(\varepsilon)$ (x and y are in the a-b plane, z is parallel to the c-axis).

2.2 High-frequency ac Conductivity

The electrical conductivity of metals and semiconductors is frequency dependent. For a metal the characteristic energy for frequency dependent conductivity is $\omega\tau \sim 1$ (or $\omega \sim 10^{14}$Hz), while for a semiconductor it is $h\omega/2\pi \sim 2E_g$. From Eq. (5) where the effective "gap" energy is ~25meV, we might expect frequency dependent conductivity near 6×10^{12} Hz for the c-axis conduction. On the other hand, several non-band transport mechanisms (including localization and variable range hopping) give frequency dependent conductivities at much lower characteristic energies, often with power law dependences such as

$$\sigma(\omega) \sim \omega^s. \qquad (6)$$

TESTARDI et al [8] have reported unusually large dielectric constants in thermally quenched (oxygen deficient) polycrystalline $YBa_2Cu_3O_y$ at very low (audio) frequencies, while REAGOR et al [9] report a strong frequency dependent conductivity in single crystal (nonsuperconducting) Eu_2CuO_4 in the microwave regime.

We have investigated the frequency dependent conductivity of $YBa_2Cu_3O_y$ single crystal specimens with different oxygen content in the frequency range 5Hz to 1GHz. Figs. 5a,b show the dc and ac (1GHz) conductivities from room temperature to below T_c. For neither the a-b plane direction nor the c-axis direction do we observe any unusual frequency dependences. This is true regardless of the oxygen content, and suggests that the effective activation energy associated with Eq. (5) is a meaningful energy scale. It also suggests that the unusual dielectric (capacitance) effect observed by TESTARDI et al is not an intrinsic effect, but is most probably due to capacitances formed at grain boundaries in polycrystalline specimens.

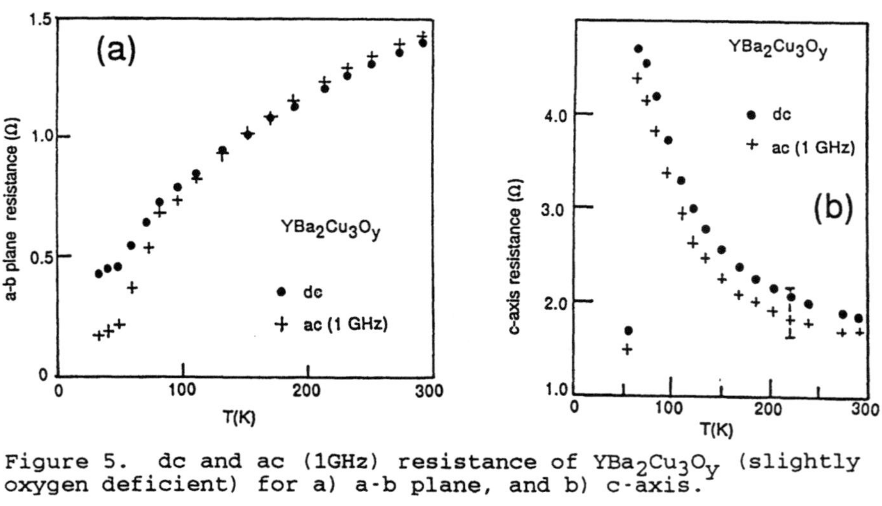

Figure 5. dc and ac (1GHz) resistance of $YBa_2Cu_3O_y$ (slightly oxygen deficient) for a) a-b plane, and b) c-axis.

2.3 Role of Interplanar Coupling and the Superconductivity Mechanism

The above results imply that conventional band transport does not describe the c-axis conduction in $YBa_2Cu_3O_7$. The important question thus arises, is the superconductivity at lower temperature driven by a novel mechanism? One of the most obvious concerns is the role of the $Cu\text{-}O_2$ planes, i.e. the dimensionality of the system. A number of models[10] take advantage of the special properties of a two dimensional system (such as density of states anomalies) to account for the high T_c's (and other unusual features such as the reduced isotope effect).

We have investigated the role of interplane coupling in $YBa_2Cu_3O_7$ by directly changing the interplane separation (through externally applied uniaxial stress) and measuring the affect on the resistivity tensor and T_c[6]. Fig. 6 shows the experimental configuration. c-axis stress (or pressure) is applied to the single crystal using a steel-sample-steel sandwich. An epoxy film electrically insulates the a-b plane surfaces of the crystal from the steel discs. The a-b plane and c-axis resistivities are determined by four-probe wire contact methods. For c-axis pressures up to 1kbar, there is no observed effect on the magnitude or temperature dependence of the a-b plane resistivity. The c-axis resistivity, on the other hand, is dramatically changed, as demonstrated in Fig. 7. The general trend is that increasing c-axis pressure (i.e. decreasing interplane separation) leads to a decrease in ρ_c. The inset to Fig. 7 shows that the functional form of the $\rho_c(T)$ curves is altered by pressure: increasing pressure tends to make the

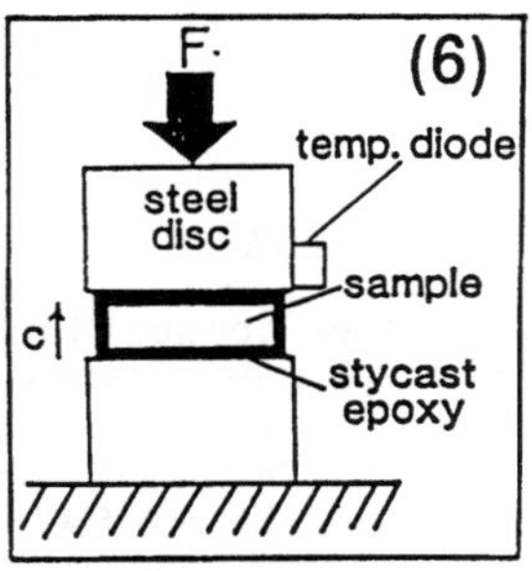

Figure 6

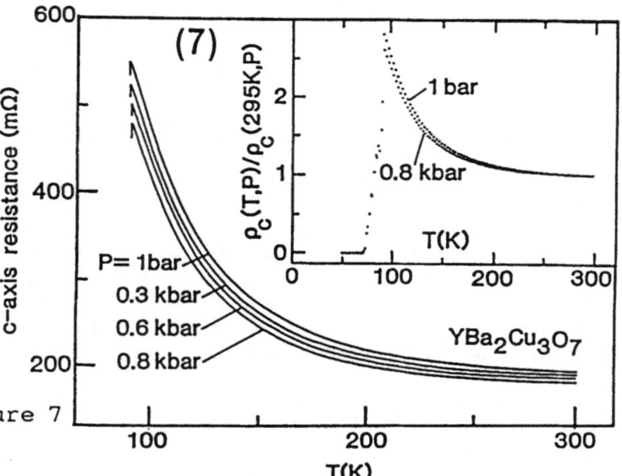

Figure 7

Figure 6. Pressure cell for applying uniaxial stress to single crystals.

Figure 7. c-axis resistance vs T for selected c-axis pressures in $YBa_2Cu_3O_7$. The inset shows normalized resistivity data for the two extreme pressures.

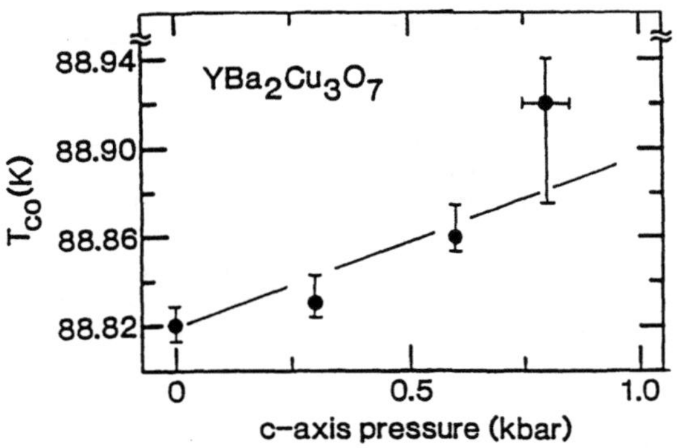

Figure 8. Superconducting onset temperature versus c-axis pressure in $YBa_2Cu_3O_7$. Decreasing interplanar spacing increases T_c.

c-axis resistivity more metallic. The most direct interpretation of this effect is that c-axis stress increases the matrix element $t_\perp$ for interplanar charge transfer. From considerations of non-band anisotropic conduction[11], one may crudely approximate

$$\sigma_c/\sigma_{a-b} \approx (t_\perp/t_{||})^2. \tag{7}$$

The data of Fig. 7 indicate $d\ln t_\perp/dP = +0.08$/kbar at 95K.

How does T_c depend (if at all) on $t_\perp$? Fig. 8 shows that T_c increases smoothly as the c-axis stress is increased. In other words, T_c increases as the electronic coupling between the planes is increased. With the measured $dT_c/dP = 0.08$K/kbar, we find $dT_c/d\ln t_\perp = 1$. This result rules out strictly two dimensional superconductivity mechanisms.

2.4 Filamentary Superconductivity and the Isotope Effect

In a homogeneous single-crystal superconductor, the transition temperature is usually well-defined. At T_c the resistance abruptly drops to zero and the specimen becomes (in low applied field) a perfect diamagnet. Sample inhomogeneities can lead to a smeared transition. Conductivity measurements always measure the path of least resistance, and hence are not always a good indicator of "bulk superconductivity"; for this the dc magnetization is in general a better probe.

Recent measurements[12] have indicated that in freshly prepared high-T_c polycrystalline specimens, T_c determined resistively is as much as 2K higher than T_c determined magnetically. The later measurement determines the bulk transition termperature for the superconducting grains. The discrepancy is evidence for filamentary superconductivity in polycrystalline samples. The fact that the filamentary T_c is higher than the bulk T_c suggests that a small part of the sample (perhaps near the grain boundaries) is actually a different material, with a volume fraction too small for x-ray detection. Alternatively, the potentially novel superconductivity mechanism may be sensitive to sample geometry and boundary effects. In this case, the filamentary superconductivity occurs in pure $YBa_2Cu_3O_7$ but only where it is truncated geometrically. We have found[12] that filamentary superconductivity in $YBa_2Cu_3O_7$ is time dependent and disappears after several months. This makes the first suggestion more probable, i.e. that the filamentary superconductivity corresponds to an unstable phase which degrades with a time constant of several months.

We have examined if the superconductivity mechanism for filamentary superconductivity is the same as for the bulk material. Isotope substitutions on the oxygen sites was performed using polycrystalline specimens of $YBa_2Cu_3O_7$. With up to 95% of the ^{16}O replaced with the ^{18}O isotope, T_c is found to be decreased for both filamentary and bulk superconductivity. The relative decrease in T_c for filamentary superconductivity was slightly but not significantly greater. Assuming the relation

$$T_c \sim M^{-\alpha} \tag{8}$$

where M is the oxygen mass, we find[12] for filamentary superconductivity $\alpha=0.028\pm.003$ and for bulk superconductivity $\alpha=0.019\pm.004$. These values are the extrapolated values, appropriate to 100% isotopic substitution. They are much smaller than the standard BCS prediction $\alpha=0.5$, and are inconsistent with three dimensional phonon-mediated pairing in general[12,13]. It seems appropriate to at least consider three dimensional theory in light of the sensitivity of T_c to interplanar coupling and the relatively high transition temperatures of the isotropic oxide superconductors Ba-K-Bi-O[14]. The calculation has not been performed assuming a dimensionally restricted phonon interaction.

3. ANISOTROPIC TRANSPORT IN Bi-Sr-Ca-Cu-O

We have investigated two distinct Bi-Sr-Ca-Cu-O single crystal structures. The first is $Bi_2Sr_2CaCu_2O_8$, a relatively well-known compound with $T_c \approx 88K$. This material shares common features with $YBa_2Cu_3O_7$. The usual crystals are mica-like platelets with the c-axis perpendicular to the untwinned plate surface. It is relatively easy to cleave the crystals in the a-b plane. We have also synthesized a new Bi-Sr-Ca-Cu-O structure[15]. The crystals grow with a long thin needle morphology, with the c-axis along the needle axis. The structure at room temperature, determined from single crystal x-ray analysis, is orthorhombic, with unit cell dimensions a=13.12Å, b=11.44Å, c=74.69Å. The c-axis dimension is less certain because of an incommensurate superstructure present. The nominal composition of the needle-like crystals was determined from SEM analysis to be approximately $Bi_{0.1}Sr_{2.2}Ca_{1.1}Cu_{6.5}O_y$, i.e. compared to standard Bi-Sr-Ca-Cu-O compounds, this material is extremely Bi poor and Cu rich. T_c onset is approximately 90K. Magnetization studies indicate bulk superconductivity below T_c.

3.1 Resistivity and Thermoelectric Power

Fig. 9a shows the resistivity of single crystal $Bi_2Sr_2CaCu_2O_8$ using the standard Montgomery method. The resistances R_1 and R_2 have the usual meaning, and reflect (but do not equal) the c-axis and a-b plane resistivities, respectively. The resistivity tensor is highly anisotropic and again the c-axis conduction is not metallic. The c-axis resistance of a $Bi_{0.1}Sr_{2.2}Ca_{1.1}Cu_{6.5}O_y$ needle is shown in Fig. 9b. There is a dramatic upturn in the resistance with decreasing temperature above T_c. We have attempted to fit the c-axis conduction of the needle crystals to formulas described above for $YBa_2Cu_3O_7$. Figs. 10a-c show respectively fits to simple activated behavior (semiconductor-like), Eq. (4)

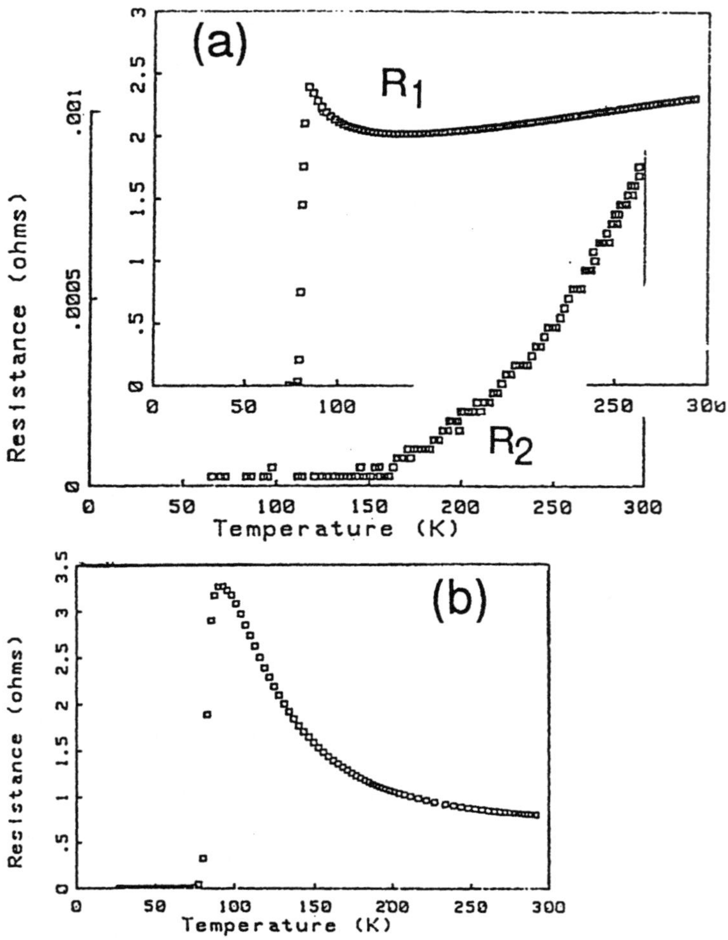

Figure 9. a) anisotropic resistances R_1 and R_2 from Montgomery method for $Bi_2Sr_2CaCu_2O_8$. b) c-axis resistance of $Bi_{0.1}Sr_{2.2}Ca_{1.1}Cu_{6.5}O_y$ needle crystals.

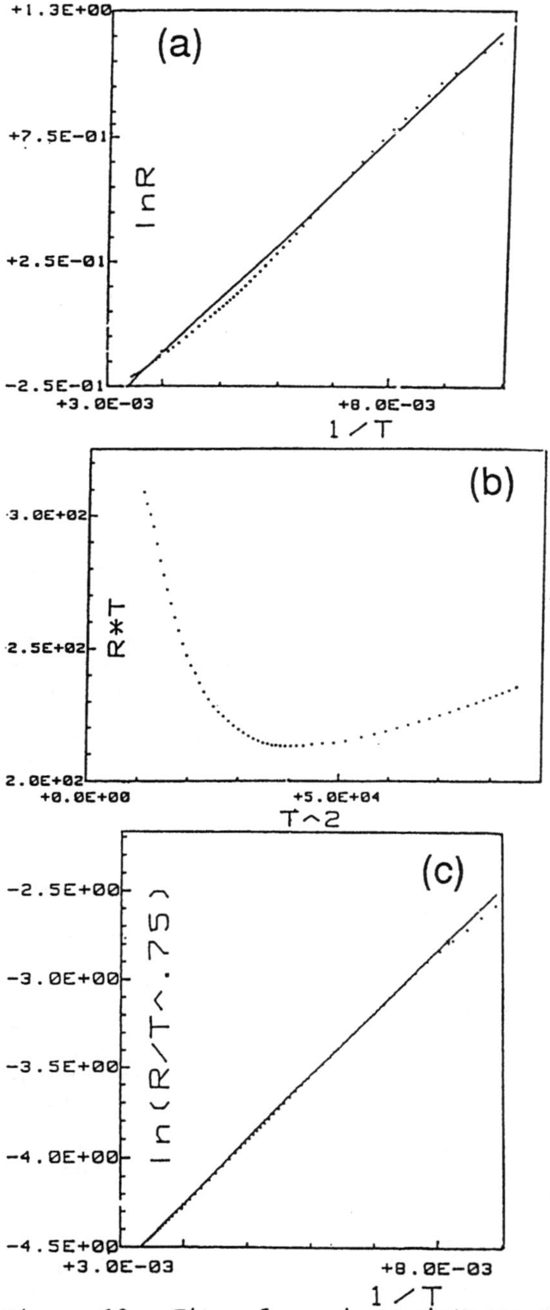

Figure 10. Fits of c-axis resistance of $Bi_{0.1}Sr_{2.2}Ca_{1.1}Cu_{6.5}O_y$ to a) activated semiconductor formula, b) RVB tunneling (Eq. 4), and c) empirical formula Eq. (5) with α=.75; ε_g=32meV.

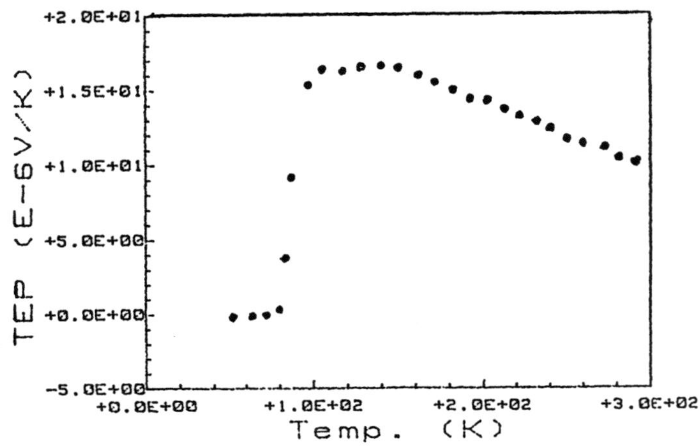

Figure 11. Thermoelectric power of $Bi_2Sr_2CaCu_2O_8$ in the a-b plane.

(appropriate to the RVB state), and the empirical expression Eq. (5). Only Eq. (5) provides an accurate fit, with α=0.75 and ε_g=32meV.

We have measured the thermoelectric power of $Bi_2Sr_2CaCu_2O_8$ single crystals in the a-b plane. Fig. 11 shows the TEP from room temperature to 50K. The rather unusual temperature dependence is very similar to that observed for the a-b plane of $YBa_2Cu_3O_7$. For $Bi_2Sr_2CaCu_2O_8$ the TEP is always positive above T_c, suggesting in the simplest interpretation positive charge carriers.

3.2 Tunneling Measurements

We have investigated the superconducting state in $Bi_2Sr_2CaCu_2O_8$ and $Bi_{0.1}Sr_{2.2}Ca_{1.1}Cu_{6.5}O_y$ single crystals by break junction tunneling measurements[16]. Both SIS and Josephson tunneling are observed. The break junctions are formed and the measurements are carried out at 4.2K. Fig. 12a shows typical and reproducible Josephson tunneling for $Bi_2Sr_2CaCu_2O_8$. From this plot we estimate an energy gap $2\Delta_0/e$ = 45mV, which leads to $2\Delta_0 = 5.9k_BT_c$. Similar values are extracted from SIS tunneling in the same material. This value corresponds to the gap in the a-b plane direction. Under certain conditions we find reproducible peak structure in dV/dI plots at regular voltage bias intervals, similar to that observed previously[17] in $YBa_2Cu_3O_7$ point contact tunnel junctions and interpreted in terms of the coulomb staircase.

We have also explored tunneling along the c-axis in $Bi_{0.1}Sr_{2.2}Ca_{1.1}Cu_{6.5}O_y$ single crystal needles. This is shown in Fig. 12b, where the dV/dI characteristics indicate an energy gap at $2\Delta_0/e$ = 37mV, or $2\Delta_0 = 4.8k_BT_c$. This is the gap in the c-axis direction for this material.

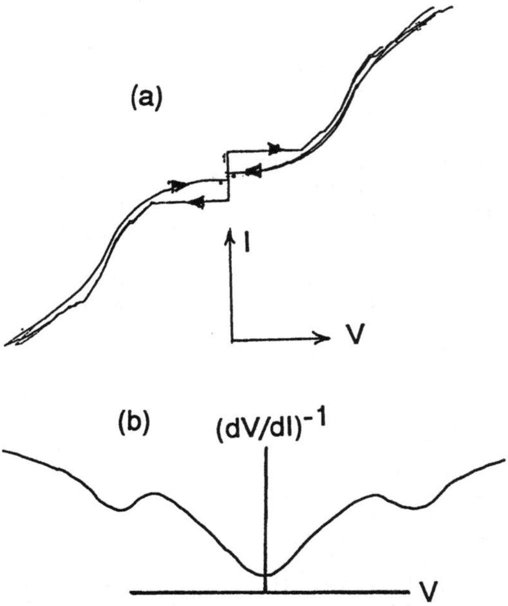

Figure 12. a) I-V characteristics of $Bi_2Sr_2CaCu_2O_8$ break junction, a-b plane, at 4.2K. Josephson tunneling is observed. b) dI/dV characteristics of $Bi_{0.1}Sr_{2.2}Ca_{1.1}Cu_{6.5}O_y$ break junction, c-axis, at 4.2K.

4. CONCLUSION

The normal state transport properties of Y-based and Bi-based superconducting oxides are unusual and suggestive of non-band transport mechanisms. The superconductivity does not appear to be confined to the copper-oxygen planes, and hence cannot be considered a strictly two dimensional effect. However, we expect the large anisotropy in the normal state to be reflected in anisotropic gap structure in the superconducting state. Reliable tunneling measurements in various crystal directions may resolve this interesting question.

We thank the following individuals for helpful interactions: T.W. Barbee III, L.C. Bourne, M.L. Cohen, C. Kim, and A. Liu. The x-ray analysis of the needle crystals was kindly provided by A. Zalkin. This research was supported in part by NSF grants DMR 83-51678 and DMR 84-00041, and by the Director, Office of Energy Research, Office of Basic Energy Sciences, Materials Sciences Division of the U.S. Department of Energy under contract No. DE-AC03-76SF00098. S. Hoen acknowledges support from the Fannie and John Hertz Foundation.

REFERENCES

1. S.W. Tozer, A.W. Kleinsasser, T. Penny, D. Kaiser, and F. Holtzberg, Phys. Rev. Lett. 59, 1768 (1987)
2. M.F. Crommie, A. Zettl, T.W. Barbee III, and M.L. Cohen, Phys. Rev. B37, 9734 (1988)
3. A.J. Heeger, in *Highly Conducting One Dimensional Solids*, ed. I.T. Devreese (Plenum, New York, 1979) p. 79
4. P.W. Anderson and Z. Zou, Phys. Rev. Lett. 60, 132 (1988)
5. S.J. Hagen, T.W. Jing, Z.Z. Wang, J. Horvath, and N.P. Ong, Phys. Rev. B37, 7928 (1988)
6. M.F. Crommie, A.Y. Liu, A. Zettl, M.L. Cohen, P. Parilla, M.F. Hundley, W.N. Creager, S. Hoen, and M.S. Sherwin, (to be published)
7. P.B. Allen, W.E. Pickett, and H. Krakauer, Phys. Rev. B37, 7482 (1988)
8. L.R. Testardi, W.G. Moulton, H. Mathias, H.K. Ng, and C.M. Rey, Phys. Rev. B37, 2324 (1988)
9. D.W. Reagor, A. Migliori, Z. Fisk, R.D. Taylor, V. Kotsubo, K.A. Martin, and R.R. Ryan, (preprint)
10. J. Labbe and J. Bok, Europhys. Lett. 3, 1225 (1987); V.Z. Kresin, Phys. Rev. B35, 8716 (1987)
11. G. Soda, D. Jerome, M. Weger, S. Alozon, J. Gallice, H. Robert, J.M. Fabre, and L. Giral, J. Physique 38, 931 (1977)
12. S. Hoen, W.N. Creager, L.C. Bourne, M.f. Crommie, T.W. Barbee III, M.L. Cohen, A. Zettl, L. Bernardez, and J. Kinney, (to be published)
13. T.W. Barbee III, M.L. Cohen, L.C. Bourne, and A. Zettl. J. Phys. C (in pres)
14. R.J. Cava, B. Batlogg, J.J. Krajewski, R. Farrow, L.W. Rupp Jr., A.E. White, K. Short, W.F. Peck, and T. Kometani, Nature 332, 814 (1988)
15. P. Pinsukanjana, M.F. Crommie, S. Hoen, and A. Zettl (to be published)
16. G. Briceno, A. Behrooz, and A. Zettl (to be published)
17. P.J.M. van Bentum, R.T.M. Smokers, and H. van Kempen, Phys. Rev. Lett. 60, 2543 (1988)

Transport Studies on High T_c Oxides

Y. Iye

The Institute for Solid State Physics, The University of Tokyo, Roppongi, Minato-ku, Tokyo 106, Japan

1. INTRODUCTION

Despite the enormous increase of our knowledge due to the recent worldwide enthusiasm for high T_c research[1], it is fair to say that we understand regrettably poorly the physics underlying the remarkable phonomenon. One of the morals clearly emerged from the efforts is that we really have to accumulate reliable data and know the normal state electronic structures before talking about the mechanism of the high T_c. We have witnessed many cases in which what appeared to be key experimental results, on which a certain class of theoretical models put their bases, were seriously questioned by more careful later studies. In this paper we present some of our recent experimental results on the transport properties of the high T_c copper oxides and related materials. Topics to be covered are (1)anisotropic superconducting and normal transport properties of single crystal $YBa_2Cu_3O_{7-\delta}$, (2)metal-insulator transition, superconductivity and magnetism in $Bi_2Sr_2(Ca_{1-x}Y_x)Cu_2O_{8+y}$, and (3)transport studies on some high T_c related cuprate materials.

2. ANISOTROPIC NORMAL TRANSPORT PROPERTIES OF $YBa_2Cu_3O_{7-\delta}$

Early experiments on high T_c materials were exclusively done on sintered polycrystalline samples. As a result of intensive effort towards improvement of crystal growth technique, increasing number of experimental data on single crystals became available. An empirical rule, which emerged from the studies of resistivity, Hall effect and thermoelectric power on single crystals, is that the transport coefficients measured in sintered polycrystalline samples basically reflect those in the basal plane of single crystals.

2.1 Resistivity

Tozer et al.[2] and Murata et al.[3], in their early work, reported quite unusual temperature dependences of anisotropic resistivity of $YBa_2Cu_3O_{7-\delta}$ single crystals. They found that while the ab-plane resistivity ρ_{ab} showed metallic T-linear behavior similar to the polycrystalline data, the c-axis resistivity was semiconductor-like. Anderson and Zou[4] fitted the ρ_c data of Tozer et al.[2] to a functional form $\rho_c = AT + B/T$ and suggested that the intrinsic behavior of ρ_c would be $\sim 1/T$. The unusual temperature dependences, $\rho_{ab} \sim T$ and $\rho_c \sim 1/T$, were claimed by Anderson and Zou[4] to be neatly explained in the holon-spinon transport scheme based on the resonating valence bond (RVB) model originally proposed by Anderson[5].

The scinario for the electrical transport in the RVB model is as follows: The charge carriers in the two-dimensional CuO_2 layer are holons. Holons are scattered by spinons, with scattering rate proportional to the number of thermally excited spinons, which is linear in T. This leads to $\rho_{ab} \sim T$. Transport along

Springer Series in Materials Science, Vol. 11
Mechanisms of High Temperature Superconductivity
Editors: H. Kamimura and A. Oshiyama
Springer-Verlag Berlin Heidelberg © 1989

the c-axis involves interlayer tunneling of charge carriers. Since a holon is an entity only meaningful within each CuO_2 layer, it has to temporarily merge with a spinon to form a real electron before tunneling to an adjacent layer. The tunneling rate is determined by the probability for a hole to find its mate, and is proportional to the number of thermally excited spinons. This mean that the conductivity along the c-axis is proportional to T. In actual experiments it is conceivable that the nominal ρ_c data contain some ρ_{ab} component due to electrode misalignment. Taking this into account, the expected temperature dependence of the c-axis resistivity is $\rho_c = AT + B/T$. Hagen et al.[6] showed the validity of this formula by plotting their data for several single crystals in the form of $\rho_c T$ vs. T^2, which came out as straight lines. It should be reminded, however, that since this plot tends to put more weight on the high temperature side, it is not a very sensitive test for the 1/T term, especially when the low temperature data are truncated by the superconducting transition.

Our data[7] on the anisotropic resistivity came out qualitatively different from the earlier reports. We devised a simple method to attach a number of electrical leads to small single crystals.[8] The right part of Fig. 1

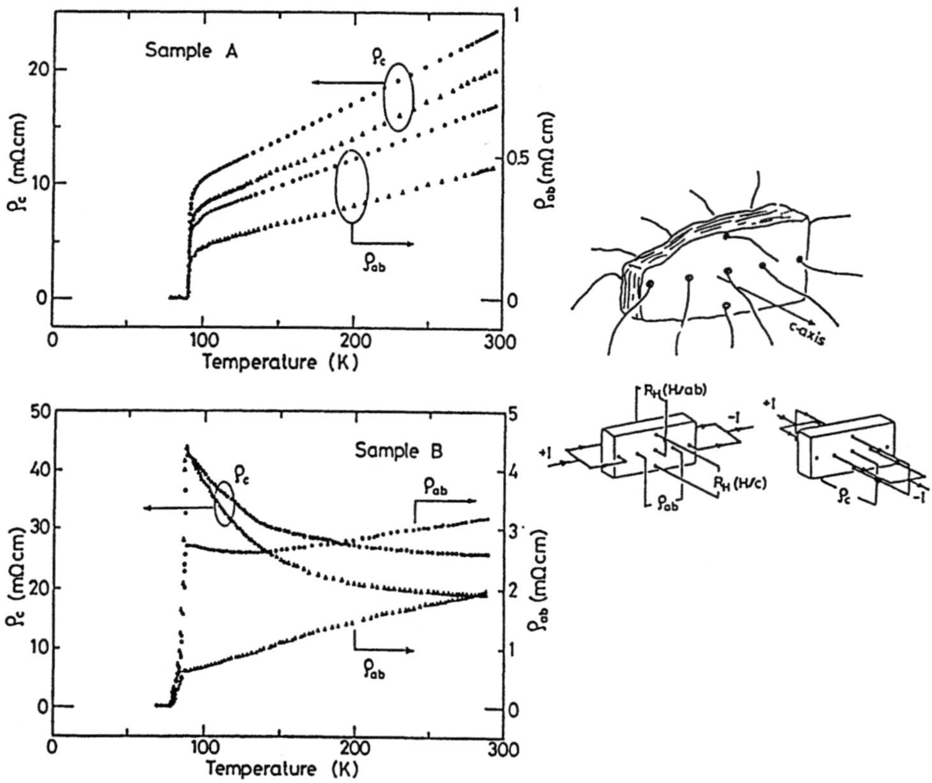

Fig. 1 Temperature dependence of ρ_{ab} and ρ_c for two single crystal samples of $YBa_2Cu_3O_{7-\delta}$ with different oxygen contents. Sample A is a fully oxidized sample, while Sample B is somewhat oxygen deficient. Note that the fully oxidized sample shows metallic ρ_{ab} and ρ_c. The right part of the figure illustrates the electrode configuration for the measurements of various transport coefficients.

illustrates an example of electrode arrangement which enables us to measure various transport coefficients using a same single crystal sample by the choice of current and voltage leads as shown in the figure. The left part of Fig. 1 shows our data on two representative samples. Sample A is a representative of our best single crystals characterized by a sharp superconducting transition ($\Delta T_c \sim 0.6$ K). Sample B is somewhat oxygen deficient as inferred from a suppressed and wider superconducting transition ($T_c^{mid}=85$ K, $\Delta T_c \sim 5$ K). The c-axis resistivity of Sample B shows a semiconductor-like temperature dependence and is qualitatively similar to those mentioned before. However, that of Sample A is metallic down to T_c. Both metallic behavior[9,10] and semiconductor-like behavior[11,12] of ρ_c are reported by other groups. Bozovic et al.[13] recently investigated optical anisotropy of $YBa_2Cu_3O_{7-\delta}$ and obtained results consistent with metalllic conduction both parallel and perpendicular to the layer plane.

The $\rho_c(T)$ issue is important because if the semiconductor-like ρ_c is intrinsic to $YBa_2Cu_3O_{7-\delta}$ and other layered high T_c cuprates, it may imply that an exotic conduction mechanism is at work in those systems. If the intrinsic ρ_c is metallic, on the other hand, we may be able to understand the electronic states in a more conventional picture, albeit greatly modified. We suspect that the semiconductor-like behavior of ρ_c may be due to oxygen deficient interior of the sample, and believe that the intrinsic ρ_c of a fully oxidized single crystal is metallic. A counterargument may be that the metallic ρ_c may arise from some sort of electrical shorting by the ρ_{ab} component due to imperfect crystallinity. Lacking a truly complete diagnostic method for the microscopic crystallinity, this problem is still open.

2.2 Hall Effect

Experiments using sintered polycrystalline samples of $YBa_2Cu_3O_{7-\delta}$ revealed the followings: (1)The Hall coefficient R_H is positive, i.e. hole-like. (2)R_H is very sensitive to oxygen stoichiometry. (3)R_H such an peculiar temperature dependence that the effective carrier density $1/eR_H$ is nearly proportional to T. These features have been also seen in single crystal samples in the configuration, H ∥ c-axis and I ∥ ab-plane.[7,12] The measurement of the Hall effect for H ∥ ab-plane was made first by Tozer et al.[2], and later by others[7,12]. Unlike the case of H ∥ c-axis, the Hall coefficient for H ∥ ab-plane is electron-like at room temperature. Figure 2 shows the temperature dependence of R_H of the same two samples as Fig.1, for the two principal field directions. It is rather surprising that in spite of the large difference in the absolute value, R_H shows similar temperature and field orientation dependences for the two samples. Forro et al.[12] reports a similar result with magnitude of R_H midway between the two data shown in Fig.2. In the language of the conventional Fermi liquid picture, the field orientation dependence of the Hall coefficient suggests a complicated Fermi surface topology.

The inserts of Fig.2 shows the linear T-dependence of $1/eR_H$ (H ∥ c-axis) noticed earlier in polycrystalline samples. An explanation for this linear T-dependence of $1/eR_H$ within a two-band model seems to require rather peculiar relationships among the carrier densities and mobilities[15,16]. Or does it indicate an exotic conduction mechanism in this system? Clayford et al.[17] propose an intimate relation between this $1/R_H \sim T$ behavior and the high T_c. The Hall coefficients of $(La_{1-x}Sr_x)_2CuO_4$ and $Bi_2Sr_2CaCu_2O_{8+y}$ also shows temperature dependences, but the dependence is much weaker than $R_H \sim 1/T$. It is noteworthy that the Hall effect in Cu-rich compound, $Y_2Ba_4Cu_8O_{20-\delta}$, is quite different from $YBa_2Cu_3O_{7-\delta}$ [16,18]. Thus, the unusual $1/R_H \sim T$ relation appears to be specific to $YBa_2Cu_3O_{7-\delta}$, and very intriguing as it certainly is, it probably

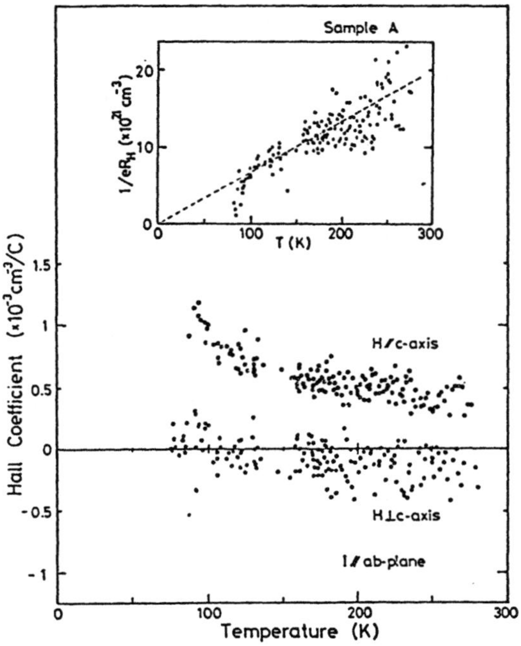

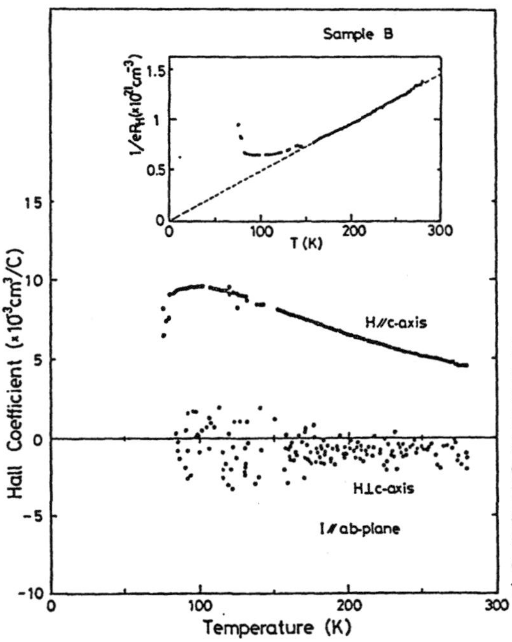

Fig. 2 Temperature dependence of the anisotropic Hall coefficient of the same two single crystal samples as Fig. 1. The Hall coefficients for H ∥ c and H ⊥ c show a marked difference in their sign and temperature dependence. The insets show the characteristic T-linear behavior of $1/eR_H$ (H ∥ c), commonly observed in the two crystals with very different carrier densities.

does not have a direct relation with the machanism of high temperature superconductivity.

3. ANISOTROPIC SUPERCONDUCTING PROPERTIES

3.1 Critical Field Anisotropy

Figure 3 shows the resistively observed superconducting transition in Sample A, for different magnetic fields applied along the two principal directions. It is immediately noted that even in a field as high as 90 kOe the onset point of transition is shifted very little, while the zero resistance point is lowered substantially. This broadening of resistive transition by magnetic field is an issue of great physical interest in its own right, and will be discussed in the next subsection. On the experimental side, it poses a practical problem in the definition of $T_c(H)$. Two commonly employed definitions are the midpoint of transition and the zero resistance point. Use of the onset point would bring in a somewhat higher degree of arbitrariness due to the rounded $\rho(T)$ curves in the onset region, which may be associated with fluctuation effect.

Figure 4 shows the temperature dependences of the upper critical fields for the two principal directions. Both the zero resistance H_{c2} and the midpoint H_{c2} are shown. The anisotropy ratio $H_{c2}^{\perp}/H_{c2}^{\parallel}$, where $\parallel$ and $\perp$ denote the field

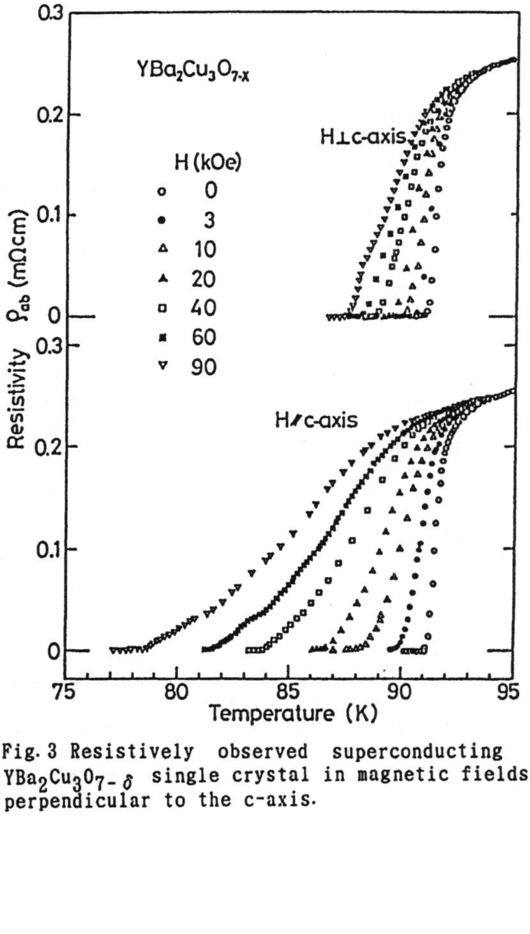

Fig. 3 Resistively observed superconducting transition in a $YBa_2Cu_3O_{7-\delta}$ single crystal in magnetic fields applied parallel and perpendicular to the c-axis.

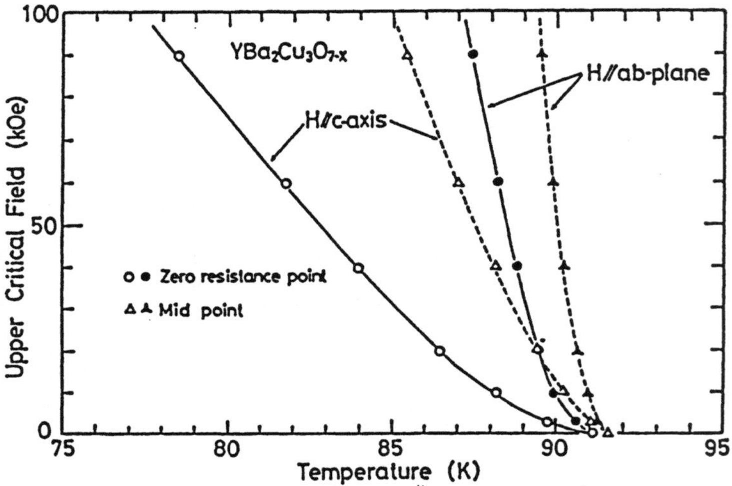

Fig. 4 Temperature dependence of $H_{c2}^{\parallel}$ and $H_{c2}^{\perp}$ of $YBa_2Cu_3O_{7-\delta}$ defined by the zero resistance points and by the midpoints.

orientation with respect to the c-axis, is about 5. The $H_{c2}(T)$ curves show characteristic upward curvature for the both field directions. The standard procedure for the estimation of coherence lengths from the values of critical field slope $-dH_{c2}^{\parallel}/dT$ and $-dH_{c2}^{\perp}/dT$ is hampered by this upward curvature. If we tentatively use values taken by drawing tangent lines to relatively straight part of the $H_{c2}(T)$ curves in higher field region, and use the Werthamer-Helfand-Hohenberg formula, we obtain $H_{c2}^{\parallel}(0)$= 550 kOe and $H_{c2}^{\perp}(0)$= 2100 kOe from the zero resistance curves and $H_{c2}^{\parallel}(0)$= 1180 kOe and $H_{c2}^{\perp}(0)$= 5100 kOe from the midpoint curves. The corresponding values of coherence lengths are ξ_{ab}= 25 A and ξ_c= 6.3 A (zero resistance), and ξ_{ab}= 17 A and ξ_c= 3.9 A (midpoint).

3.2 Behavior Associated with Short Coherence Length

The coherence length comparable to the unit cell dimensions is a key feature of the high T_c superconductors and makes the phenomenology of the high temperature superconductivity very different from the conventional one. As can be readily seen by recalling the BCS coherence length formula, $\xi = hv_F/\pi\Delta$, it is a direct consequence of the high transition temperature and low carrier density. It also reflects a highly local nature of the superconductivity in this class of materials. The value of $\xi_c(0)$ in comparison with the CuO_2 layer separation is particularly important because it determines the intrinsic dimensionality of superconductivity. The estimate of $\xi_c(0)$ ranges from ~2 to ~7 A, depending on the definition of H_{c2}. This range of $\xi_c(0)$ is tantalizingly merginal between the two- and three-dimensional behavior.

There are three major lines of interpretation for the broadening of the resistive transition by magnetic fields. The first one is to attribute it simply to sample inhomogeneity, or weak links with distributed strength. While these undoubtedly play a role in low quality crystals and in ceramics, the agreement of experimental data on high quality single crystals suggests the phenomenon has more intrinsic origin. The second and the third approaches invoke fluctuation and giant flux creep, respectively. Oh et al. [19] suggested that the upward curvature of the H_{c2} curve was a manifestation of critical fluctuation effect. However, this seems to run into a conceptual difficulty because this would mean that the critical fluctuaion regime is much wider below T_c than above T_c. Tsuneto et al. [20] recently persued the fluctuation effect on the resistive transition in magnetic fields by taking account of the renormalization effect due to interaction between fluctuations and showed that the higher temperature part of the resistive transition could be fitted by suitable choice of parameters. Tinkham[21] explained the broadening of resistive transition in terms of the giant flux creep model originally proposed by Yeshurun and Malozemoff[22]. The effect consistent with this model (and even more drastic than in $YBa_2Cu_3O_{7-\delta}$) is recently reported by Palstra et al. [23] for $Bi_2Sr_2CaCu_2O_{8+y}$ crystal. A critical test discriminating among these models is to study the dependence on the direction of the transport current with respect to the field, because the flux creep depends on the $J \times B$ Lorentz force. A preliminary measurement with $I \parallel H \parallel$ c-axis yielded behavior quite similar to the case of $I \perp H \parallel$ c-axis. Whether this poses a serious question to the flux creep model, or simply indicates that the supercurrent flows in tortuous percolative paths, remains to be clarified by further studies.

4. TRANSPORT STUDIES ON HIGH T_c RELATED CUPRATE MATERIALS

4.1 Metal-Insulator Transition in $Bi_2Sr_2Ca_{1-x}Y_xCu_2O_{8+y}$ System

The bismuth containing high T_c cuprates discovered by Maeda et al. [24], together with the thallium compounds discovered by Sheng and Harmann[25], offer a unique oppotunity to study the systematics of high temperature superconductivity with varying number of CuO_2 layers. In cases of $(La_{1-x}Sr_x)_2CuO_4$ and $YBa_2Cu_3O_{7-\delta}$, it has been clarified that the high T_c phase is located near the antiferromagnetic insulator phase. It is of crucial importance to see if the same is true for the new Bi and Tl systems. We found that the 80K superconducting Bi compound, $Bi_2Sr_2CaCu_2O_{8+y}$ can be modified by substituting trivalent Y for divalent Ca[26].

Figure 5 shows the temperature dependences of the resistivity and the Hall coefficient for a series of $Bi_2Sr_2Ca_{1-x}Y_xCu_2O_{8+y}$ compounds[27]. Similar results are reported by Yoshizaki et al. [28] and by Ando et al. [29] The resistively determined T_c vanishes at $x \sim 0.5$, where a metal-to-insulator transition takes place. Figure 6 shows the carrier density estimated from the Hall coefficient assuming single band conduction, together with that determined by chemical analysis assuming the above chemical formula. The dashed curve in the figure shows the relation $n \sim \alpha(1-x)$, corresponding to a linear variation of the carrier density with the Y-substitution. The mobile carrier density as given by the Hall effect changes more rapidly (but continuously) with x. It has been noticed that T_c is slightly higher in compounds with small amount of Y than the all Ca (x=0) compound. Whether this is intrinsic or is simply due to difference in the sample preparation conditions must await for further studies.

Recently, Nishida et al. [30] found by a muon rotation experiment that the x=1 compound, $Bi_2Sr_2YCu_2O_{8+y}$, has a static magnetic order, presumably an antiferromagnetic one, up to ~400 K. The Neel temperature decreases to ~20 K for the x=0.3 compound. Figure 7 shows schematic phase diagrams for the three high T_c copper oxide systems. It is noteworthy that the $Bi_2Sr_2Ca_{1-x}Y_xCu_2O_{8+y}$ system share a common feature with $(La_{1-x}Sr_x)_2CuO_4$ and $YBa_2Cu_3O_{7-\delta}$. The universaly ovserved close relationship between the superconducting phase and the antiferromagnetic phase presents a strong circumstantial evidence in favor of the magnetic mechanisms for the high temperature superconductivity.

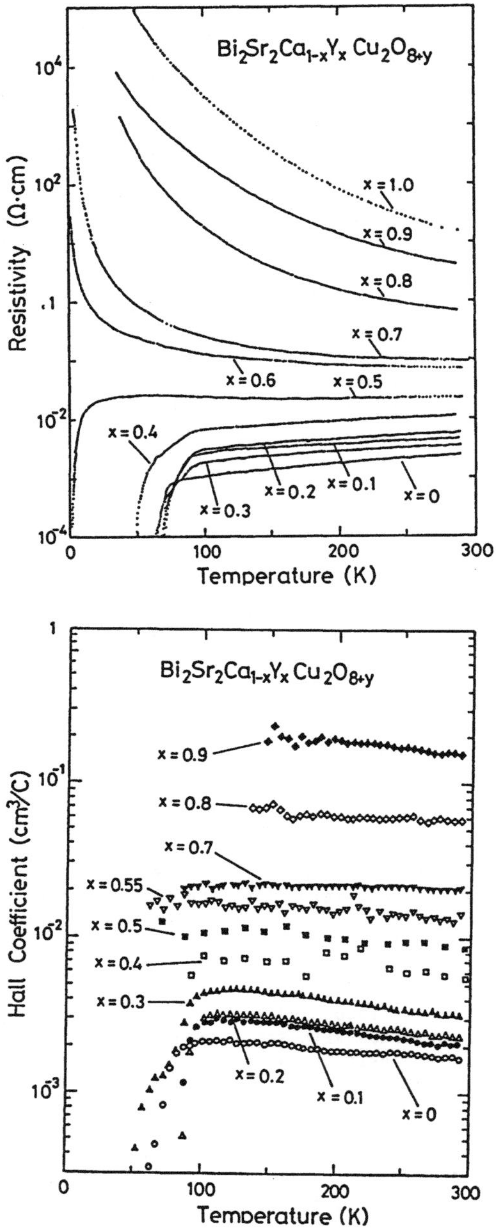

Fig. 5 Temperature dependence of the resistivity and the Hall coefficient for a series of $Bi_2Sr_2Ca_{1-x}Y_xCu_2O_{8+y}$ compounds with different x.

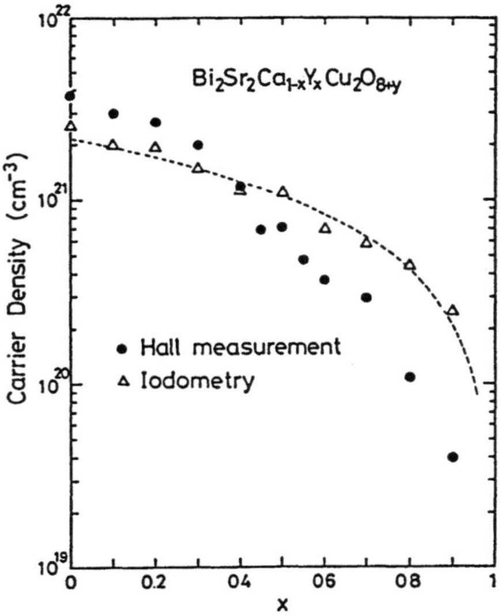

Fig. 6 Carrier concentration change with x in the $Bi_2Sr_2Ca_{1-x}Y_xCu_2O_{8+y}$ system. Solid circles and open triangles represent the hole densities deduced from the Hall measurement and from the chemical analysis, respectively. The dashed curve corresponds to the dependence linear in (1-x).

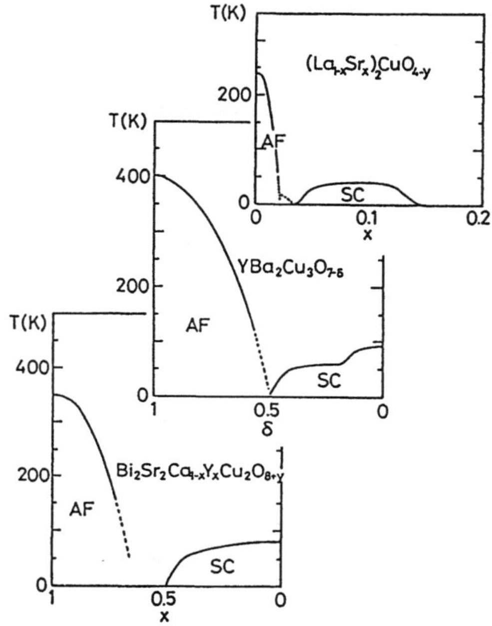

Fig. 7 Schematic phase diagram for the three high T_c copper oxide systems, sharing the common existence of the antiferromagnetic phase in the vicinity of the high T_c phase.

4.2 Normal Metallic La-Sr-Cu Oxides

Notable features commonly observed in high T_c copper oxides, besides the above mentioned close relation with the magnetic phase, are the p-type conduction and the quasi-two dimensionality. In order to gain further insight into these issues, it is interesting to study systems with similar structures[31, 32]. The high T_c cuprate, $(La_{1-x}Sr_x)_2CuO_4$, has the cation ratio (La, Sr):Cu = 2:1, and has the K_2NiF_4 structure. There exist compounds with the cation ratio (La, Sr):Cu = 1.5:1 and = 1:1. The former, $La_{1.9}Sr_{1.1}Cu_2O_{6+y}$, has double CuO_2 layers similar to those in $YBa_2Cu_3O_{7-\delta}$. The latter is denoted according to Michel et al. [31] as $La_{8-x}Sr_xCu_8O_{20-y}$, and has a three-dimensional Cu-O network.

Figure 8 shows the temperature dependences of the resistivity and the Hall coefficient in $La_{1.9}Sr_{1.1}Cu_2O_{6+y}$ and $La_{8-x}Sr_xCu_8O_{20-y}$ (x=1.6 and 2.13). The first thing to note is that the values of resistivity are comparable to those of the high T_c cuprates. Namely, they are well metallic but do not show superconductivity down to 0.3 K. The sign of the Hall coefficient is positive for the 2D systems, $La_{1.9}Sr_{1.1}Cu_2O_{6+y}$ and its kin $La_{1.9}Ca_{1.1}Cu_2O_6$. The Hall coefficient in the 3D systems, $La_{8-x}Sr_xCu_8O_{20-y}$ (x=1.6 and 2.13), turns out to be negative.

In the low temperature region, the resistivity of the $La_{1.9}Sr_{1.1}Cu_2O_{6+y}$ compound shows an upturn reminiscent of the weak localization effect. In fact, a

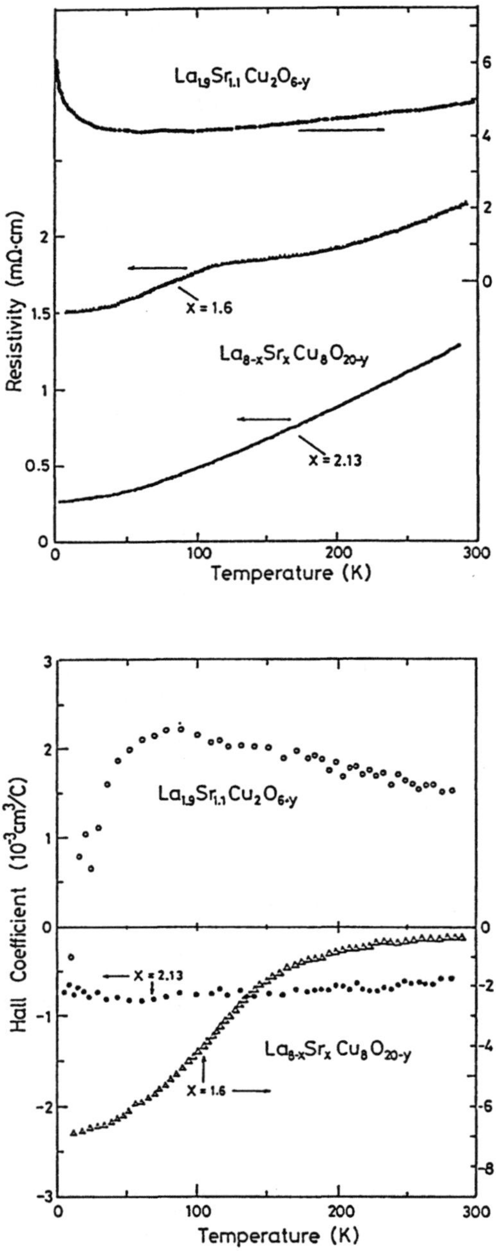

Fig. 8 Temperature dependence of the resistivity and the Hall coefficient in $La_{1.9}Sr_{1.1}Cu_2O_{6+y}$ an $La_{8-x}Sr_xCu_8O_{20-y}$ (x=1.6 and 2.13). Note that they show no trace of the superconducting $(La_{1-x}Sr_x)_2CuO_4$ phase. Anomalies are seen at ~130 K in the x=1.6 compound.

σ versus log T plot yields a straight line over a fairly wide temperature range. The resistivity curve for the $La_{8-x}Sr_xCu_8O_{20-y}$ (x=1.6) compound shows an anomaly at ~130 K, which is also reflected in the Hall coefficient curve. A susceptibility anomaly concomittant with the above-mentioned transport anomaly has been observed in the x=1.6 compound. Although the nature of this anomaly is not clear at the moment, a similarity with the magnetic phase transition in La_2CuO_4 may be worth pointing out.

CONCLUSIONS

We conclude this paper by addressing the following questions from an empirical viewpoint:

(1) Fermi liquid or not? --- As far as the transport data are concerned, there appears to be yet no positive evidence strong enough to force us to give up the Fermi liquid picture.

(2) 2D or anisotropic 3D? --- On the basis of the resistivity anisotropy and the critical field anisotropy data, $YBa_2Cu_3O_{7-\delta}$ appears more like a anisotropic 3D system than a 2D system.

(3) Is the magnetism intimately related with the high T_c? --- Judging from the universality of the phase diagram, it seem highly likely that the magnetic mechanism plays the essential part in the high temperature superconductivity.

(4) Are the p-type conduction and the quasi-2D sturucture necessary for the oc curence of the high T_c? --- Within the limited number of systems so far studied, it appears to be the case. In this context, the absense of superconductivity in the $La_{1.9}Sr_{1.1}Cu_2O_{6+x}$ system is extremely interesting.

ACKNOWLEDGEMENTS

The author wishes to thank the coauthors of refs. [7], [8], [14], [26], [27] and [30] for exciting collaborations. In particular, contributions of T. Tamegai, H. Takeya and H. Takei are invaluable. This work was supported by the Grant-in-Aid of Scientific Research on Priority Areas "Mechanism of Superconductivity" from the Ministry of Education, Science and Culture, Japan.

REFERENCES

1. For recent development, see for example, Proc. Int. Conf. on High Tempera ture Superconductors and Materials and Mechanisms of Superconductivity, (Interlaken, 1987), Physica C153-155.
2. S. W. Tozer, A. W. Kleinsasser, T. Penny, D. Kaiser and F. Holtzberg: Phys. Rev. Lett. 59, 1768 (1987).
3. K. Murata, K. Hayashi, Y. Honda, M. Tokumoto, H. Ihara, M. Hirabayashi, N. Terada and Y. Kimura: Jpn. J. Appl. Phys. 26, L1941 (1987).
4. P. W. Anderson and Z. Zou: Phys. Rev. Lett. 60, 132 (1988).
5. P. W. Anderson: Science 235, 1196 (1987).
6. S. J. Hagen, T. W. Jing, Z. Z. Wang J. Horvath and N. P. Ong: Phys. Rev. B37, 7928 (1988).
7. Y. Iye, T. Tamegai, T. Sakakibara, T. Goto, N. Miura, H. Takeya and H. Takei: Physica C153-155, 26 (1988).
8. Y. Iye, T. Tamegai, H. Takeya and H. Takei: Jpn. J. Appl. Phys. 27, L658 (1987).
9. L. Ya. Vinnikov, G. A. Emelchenko, P. A. Kononovich, Yu. A. Ossipian, I. F. Schegolev, L. I. Buravov and V. N. Laukhin: Physica C153-155, 1359 (1988).
10. M. Oda, Y. Hidaka, M. Suzuki and T. Murakami: Phys. Rev. B, to be puplished.
11. M. F. Crommie, A. Zettl, T. W. Barbee III and M. L. Cohen: Phys. Rev. B37, 9734 (1988).

12. L. Forro, M. Raki, C. Ayache, P. C. E. Stamp, J. Y. Henry and J. Rossat-Mignod: Physica C153-155, 1357 (1988).
13. I. Bozovic, K. Char, S. J. B. Yoo, A. Kapitulnik, M. R. Beasley, T. H. Geballe, Z. Z. Wang, S. Hagen, N. P. Ong, D. E. Aspnes and M. K. Kelly, Phys. Rev. B38 (1988) 5077.
14. Y. Iye, T. Tamegai, H. Takeya and H. Takei: submitted to Jpn. J. Appl. Phys.
15. A. Davidson, P. Santhanam, A. Palevski and M. J. Brady: Phys. Rev. B38, 2828 (1988).
16. H. L. Stormer, A. F. Levi, K. W. Baldwin, M. Anzlowar and G. S. Boebinger: Phys. Rev. B38, 2472 (1988).
17. J. Clayhold, N. P. Ong, Z. Z. Wang, J. M. Tarascon, P. Barboux, preprint.
18. K. Char, M. Lee, R. W. Barton, A. F. Marshall, I. Bozovic, R. H. Hammond, M. R. Beasley, T. H. Geballe, A. Kapitulnik and S. S. Laderman: Phys. Rev. B38, 834 (1988).
19. B. Oh, K. Char, A. D. Kent, M. Naito, M. R. Beasley, T. H. Geballe, R. H. Hammond and A. Kapitulnik: Phys. Rev. B37, 7861 (1988).
20. T. Tsuneto: J. Phys. Soc. Jpn. 57 (1988) in press; R. Ikeda, T. Ohmi and T. Tsuneto, preprint.
21. M. Tinkham: Phys. Rev. Lett. 61, 1658 (1988).
22. Y. Yeshurun and A. P. Malozemoff: Phys. Rev. Lett. 60, 2202 (1988).
23. T. T. M. Palstra, B. Batlogg, L. F. Schneemeyer and J. V. Waszczak: Phys. Rev. Lett. 61 (1988) 1662.
24. H. Maeda, Y. Tanaka, M. Fukutomi and T. Asano: Jpn. J. Appl. Phys. 27 (1988) L209.
25. Z. Z. Sheng and A. M. Hermann: Nature 332, 55 (1988).
26. T. Tamegai, A. Watanabe, K. Koga, I. Oguro and Y. Iye: Jpn. J. Appl. Phys. 27 (1988) L1074.
27. T. Tamegai, K. Koga, K. Suzuki, M. Ichihara, F. Sakai and Y. Iye: Jpn. J. Appl. Phys., submitted.
28. R. Yoshizaki, Y. Saito, Y. Abe and H. Ikeda: Physica C152 (1988) 408.
29. Y. Ando, K. Fukuda, S. Kondoh, M. Sera, M. Onoda and M. Sato: Solid State Commun. 67 (1988) 815.
30. N. Nishida, H. Miyatake, S. Okuma, T. Tamegai, Y. Iye, R. Yoshizaki, K. Nishiyama and K. Nagamine: Physica C156 (1988) 625.
31. C. Michel, L. Er-Rakho and B. Raveau: J. Phys. Chem. Solids 49 (1988) 451.
32. J. B. Torrance, Y. Tokura, A. Nazzal and S. S. P. Parkin: Phys. Rev. Lett. 60 (1988) 542.

III.5 Thermal

Recent Experimental Studies on High-T_c Oxides at IMS

M. Sato[1], *M. Sera*[1], *S. Shamoto*[1], *M. Onoda*[1], *S. Kondoh*[2], *K. Fukuda*[1], *and Y. Ando*[3]

[1]Institute for Molecular Science, Myodaiji, Okazaki 444, Japan
[2]Permanent Address, Research and Development Division, Asahi Glass Co., Ltd., Hazawa, Kanagawa-Ku, Yokohama 221, Japan
[3]Permanent Address, Nippon Soken, Inc., Shimohasumi, Nishio 445, Japan

1. Introduction

At the very beginning of the studies on high-T_c oxides, results of our tunneling experiments /1/ suggested that the rough features of the high-T_c superconductivity could still be described by the BCS's universal theory. The superconducting coherence lengths of $(La_{1-x}M_x)_2CuO_4$ (M=Sr and Ba) crystals estimated from the upper critical magnetic field /2/ also suggested that the mean field description should at least approximately be valid even for the surprisingly high value of their transition temperatures. It meant that the superconducting behaviors did not essentially depend on the detailed origin(s) of the electron-electron pairing force and therefore did not easily give the useful clue to elucidate the mechanism of the superconductivity. Then, we felt it quite important to carry out detailed and careful studies not only on the superconducting state but also on the normal state. Much effort to prepare large single crystals of $(La_{1-x}Sr_x)_2CuO_4$ and $YBa_2Cu_3O_{7-\delta}$ has been made /3,4/ and the obtained crystals have been used by ourselves or in collaboration with many other groups in various kinds of experiments, transport studies /5/, Raman scattering /6/, neutron scattering /7-9/ and so on. However, for certain experiments where the oxygen number should be well controlled, powder specimens are realized to be much more suitable than single crystal specimens. For example, the low temperature specific heat was measured for Bi-Sr-Ca-Cu-O powder specimen which has very large Meissner volume fraction /10/.

Search for new oxide superconductors was also important at least at the early stage of the studies to extract the basic characteristics of the high-T_c oxides. We discovered the Ln-Ba-Cu-O with magnetic lanthanide atoms Ln /11/. Superconductivity in Tl-Ba-Cu-O was also announced by the present author's group first /12/. For the understanding of the structural characteristics of Bi-containing systems, four circle X-ray diffraction studies have been carried out on $Bi_2Sr_2CuO_y$ /13/, result of which clarifies the origin of the long period structure of Bi-Sr-Ca-Cu-O system.

Now, it seems rather well understood experimentally that the superconducting behaviors are, as we felt, not much different from those expected by the BCS mean field theory even though the phonon exchange is not considered to be a main mechanism of the superconductivity. To clarify the origin of the superconductivity, it seems to be quite important to understand the natures of the O2p hole system which strongly interacts with the 3d spins at Cu^{2+} sites. In the present article, we discuss the low temperature specific heat of the high-T_c oxides. The thermoelectric powers studied on single crystals of $(La_{1-x}Sr_x)_2CuO_4$ and $YBa_2Cu_3O_{7-\delta}$ are also discussed as one of the important quantities to understand the O2p hole system. Brief comments on the nature of the electrons in perovskite superconductor $Ba_{1-x}K_xBiO_3$ are also given.

2. Low Temperature Specific Heat of $Bi_4Sr_3Ca_3Cu_4O_y$

Presence of the T-linear term in the low temperature specific heat was one of the most remarkable things in the early stage of the studies on high-T_c oxides /14/. If the T-linear specific heat γT is really intrinsic, it suggests that the superconducting properties of high-T_c oxides are quite different from those expected by the BCS theory. However, contributions from the possible impurity phases and/or the normal phase which might be caused by the partial oxygen deficiencies in the specimens should be carefully avoided to conclude that the term really exists. We adopted Bi-Sr-Ca-Cu-O sintered pellets in the specific heat measurement /10/, because oxygen deficiencies are not easily introduced in this oxide system. The nominal chemical formula was

$Bi_4Sr_3Ca_3Cu_4O_y$. The Meissner volume fraction was about 80% at ~5K. The EPMA showed that only a very small amount of impurity phases of $(Ca,Sr)_2CuO_y$, $Bi_2Sr_2CuO_y$ and CuO existed in the major superconducting phase of $Bi_4Sr_3Ca_3Cu_4O_y$. Figure 1 shows the C/T vs. T^2 plot of the observed results. Although small deviation from the linear behavior is observed at low temperatures which may be explained by the possible existence of the lattice imperfections, the straight extrapolation to T=0 clearly indicates the absence of the T-linear component in the present high-T_c oxide. To the authors' knowledge, it was a first experimental confirmation of the non-existence of the T-linear term in the low temperature specific heat of high-T_c oxides. Since then, many groups have gotten the same conclusion up to now /15, 16/. The absence of the T-linear term seems to be common to all other high-T_c oxides. As one of the origins of the experimentally observed T-linear term of $YBa_2Cu_3O_{7-\delta}$, KUENTZLER et al. /17/ proposed the contributions from the impurity phases based on their observation of the large specific heat of $Y_2Cu_2O_5$ and $BaCuO_2$ at low temperatures. Fortunately, Bi-Sr-Ca-Cu-O system does not contain these impurity phases.

The x-dependence of γ in $(La_{1-x}Sr_x)_2CuO_4$ and $(La_{1-x}Ba_x)_2CuO_4$ has been studied by KUMAGAI et al. /18, 19/. Its complicated behavior observed for $(La_{1-x}Ba_x)_2CuO_4$ may be caused by the anomalous double peak structure of the T_c vs. x curve found by MOODENBAUGH et al. /20/. KUMAGAI et al. /21/ observed similar behavior for their own specimens. The present authors have found

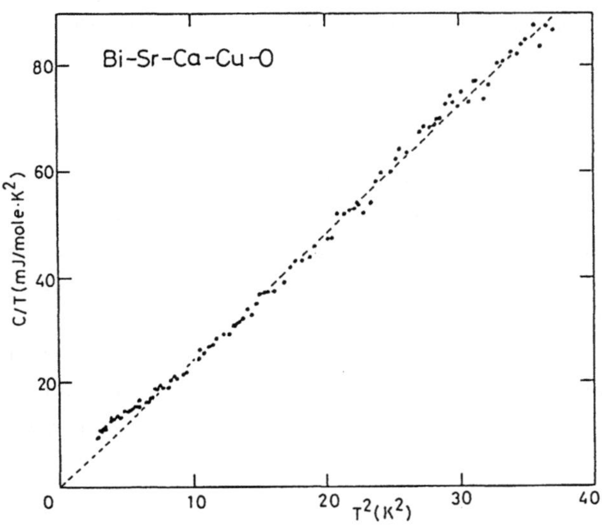

Fig. 1 The C/T of a sintered pellet with a nominal formula of $Bi_4Sr_3Ca_3Cu_4O_y$ is plotted as a function of T^2

striking anomalies in the temperature dependence of the thermoelectric power S and the Hall coefficient R_H of sintered pellets of $(La_{0.94}Ba_{0.06})_2CuO_4$ prepared by Kumagai's group, which seem to be attributed to the structural transition in the compound. The depression of the superconducting transition temperature T_c by the structural change produces the dip of T_c vs. x curve near x=0.06. Therefore, the T-x phase diagram of $(La_{1-x}Ba_x)_2CuO_4$ should be modified and the new phase diagram can explain the observed γ vs. x curve of the compound.

3. Thermoelectric Power of $(La_{1-x}Sr_x)_2CuO_4$ and $YBa_2Cu_3O_{7-\delta}$ Single Crystals

Transport properties of the high-T_c oxides have been known to behave in an anomalous way. For example, the T-linear dependence of the resistivity ρ=AT in the wide temperature region has been discussed in the framework of the RVB theory. However, we now know that there does not exist the intrinsic T-linear component in the low temperature specific heat which is caused, according to the theory, by the same origin as that of the T-linear resistivity. Therefore the behavior of the resistivity may have a different physical explanation. We have to be very careful if it is an intrinsic one or not, because the coefficient A has quite large sample dependence which suggests the fact that the existence of the grain boundaries in the specimens may be crucial to the behavior of the ρ vs. T curve. In contrast to the case of the resistivity, the thermoelectric power S(T) has been found to exhibit an intrinsic behavior as will be shown in the next section. It is possibly due to the fact that in the measurement of S(T), the current is not applied, which seems to eliminate the boundary effects.

Figure 2 shows the thermoelectric powers of $YBa_2Cu_3O_{6.9}$ single crystals /5/. The T-dependence of the resistivities along the direction within and perpendicular to the CuO_2 plane is also shown by the broken lines. The anisotropic nature is quite evident. Roughly speaking, S is linear in T along the direction perpendicular to the plane for both compounds of $(La_{1-x}Sr_x)_2CuO_4$ and $YBa_2Cu_3O_{7-\delta}$. In Figs. 3a and 3b, the thermoelectric powers S(T) of $(La_{1-x}Sr_x)_2CuO_4$ and $YBa_2Cu_3O_{7-\delta}$, respectively, are shown for various values of x or δ. Generally, as the carrier concentration increases, the S vs. T curve shifts downwards and often changes its sign as a function of T, an example of which is shown in Fig. 4 for a sintered pellet with a nominal formula of $Bi_4Sr_3Ca_3Cu_4O_y$.

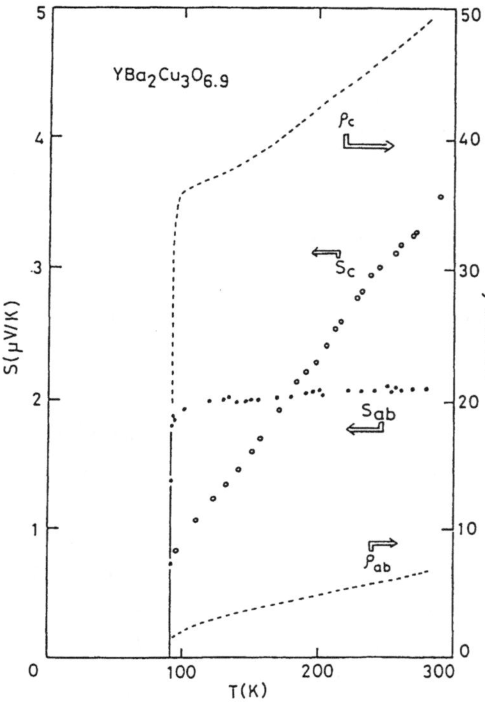

Fig. 2 Thermoelectric powers of $YBa_2Cu_3O_{6.9}$ are shown along two directions within (S_{ab}) an perpendicular to (S_c) the CuO_2 planes. Broken lines indicate the resistivities

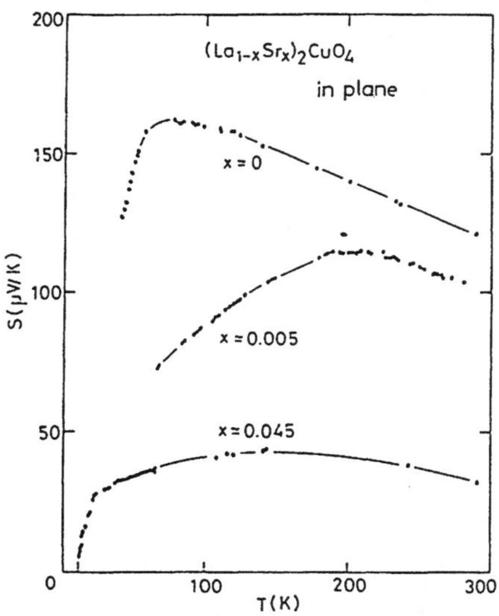

Fig. 3a Thermoelectric powers of $(La_{1-x}Sr_x)_2CuO_4$ within the CuO_2 plane for various values of x. The values of x were determined by EPMA

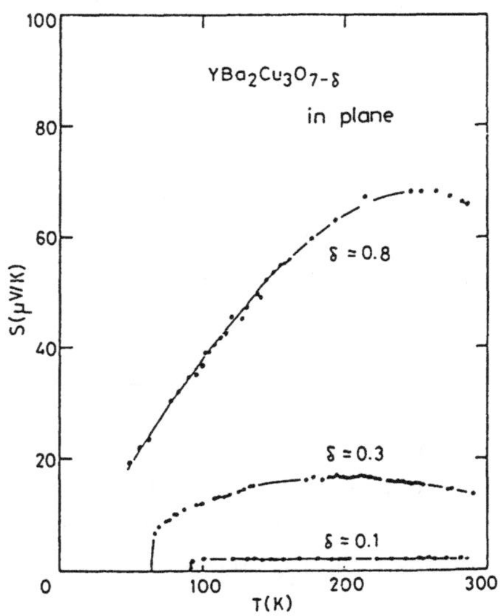

Fig. 3b Thermoelectric powers of $YBa_2Cu_3O_{7-\delta}$ within the CuO_2 plane for various values of δ. The values of δ were determined from the lattice parameters

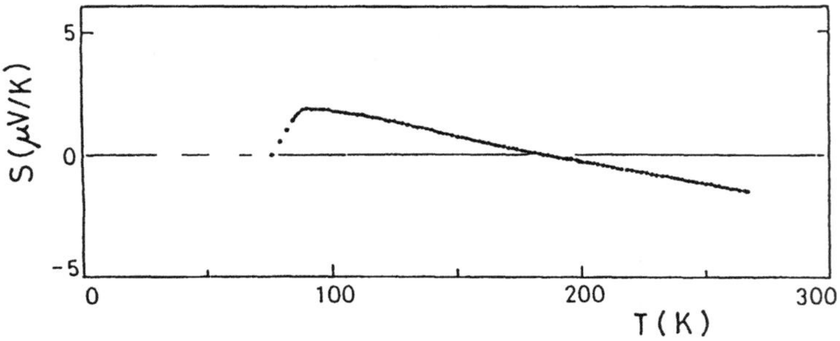

Fig. 4 Thermoelectric power of a sintered pellet with nominal formula of $Bi_4Sr_3Ca_3Cu_4O_y$

As discussed in ref. 5, these behaviors of S(T) are quite characteristic but cannot consistently be explained so easily by the mechanisms proposed by many authors previously. As is well known, the thermoelectric power is described as $S \propto (\partial \ln\sigma/\partial\varepsilon)_{\varepsilon=\varepsilon_F}$, where $\sigma = N(\varepsilon)e^2 D(\varepsilon)$ with the electronic density of states $N(\varepsilon)$ and the electron diffusion constant $D \propto v^2\tau(\varepsilon)$. It seems to be quite difficult to get the observed behaviors from the term of $\partial v^2 N(\varepsilon)/\partial\varepsilon$. Therefore, we may have to consider the main contribution of $\partial\tau(\varepsilon)/\partial\varepsilon$ to explain the observed characteristic behaviors of S(T).

It is tempting to attribute the S vs. T behavior to the magnetic scattering of O2p holes due to the Cu3d spins. If we use the possible analogy to Kondo lattice, T_K is expected to be quite high, because the interaction between the conduction holes and 3d spins may be very strong. Then, we may be able to expect the anomalous behavior of S vs. T curves in the observed temperature region. Here, we just propose the possible similarity of the models. The characteristic behaviors of S(T) may give us certain clue to understand the nature of the system with O2p conducting holes and Cu3d spins.

4. Electronic Nature of $Ba_{1-x}K_xBiO_3$

In the sense that very high-T_c value of about 30K was found in a material without Cu atom, the superconductivity of $Ba_{1-x}K_xBiO_3$ attracts much attention /22, 23/. It is also interesting that it is a three dimensional conductor with rather high-T_c. We have also prepared $Ba_{1-x}K_xBiO_3$ specimens /24/ by heating the mixtures of BaO, KO_2 and Bi_2O_3 in Ag tube with a condition of excess KO_2 as described by CAVA et al. /23/. The as-sintered specimens had dark blue color and some of them exhibited superconductivity with a volume

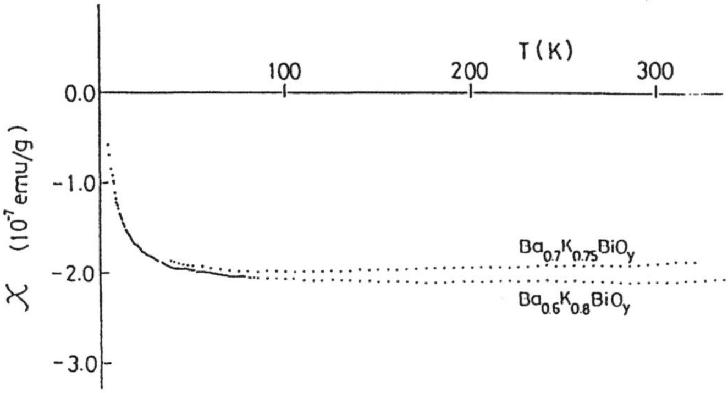

Fig. 5 Magnetic susceptibility of $(Ba,K)BiO_3$ specimens. Details are in the text

fraction of about 10%. The thermal treatment in oxygen atmosphere did not improve the superconductivity at all. Figure 5 show the temperature dependence of the magnetic susceptibility of two specimens with the nominal formulae of $Ba_{0.7}K_{0.75}BiO_y$ and $Ba_{0.6}K_{0.8}BiO_y$ with the Meissner volume fractions of 9.1% and 0.03%, respectively. As the ion-core diamagnetism is estimated to be about 2.35×10^{-7}emu/g, the paramagnetic contribution to the susceptibility is quite small for both specimens irrespective of the magnitudes of the Meissner volume fractions. This suggests that the order of the paramagnetism of the conduction electron system seem to be the same as that of $Ba_{1-x}Pb_xBiO_3$.

Figure 6 shows the temperature dependence of the thermoelectric power S of the specimen with the Meissner volume fraction of 9.1% /5/. Although the temperature dependence of the resistivity ρ has serious grain boundary effects as can be seen in the inserted figure, S(T) seems to be insensitive to the effects as is already mentioned in the previous section. This clearly indicates that S(T) is a more reliable quantity than ρ(T). Though an anomaly appears near 50K possibly due to the phonon dragg effect, the over all behavior has the simple T-linear dependence expected for usual metals.

From the EPMA, the actual chemical formula of the specimen used in the measurement is found to be $Ba_{0.77}K_{0.23}BiO_3$. The carrier number is, therefore, estimated to be 3.1×10^{21}el/cm^3. Then, using the relations $S=\pi^2k_B{}^2T/2eE_F$ and $E_F=\hbar^2(3\pi^2n)^{2/3}/2m^*$, we estimate m^* as 0.3m, m being the free electron mass. Then, both the magnetic susceptibility and the thermoelectric power indicate

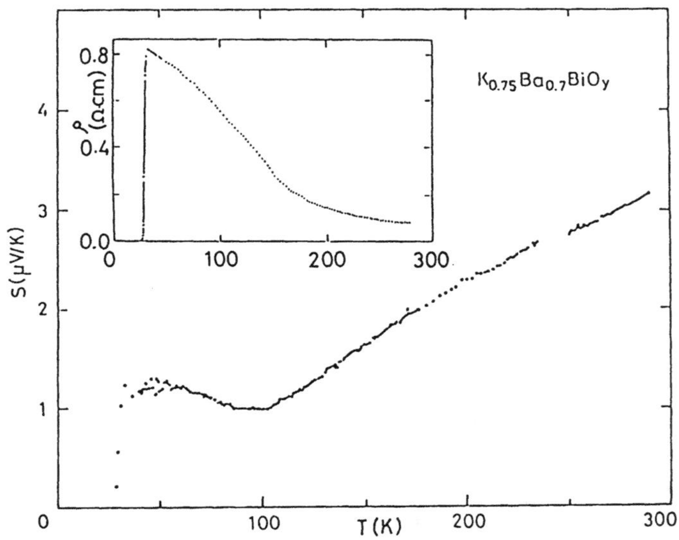

Fig. 6 Thermoelectric power of a (Ba,K)BiO_3 sintered specimen. In contrast to the case of the high-T_c oxides with CuO_2 planes, S has rather normal behavior. Effects of grain boundaries on S seem to be negligible while ρ is found to have a significant effect of the boundaries (see insertion)

that the electronic nature should be much different from that of the high-T_c oxides with CuO_2 planes, in which the holes seems to have very strong interaction with spins of Cu^{2+} ions. Quite recently, μSR experiment by UEMURA et al. /25/ reported the absence of the magnetic order in (Ba,K)BiO_3 system. Rather large isotope effect was observed by two groups /26, 27/. All of these results indicate that the pairing mechanisms related with spins, which are being developed in the discussion of the superconductivity of the system with CuO_2 planes should be excluded to explain, at least, the superconductivity of the present perovskite (Ba,K)BiO_3 system.

Acknowledgements

The present work was supported by a Grant in Aid for Scientific Research from the Japanese Ministry of Education, Science and Culture.

References

1. T. Ekino, J. Akimitsu, M. Sato and S. Hosoya: Solid State Commun. 62 535 (1987)
2. S. Shamoto, M. Onoda, M. Sato and S. Hosoya: Solid State Commun. 62 479 (1987)

3. M. Sato: Physica C153-155 38 (1988)
4. S. Shamoto: Solid State Commun. 66 1151 (1988)
5. M. Sera, S. Shamoto and M. Sato: Solid State Commun. 68 649 (1988)
6. S. Sugai, S. Shamoto and M. Sato: Phys. Rev. B
7. M. Sato, S. Shamoto, J.M. Tranquada, G. Shirane and K. Keimer: Phys. Rev. Letters 61 1317 (1988)
8. M.A. Kastner, R.J. Birgeneau, T.R. Thurston, P.J. Picone, M. Sato, K. Fukuda, S. Shamoto, Y. Endoh, Y. Yamada and G. Shirane: Phys. Rev. B
9. R.J. Birgeneau, D.R. Gabbe, H.P. Jenssen, M.A. Kastner, P.J. Picone, T.R. Thurston, G. Shirane, Y. Endoh, M. Sato, K. Yamada, Hidaka, M. Oda, Y. Enomoto, M. Suzuki and T. Murakami: Phys. Rev. B
10. M. Sera, S. Kondoh, K. Fukuda and M. Sato: Solid State Commun. 66 1101 (1988)
11. S. Hosoya, S. Shamoto, M. Onoda and M. Sato: Jpn. J. Appl. Phys. 26 L325 (1987) and 26 L456 (1987)
12. S. Kondoh, Y. Ando, M. Onoda and M. Sato: Solid State Commun. 65 1329 (1988)
13. M. Onoda and M. Sato: Solid State Commun. 67 799 (1988)
14. For example, N.E. Phillips, R.A. Fisher, S.E. Lacy, C. Marcenat, J.A. Olsen, W.K. Ham, A.M. Stacy: Jpn. J. Appl. Phys. 26 1115 (1987)
15. K. Kumagai and Y. Nakamura: Physica C
16. R.A. Fisher, S. Kim, S.E. Lacy, N.E. Phillips, D.E. Morris, A.G. Markelz, J.Y.T. Wei and D.S. Ginley:
17. R. Kuentzler, Y. Dossmann, S. Vilminot and S. EL Hadigui: Solid State Commun. 65 1529 (1988)
18. K. Kumagai, Y. Nakamichi, I. Watanabe, Y. Nakamura, H. Nakajima, N. Wada and P. Lederer: Phys. Rev. Letters 60 724 (1988)
19. K. Kumagai, I. Watanabe, Y. Nakamura, H. Aoki, Y. Nokamichi and H. Nakajima: Proc. of 4th Symposium on Magnetism and Magnetic Materials, Taiwan 1988
20. A.R. Moodenbaugh, Y. Xu, M. Suenaga, T.J. Folkerts and R.N. Shelton: Phys. Rev. B38 4596 (1988)
21. K. Kumagai: private communication
22. L.F. Mattheiss, E.M. Gyorgy and D.W. Johnson, Jr: Phys. Rev. B37 3745 (1988)
23. R.J. Cava, B. Batlogg, J.J. Krajewski, R.C. Farrow, L.W. Rupp, Jr, A.E. White, K.T. Short, W.F. Peck, Jr, T.Y. Kometani: Nature 332 814 (1988)
24. S. Kondoh, M. Sera, K. Fukuda, Y. Anodo and M. Sato: Solid State Commun. 67 879 (1988)
25. Y.J. UEMURA, B.J. Sternlieb, D.E. Cox, J.H. Brewer, R. Kadono, J.R. Kempton, R.F. Kiefl, S.R. Kreitzman, G.M. Luke, P. Mulhern, T. Riseman, D.L. Williams, W.J. Kossler, X.H. Yu, C.E. Stronach, M.A. Subramanian, J. Gopalakrishnan and A.W. Sleight: Nature 335 151 (1988)
26. D.G. Hinks, D.R. Richards, B. Dabrowski, D.T. Marx and A.W. Mitchell: Nature 335 419 (1988)
27. B. Batlogg, R.J. Cava, L.W. Rupp, Jr., A.M. Mujsce, J.J. Krajewski, J.P. Remeika, W.F. Peck, Jr., A.S. Cooper and G.P. Espinosa: Phys. Rev. Letters 61 1670 (1988)

After writing this paper, I learned the result obtained by J.D. Axe et al. /28/. Their preprint discussed the structure of $(La_{0.95}Ba_{0.05})_2CuO_4$ below the structural transition at about 60K.

28. J.D. Axe, D.E. Cox, K. Mohanty, H. Moudden, A.R. Moodenbaugh, Y. Xu and T.R. Thurston IBM J. Rev. Dev.

Substitution Effects in High-T_c Superconductive Oxides

T. Fujita

Department of Physics, Hiroshima University, Hiroshima 730, Japan

Experimental work on the effects of substitution for copper in high-T_c oxide superconductors $(La,M)_2CuO_4$ and $YBa_2Cu_3O_7$ are reviewed. Substitution of foreign metal elements for copper always results in depression of superconductivity in these oxides. Not only the transition temperature T_c but also superconducting volume estimated by diamagnetic susceptibility decrease with increasing the concentration of elements irrespective of magnetic or non-magnetic substitution. The experimental results suggest that the depression of superconductivity is not simply ascribed to the magnetic pair-breaking mechanism alone. Discussion is focussed mainly on structure, superconductivity, transport properties and magnetism.

1. INTRODUCTION

Oxide superconductors with high transition temperature T_c have rather complex structures which include a variety of elements. In order to understand the high-T_c superconductivity in these oxides, it is of special importance to specify what role each of the elements plays. Especially copper and oxygen must be the key elements because most high-T_c oxide superconductors contain layers of these two elements in common. One exception reported so far is a 30 K superconductor of $(Ba,K)BiO_3$ which has a cubic perovskite structure and does not contain copper[1]. However, it is doubtful at present whether this cubic compound and the other copper oxides can be classified into a same group of high-T_c superconductors.

The first family of superconductive copper oxides, $(La_{1-x}M_x)_2CuO_4$ with M = Ba, Sr or Ca[2,3], show the maximum T_c of $\sim$40 K around $x = 0.075$, and their structure involves only one crystalline site for Cu, which are octahedrally coordinated by six oxygen atoms and arranged so as to form quasi-two-dimensional networks parallel to the crystallographic **ab** plane. In the second family including the orthorhombic $YBa_2Cu_3O_7$[4] with $T_c \sim 90$ K, there exist two inequivalent sites for Cu. The Cu1 site, which is located between two Ba layers, is surrounded by four oxygen atoms in square coordination. The adjacent squares share oxygen atoms at the corner and form a linear Cu-O chains parallel to the b axis in case of O_7

Springer Series in Materials Science, Vol. 11
Mechanisms of High Temperature Superconductivity
Editors: H. Kamimura and A. Oshiyama

stoichiometry. The ordered Cu-O chains are responsible for the orthorhombicity of $YBa_2Cu_3O_7$. The Cu2 site located between Y and Ba layers has pyramidal coordination of oxygen. These pyramids share the corners of their bases to form a quasi-two-dimensional Cu-O sheet. There were some arguments in the early stage of investigation claiming that the Cu1-O chains are responsible for the substantially higher T_c of $YBa_2Cu_3O_7$ compared with that of the first family. By analyzing the effect of substitution for Cu, we succeeded in finding one of the earliest evidences[5] against this speculation, and proposed that the long-range ordering of oxygen which forms linear Cu1-O chains is not essential for the high-T_c superconductivity and that the Cu2-O planes mainly carry superconducting current in this oxide. This view was strongly supported by the discovery of the third family of superconductive copper oxides[6,7].

The third family containing Bi or Tl involves several kinds of structural variation forming a series of compounds represented with the formula of $(BiO)_2Sr_2Ca_{n-1}Cu_nO_{2n+2}$ and $(TlO)_2Ba_2Ca_{n-1}Cu_nO_{2n+2}$. In the simple oxides with n = 1, a single-layered plane of Cu with octahedral coordination is sandwiched between Bi-O or Tl-O layers. These oxides with $T_c = 6 \sim 20$ K are very similar to the first family of superconductive copper oxides. The oxides with n = 2 and with $T_c \sim 80$ K (Bi) and 100 K (Tl) have a double-layered structure in which each Cu is surrounded by five oxygen atoms in pyramidal coordination while no linear chains of Cu-O exist. The highest T_c of ~ 110 K (Bi) or ~ 120 K (Tl) has been achieved in the oxides with n = 3 which contain an extra Cu-O sheet with square coordination in addition to the double-layered structure for n = 2.

Now that the Cu-O planes are believed to carry an important implication for the superconductivity in all these copper oxides, the systematic study of the effects of substitution for copper by various elements should be an effective approach to the mechanism of high-T_c superconductivity. This paper reviews the experiments on the effects of substitution for copper in $YBa_2Cu_3O_7$ as well as in $(La,M)_2CuO_4$ focussing on crystalline structure, oxygen content, superconductivity, carrier concentration, transport properties, magnetism.

2. STRUCTURE

Powder X-ray diffraction[8] indicates that the insulating La_2CuO_4 has an orthorhombic structure at room temperature. When the sample is heated up, however, the structure becomes tetragonal. The orthorhombic to tetragonal transition is also found when the oxide is doped with M = Ba, Sr or Ca. As the doping concentration x in $(La_{1-x}M_x)_2CuO_4$, the distortion temperature T_d is gradually lowered from ~ 530 K for x = 0 through room temperature around x = 0.04 toward zero above x = 0.1. Further substitution of metallic elements for Cu has also been investigated in this family. Both for Ni and Zn substitution[9], the tetragonal lattice parameter **a** decreases but the parameter **c** increases with increasing y in $La_{1.85}Sr_{0.15}(Cu_{1-y}D_y)O_4$ at room temperature. However, the volume change is different between Ni and Zn substitution resulting in reduction for Ni and increase for Zn. This may be simply explained by difference in ionic size between Ni^{2+}, Cu^{2+} and Zn^{2+}.

In $YBa_2Cu_3O_{7-\delta}$, the oxygen content $7-\delta$ is one of the key parameters which control the properties of this oxide. The structure of the samples with $\delta \sim 0$ is orthorhombic at room temperature, whereas it becomes

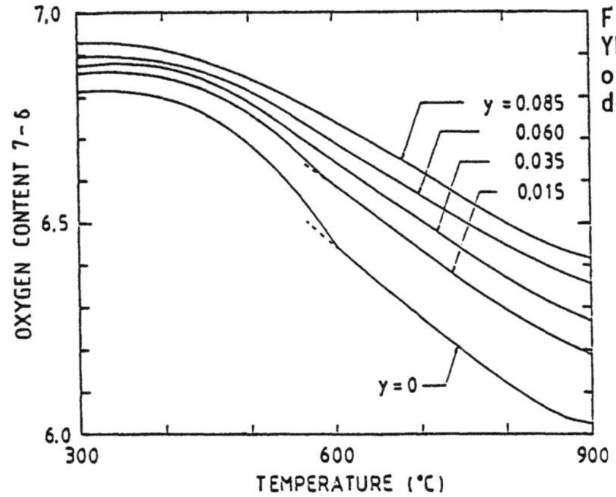

Fig. 1. Oxygen content $7-\delta$ in $YBa_2(Cu_{1-y}Fe_y)_3O_{7-\delta}$ as a function of temperature, which were determined by TGA and iodometry.

tetragonal when the samples are heated. This structural change from orthorhombic to tetragonal occurs at T_d = 940 K in air. As shown in Fig.1, thermogravimetric analysis (TGA), suggests that T_d = 940 K is the temperature at which the oxygen content $7-\delta$ decrease down to 6.3[10,11]. The value is consistent with the critical oxygen content for the orthorhombic to tetragonal phase boundary observed at room temperature in the oxygen-deficient samples quenched from a high temperature phase. The fact in turn suggests that the critical oxygen content is nearly temperature-independent and the structural change at T_d is not induced only by temperature but dominantly by the release of a large amount of oxygen.

A similar transition from the orthorhombic to tetragonal phase is also induced by substitution of metallic elements D = Fe, Co, Ga and Al[5,9,12] in $YBa_2(Cu_{1-y}D_y)_3O_{7-\delta}$. Figure 2 shows the variation of lattice parameters determined at room temperature by powder X-ray diffraction for samples with D = Fe or Co. As either Fe or Co concentration increases, the parameters **a** and **b** join at x = 0.02 $\sim$0.03 suggesting an orthorhombic to tetragonal transition. The variation of the unit-cell volume[10] is also plotted in Fig.2. The increase in volume can not be interpreted in terms of ionic size because the ionic radius of Fe and Co cations is expected to be smaller than Cu^{2+} radius of 0.72 A (see Table 1). The lattice expansion suggests the enhancement of oxygen content $7-\delta$. Figure 1 illustrates the variation of oxygen content as a function of temperature for various y in $YBa_2(Cu_{1-y}Fe_y)_3O_{7-\delta}$, which was determined by TGA in combination with iodometric titration[11]. Both for Fe and Co substitution, Fig.3 shows more clearly the change in oxygen content $\Delta\delta = \delta(0)-\delta(y)$ at 810°C, at which oxygen is expected to be removed completely from Cu1 layers in the undoped $YBa_2Cu_3O_{7-\delta}$. The initial slope of $\Delta\delta$ against y is surprisingly large for Fe substitution suggesting nearly 4 oxygen atoms trapped around each Fe atom even at this temperature[10]. Since there is no site for four additional oxygens around Cu2 site, the above result implies that Fe atoms preferentially occupy the Cu1 site for low level of substitution. On the other hand, $\Delta\delta$ increases linearly with y for Co substitution at the rate of one additional oxygen per Co.

In Fig.4, the orthorhombic to tetragonal phase boundary[10] is plotted for Fe substitution in the temperature-concentration plane. The critical

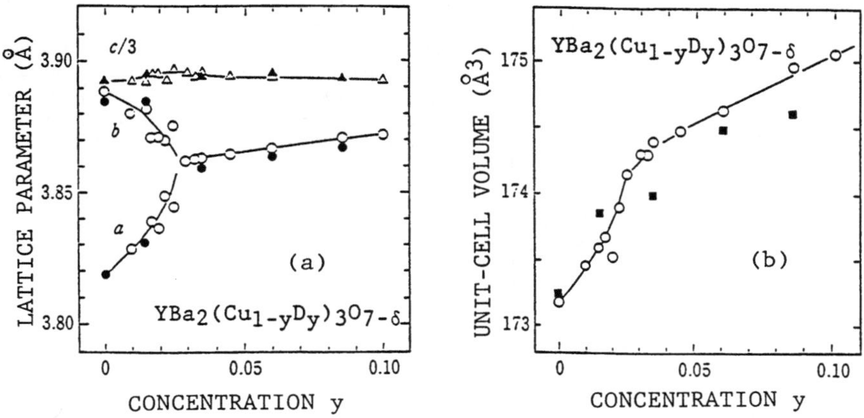

Fig. 2. (a) Lattice parameters and (b) unit-cell volume at room temperature for $YBa_2(Cu_{1-y}D_y)_3O_{7-\delta}$ with D = Fe (close symbols) and Co (open symbols).

Table 1. Parameters for the substituted elements D in $YBa_2(Cu_{1-y}D_y)_3O_{7-\delta}$. y_1 is the occupancy for the Cu1 site and y_2 for the Cu2 sites. $3y=y_1+2y_2$.

D	Fe	Co	Ni	Zn
valence	+3 or +4	+3	+2	+2
ionic radius	0.64 A	0.63 A	0.69 A	0.74 A
y	0.05	0.1	0.067	0.1
y_1	0.1	0.03	0	0.2
y_2	0.03	0	0.1	0.05
ref.	[13]	[17]	[18]	[18]

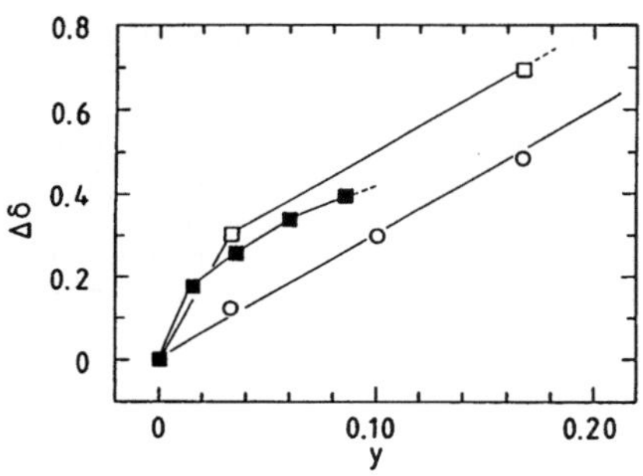

Fig. 3. Enhancement of oxygen content, $\Delta\delta = \delta(0)-\delta(y)$, with y at 810 °C for $YBa_2(Cu_{1-y}D_y)_3O_{7-\delta}$. Closed squares are the present results for D=Fe. Open symbols are the re-analysed data from ref. 12.

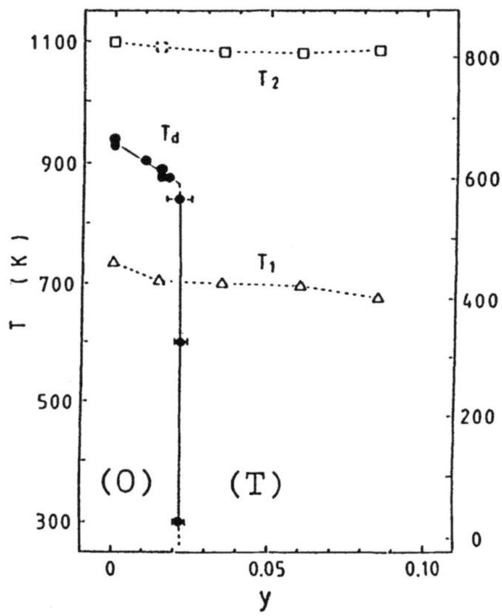

Fig. 4. The orthorhombic and tetragonal phases of $YBa_2(Cu_{1-y}Fe_y)_3O_{7-\delta}$ in the T-y plane. The phase boundary was determined by X-ray diffraction in air. T_1(triangles) and T_2(squares) are the temperatures at which inflections occur in the variation of the lattice parameter **c**.

concentration y_c is nearly temperature-independent for substitution-induced transition. The tetragonality of $YBa_2(Cu_{0.93}Fe_{0.05})_3O_{7-\delta}$ has been confirmed down to 3 K by powder neutron diffraction[13]. The structural transition can be explained as a consequence of competition between the preference of Fe for octahedral coordination and the oxygen correlation to form Cu-O chains. High resolution electron microscopy and neutron diffraction[14,15] revealed the existence of microdomains containing short range ordering of Cu1-O. Even inside the microdomains, however, the orthorhombic distortion b-a of the local lattice is expected to be small so as to minimize the total free energy including the strain energy for lattice matching.

The lattice parameter **c** plotted against temperature exhibits two clear inflections at T_1 and T_2[11]. T_1 is a measure for the temperature at which oxygen starts to be released and T_2 for the temperature at which all removable oxygen atoms are lost from Cu1 layers. These two temperatures are also shown in Fig.4. Since Fe atoms prefer a higher oxidized state, heat treatment in reduced oxygen atmosphere would give rise to modification also in structure[16]. If tetragonal $YBa_2Cu_{2.9}Fe_{0.1}O_{7-\delta}$ is heated in nitrogen atmosphere at 920°C for 20 h and then annealed in flowing 1 atm oxygen at 400°C for 5 h, the structure of the sample is found to become orthorhombic with $7-\delta$ = 6.92 at room temperature[10]. The orthorhombic structure maintains until the sample is heat-treated again at high temperatures over 700°C in air.

The substitution for Cu by divalent Ni and Zn ions retains the orthorhombic structure at least up to y = 0.1 with a slight decrease (Ni) or increase (Zn) in unit-cell volume[9], as is the case for the substitution in $(La,Sr)_2CuO_4$. Various properties including the site preference, ionic radius, etc. are summarized for some elements substituted in YBa_2CuO_7 in Table 1.

3. SUPERCONDUCTIVITY

Substitution of foreign elements for copper always results in depression of superconductivity in the copper oxides[5,17]. In Fig.5, a typical example of resistive transition is shown for a series of samples $YBa_2(Cu_{1-y}Co_y)_3O_{7-\delta}$. We determined T_c by the midpoint of the transition. This value of T_c coincides well with the onset temperature of the superconducting diamagnetism. In Fig.6, the variation of T_c is plotted as a function of concentration y for the samples of $YBa_2(Cu_{1-y}Co_y)_3O_{7-\delta}$ and $YBa_2(Cu_{1-y}Fe_y)_3O_{7-\delta}$ all of which were identically heat-treated in air at various stages of 900, 920, 550 and finally 380°C totally for about one week. The T_c vs y curves are almost identical for both the Fe and Co substitutions and exhibit no discontinuous change across the orthorhombic to tetragonal transition, at which the Cu1-O chains are expected to be strongly disturbed, although a slight change seems to exist in the slope dT_c/dy. This was the first evidence which we proposed in our early paper for the quasi-two-dimensional superconductivity in the Cu2-O planes instead of the quasi-one-dimensional superconductivity in the Cu1-O chains. As shown in Fig. 6, the diamagnetic components of the ac susceptibility measured at 4.2 K decrease continuously with increasing

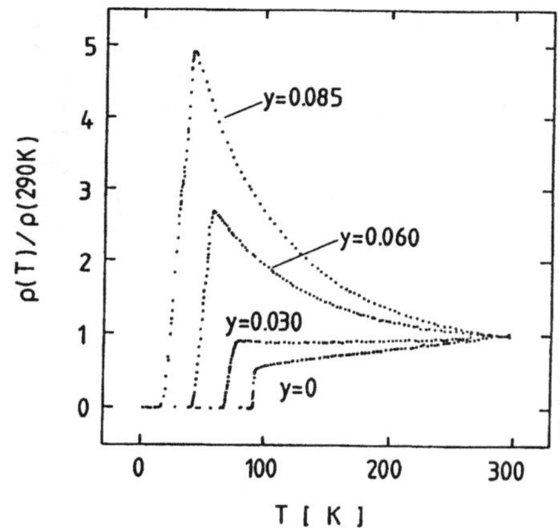

Fig. 5. Temperature dependence of the resistivity of $YBa_2(Cu_{1-y}Co_y)_3O_{7-\delta}$.

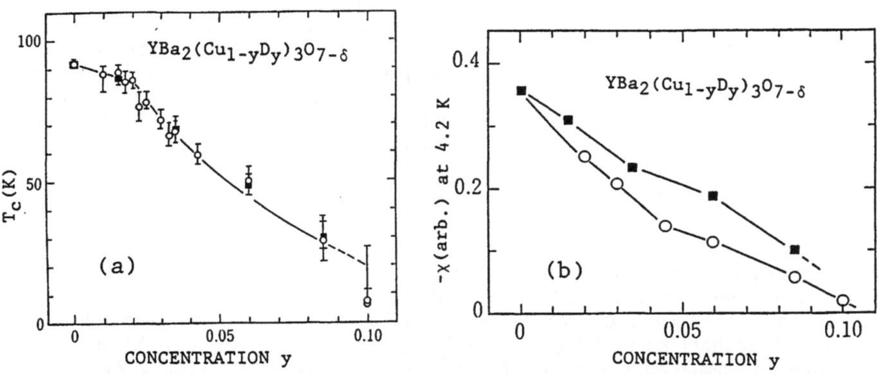

Fig. 6. Concentration dependences of (a)T_c and (b)diamagnetism in $YBa_2(Cu_{1-y}D_y)_3O_{7-\delta}$ with D = Fe(closed symbols) and Co(open symbols).

concentration of the substitutes. This can be another support for the plane superconductivity. However, the approximately linear reduction of diamagnetism also tempts us to speculate that the substitution destroys the superconductivity locally, while the sample remains superconducting as a whole because of short coherence length ξ. Further investigation is needed to make clear on this point.

Figure 7 shows the initial slope dT_c/dy of decrease in T_c for various element substitutions in $YBa_2(Cu_{1-y}D_y)_3O_{7-\delta}$ as well as in $La_{1.85}Ba_{0.15}(Cu_{1-y}D_y)O_4$[9]. There is no convincing interpretation for the general behavior of dT_c/dy. We did not succeed in finding any systematic correlation of T_c with the magnetic or non-magnetic substitution and the orthorhombic or tetragonal structure. In $La_{1.85}Ba_{0.15}(Cu_{1-y}D_y)O_4$, Fe substitution exhibits a different result from the other substitution. Non-magnetic Zn and Ga are effective in depressing T_c compared with Co and Ni. Also in $YBa_2(Cu_{1-y}D_y)_3O_{7-\delta}$, non-magnetic Zn and Ga substitutions depress T_c more effectively than Fe, Co and Ni substitutions, suggesting that the magnetic pair-breaking mechanism is not dominant. In Fig. 7, open triangles are the result for the samples which were well annealed in air at low temperatures between 350 and 380 K. Open circles are the data for the corresponding samples prepared without low-temperature annealing. Thus the substitution especially by high-valent elements appears to provide a sensitive dependence on the oxygen content. This is possibly a consequence of substitution in Cu1 and available room for extra oxygen in this crystalline structure. In case of $(La_{1-x}M_x)_2Cu_{1-y}D_yO_4$, however, all Cu sites are equivalent and no room for extra oxygen exists. Substitution of Fe, which particularly demands extra oxygen around it, reduces the effective number of oxygen which is supposed to provide carriers to the Cu-O networks. The effect can be relaxed in $YBa_2(Cu_{1-y}Fe_y)_3O_7$ by introducing extra oxygen. Similar argument will be applicable to the results for Co and Ga substitutions.

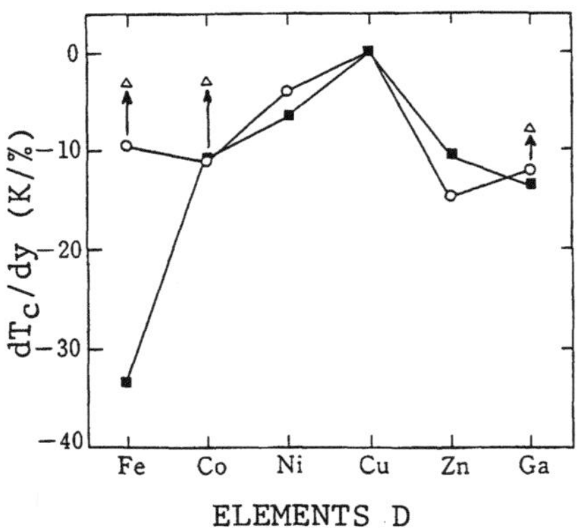

Fig. 7. Initial slope of T_c-depression, dT_c/dy, for various substitutions D in $YBa_2(Cu_{1-y}D_y)_3O_{7-\delta}$ (open circles: without low temperature annealing, open triangles: after annealing in air) and $La_{1.85}Ba_{0.15}(Cu_{1-y}D_y)O_4$ (closed squares).

The pressure dependence of T_c seems to have a strong correlation with the crystalline structure[20]. The rate of pressure enhancement of the transition temperature dT_c/dP is about 0.15 K/kbar for the air-annealed samples of orthorhombic $YBa_2(Cu_{1-y}D_y)_3O_{7-\delta}$ with D = Fe, Co, Ni and Zn. This rate is comparable with $dT_c/dP \simeq 0.17$ for $La_{1.85}Ba_{0.15}CuO_4$[21] which is orthorhombic at low temperatures. However, the rate is about four times larger, i.e. $dT_c/dP \sim 0.6$ K/kbar, for the substitution-induced tetragonal $YBa_2(Cu_{1-y}D_y)_3O_{7-\delta}$ with D = Fe and Co. At present, it is an open question whether the result reflects the direct difference in structure or the resultant difference in compressibility.

4. TRANSPORT PROPERTIES AND MAGNETISM IN THE NORMAL STATE

A striking feature of the superconductive copper oxides with high quality is the linear dependence of normal-state resistivity on temperature[22]. As foreign elements are substituted in Cu sites, however, the normal-state resistivity of the samples generally increases and the upturn in resistivity grows remarkably at low temperatures. As a result, the slope of resistance-temperature curves changes in sign from metal-like to semiconductor-like behavior as demonstrated in Fig. 5. This change may be interpreted in terms of localization. Figure 8 shows an example of the thermoelectric power S for $YBa_2(Cu_{1-y}Fe_y)_3O_{7-\delta}$. In the pure sample with y = 0, the observed value of S is very small and almost temperature-independent. As Fe content increases, S increases significantly. Similar behavior is also observed when we measure the Hall coefficient R_H as shown in Fig. 9. These results indicate that the positive carriers predominantly contribute to the transport properties and the carrier concentration decreases continuously with substitution. This is consistent with the preceding argument that the effective number of free carriers is reduced by trapping of oxygen around high-valent Fe and/or by localization.

The magnetic susceptibility of $La_{1.85}Sr_{0.15}CuO_4$ has a remarkable feature in the normal sate, i.e. a linear but weak temperature-dependence over a wide range of temperature. The effect of substitution on the susceptibility has also been investigated for $La_{1.85}Sr_{0.15}Cu_{1-y}D_yO_4$[8]. Room-temperature susceptibility increases with Ni substitution, whereas

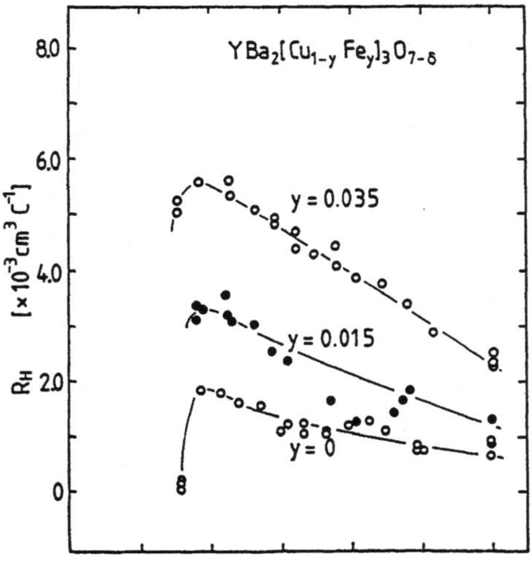

Fig. 8. Thermoelectric power S of $YBa_2(Cu_{1-y}Fe_y)_3O_{7-\delta}$.

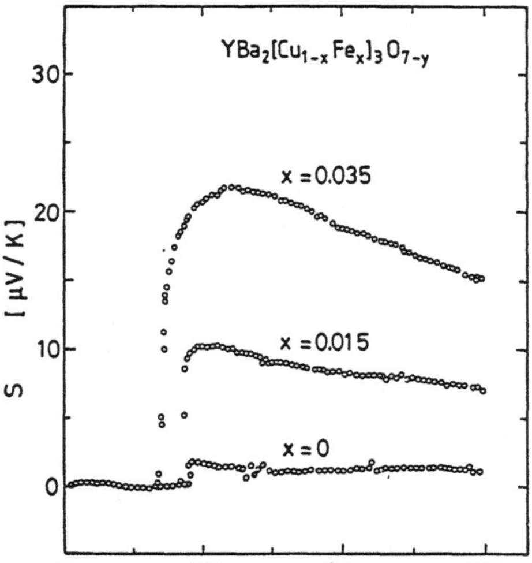

Fig. 9. Hall coefficient R_H of $YBa_2(Cu_{1-y}Fe_y)_3O_{7-\delta}$.

it decreases with Zn substitution. In both cases, an increase of Curie-Weiss type is observed at low temperatures and fitted by the formula $\chi = C/(T-\Theta)+\chi_0$. The result suggests that foreign element substitution induces small magnetic moment on the neighboring Cu sites. Essentially similar behavior is observed also in $YBa_2(Cu_{1-y}D_y)_3O_{7-\delta}$. By fitting the observed susceptibility to the formula of Curie-Weiss type, the effective magnetic moments of 4.03 μ_B/Fe and 4.08 μ_B/Co are estimated for slight substitution of Fe and Co, respectively[23]. These values appear to decrease with increasing concentration y. Mössbouer spectrum[24] shows a hyperfine splitting at low temperatures. This suggests not only the existence of magnetic moment on Fe, but also the possibility of coexistence of superconductivity and induced magnetic order. Neutron experiment[25] of oxygen deficient $YBa_2(CU_{1-y}Co_y)_3O_{7-\delta}$ revealed that Co substitution not only modifies the original antiferromagnetism on the Cu2 sites, but also induces an ordered magnetic on the Cu1 sites. Since the the high-T_c superconductivity in copper oxides is supposed to have some kind of affinity with magnetism, further contribution is expected from substitution experiments.

ACKNOWLEDGEMENTS

The author wish to thank his collaborators, especially Dr.Y.Maeno, Prof.J.Sakurai, Prof.K.Okuda, Prof.A.Ito, Dr.M.Kurisu and Dr.T.Fujiwara for their important contributions and valuable discussion. Thanks are also due to his students Y.Aoki and T.Tomita for their assistance in experiments and to Dr.K.Satoh and T.Nojima for their assistance in preparing the manuscript.

REFERENCES

1. R.J.Cava, B.Batlogg, J.J.Krajewski, R.C.Farrow, L.W.Rupp.Jr., A.E.White, K.T.Short, W.F.Peck.Jr. and T.Y.Kometani: Nature 332 (1988) 814.
2. J.G.Bednorz and K.A.Müller: Z. Phys. B64 (1986) 189.
3. K.Kishio, K.Kitazawa, S.Kanbe, I.Yasuda, N.Sugii, H.Takagi, S.Uchida, K.Fueki and S.Tanaka: Chem. Lett. (1987) 429.
4. M.K.Wu, J.R.Ashburn, C.J.Torng, P.H.Hor, R.L.Meng, L.Gao, Z.J.Huang, Y.Q.Wang and C.W.Chu: Phys. Rev. Lett. 58 (1987) 908.
5. Y.Maeno, T.Tomita, M.Kyogoku, S.Awaji, Y.Aoki, K.Hoshino, A.Minami, and T.Fujita: Nature 328 (1987) 512.
6. H.Maeda, Y.Tanaka, M.Fukutomi and T.Asano: Jpn. J. Appl. Phys. 27 (1988) L209.
7. Z.Z.Sheng and A.M.Hermann: Nature 332 (1988) 55.
8. T.Fujita, Y.Aoki, Y.Maeno, J.Sakurai, H.Fukuba and H.Fujii: Jpn. J. Appl. Phys. 26 (1987) L368.
9. J.M.Tarascon, L.H.Greene, P.Barboux, W.R.McKinnon, G.W.Hull, T.P.Orlando, K.A.Delin, S.Foner and E.J.McNiff.Jr.: Phys. Rew. B36 (1988) 8393.
10. Y.Maeno, M.Kato, Y.Aoki and T.Fujita: Jpn. J. Appl. Phys. 26 (1987) L1982. Y.Maeno and T.Fujita: Physica C 153-155 (1988) 1105.
11. T.Fujiwara, H.Nakata, T.Kumamaru, Y.Maeno, Y.Aoki and T.Fujita: Chem. Lett. (1988) 1527.
12. J.M.Tarascon, P.Barboux, P.F.Miceli, L.H.Greene, G.W.Hull, M.Eibschutz and S.A.Sunshine: Phys. Rev.B37 (1988) 7458.
13. G.Roth, G.Heger, B.Renker, J.Pannetier, V.Caignaert, M.Hervieu and B.Raveau: Z. Phys. B71 (1988) 43.

14. M.Takano, Z.Hiroi, H.Mazaki, Y.Bando, Y.Takeda and R.Kanno: Physica C153-155 (1988) 860.
15. P.Bordet, J.L.Hodeau, P.Strobel, M.Marezio and A.Santoro: Solid State Comm. 66 (1988) 435.
16. E.Takayama-Muromachi, Y.Uchida and K.Kato: Jpn. J. Appl. Phys. 27 (1988) L2087.
17. T.Kajitani, K.Kusaba, M.Kikuchi, Y.Syono and M.Hirabayashi: Jpn. J. Appl. Phys. 26 (1987) L1727.
18. T.Kajitani, K.Kusaba, M.Kikuchi, Y.Syono and M.Hirabayashi: Jpn. J. Appl. Phys. 27 (1988) L354.
19. G.Xiao, F.H.Streitz, A.Gavrin, Y.W.Du and C.L.Chien: Phys. Rev. B35 (1987) 8782.
20. M.Kurisu, K.Kumagai, Y.Maeno and T.Fujita: Physica C152 (1988) 339.
21. M.Kurisu, H.Kadomatsu, H.Fujiwara, Y.Maeno and T.Fujita: Jpn. J. Appl. Phys. 26 (1987) L361.
22. M.Gurvitch and A.T.Fiory: Novel Superconductivity ed. by S.A.Wolf and V.Z.Kresin, (Plenum Press,1987) p663.
23. S.Noguchi, J.Inoue, K.Okuda, Y.Maeno and T.Fujita: Jpn. J. Appl. Phys. 27 (1988) L390.
24. T.Tamaki, T.Komai, A.Ito, Y.Maeno and T.Fujita: Solid State Commun. 65 (1988) 43.
25. P.F.Miceli, J.M.Tarascon, L.H.Greene, P.Barboux, M.Giroud, D.A.Neumann, J.J.Rhyne, L.F.Schneemeyer and J.V.Waszczak: Preprint.

Strong Dependence of T_c on Hole Concentration in CuO_2 Sheets

Y. Tokura[1;2], *J.B. Torrance*[2], *A.I. Nazzal*[2], *H. Takagi*[3], *and S. Uchida*[3]

[1]Department of Physics, University of Tokyo, Tokyo 113, Japan
[2]IBM Research Devision, Almaden Research Center, CA 95120-6039, USA
[3]Department of Applied Physics, University of Tokyo, Tokyo 113, Japan

We review here recent experimental investigations on the dependence of T_c on hole concentration in High T_c cuprates, $La_{2-x}Sr_xCuO_4$, $YBa_2Cu_3O_y$ - like compounds and also the new compound $Nd_{2-x-y}Ce_xSr_yCuO_4$. In all these compounds, the hole concentration in the sheet of Cu-O pyramids or octahedra is the major experimental variable closely correlated with T_c.

1. Introduction

Since the discovery of high temperature superconductivity in cuprate compounds[1], extensive studies have been carried out to elucidate why those show High T_c or what is crucial for High T_c. There are widely diverse ideas so far proposed , yet there is, at least, a general consensus that introduction of holes into CuO_2 sheets are indispensable for appearance of High T_c. Here, we would like to review recent experimental results on the dependence of T_c on hole concentration (p) per [CuO_2] unit in representative High T_c cuprates. The first topic is on the strong correlation between T_c and p in the possibly simplest compound $La_{2-x}Sr_xCuO_4$. In particular, we summarize the recent investigation [2] over an extended range of p, which has revealed anomalous disappearance of superconductivity beyond $p \sim 0.3$. A similar T_c-p correlation has been studied also in YBa_2Cu_3Oy-like systems. The critical dependence of superconducting properties of $YBa_2Cu_3O_y$ on the oxygen content y, has been investigated by a number of researchers. By decreasing the oxygen content, the average hole concentration p_{av} in the sample can be continuously changed. At the same time, however, there are important modifications to the structure, which contains chains near y=7, but which are no longer present for y=6. Here, we describe the way to decouple the contributions of y and p, and present a broader perspective on the electronic phase diagram in the p-y space[3]. As the final topic, we present recent experimental results on the new superconductor $(Nd,Ce,Sr)_2CuO_4$, which shows the single Cu-O layer structure with the network of pyramids. In these materials, hole concentration p, which is again the major experimental variable, is critically dependent on the relative concentration of tetravalent (Ce^{4+}) and divalent (Sr^{2+}) cations to the host trivalent(Nd^{3+}) cations as well as on oxygen deficiencies.[4]

2. Dependence of T_c on hole concentration in $La_{2-x}Sr_xCuO_4$

Samples of $La_{2-x}Sr_xCuO_{4-\delta}$ have previously shown a maximum concentration of p=0.15-0.20 holes per [CuO_2] unit, because increasing x > 0.15-0.20 normally induces compensating oxygen vacancies. The oxygen vacancies

Springer Series in Materials Science, Vol. 11
Mechanisms of High Temperature Superconductivity
Editors: H. Kamimura and A. Oshiyama
Springer-Verlag Berlin Heidelberg © 1989

greatly complicate the properties of this potentially simplest system : for example, the sample with x > 0.15 often shows degraded superconducting behavior or evidence of phase separation. To investigate the properties in a wider range of p, the formation of oxygen vacancies should be inhibited in the sample with higher x. Recently, Torrance et al.[2] have found the way to prepare the high-x sample with no oxygen vacancies. By annealing a set of samples at 600°C in 100 bars of oxygen pressure, one can fill the oxygen vacancies, which are inevitablly introduced in the process of firing the samples with x>0.15 at temperatures above 1000° C, and hence extend the range of accessible hole concentrations, up to p=0.4.

In Fig.1, the measured hole concentration is shown versus x for a series of samples from Shafer et al.[5], Nguyen et al[6]., and Torrance et al.[2] (1bar and 100bar O_2). In those experiments, the hole concentration p is measured directly by determining the charge on the [Cu-O] unit with use of an iodometric titration technique[7]. In oxygen deficient samples, the general formula is given by $La_{2-x}Sr_xCuO_{4-\delta}$, where the hole concentration in the CuO_2 sheet is given by conservation of charge: $p=x-2\delta$.
The dashed line in Fig.1 shows the relation between p and x with no vacancies (δ=0). As seen in Fig.1, the hole concentration increases as the samples are doped with higher concentration of Sr until the onset of oxygen vacancies (marked by arrows in each case in the figure). With the procedure of annealing the samples at 600°C in 100bars of oxygen, the range of hole concentration has clearly increased up to p=+0.40.[8]

In Fig.2 we show the dependence of T_c on hole concentration, which was first presented by Torrance et al.[2]. For low hole concentrations are plotted the Meissner data by van Dover et al.[9] and resistance data of Shafer et al.[5] (when resitivity is equal to 10% of value just above T_c). The Meissner data in ref.[2] on the samples annealed in 100bars of oxygen

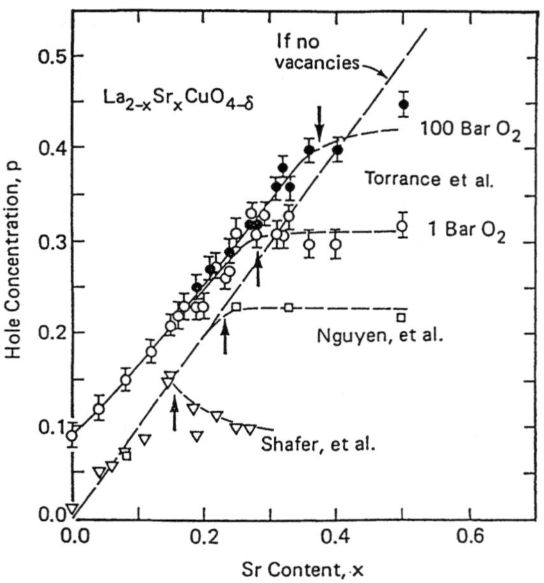

Fig.1. The measured hole concentration, p, vs the Sr content for four sets of samples.

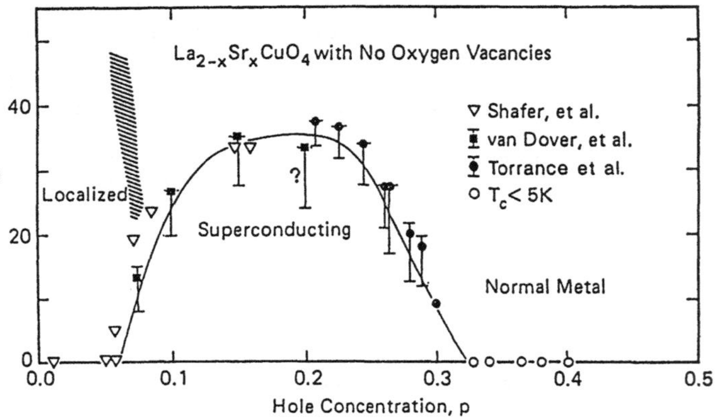

Fig.2. The dependence of T_c on hole concentration in $La_{2-x}Sr_xCuO_4$.

are plotted for high hole concentration above p=0.2. Meissner data in samples of $La_{2-x}Sr_xCuO_4$ usually show the broad width of the transition. This feature is attributed, not to poor sample quality, but rather to a distribution of local Sr concentrations. For this reason, the onset temperature is defined as the temperature when the Meissner susceptibility (χ_M) has increased to $-0.05/4\pi$. To represent the distribution of local T_c, the bar is plotted in Fig.2, showing the temperature width over which χ_M is between 20% and 80% of its observed low-temperature value.

The most striking feature seen in Fig.2 is the anomalous disappearance of superconductivity above p=0.32, even though the sample is more conducting. Evidently, High T_c phenomena are found in a somewhat narrow window of hole concentration. In particular, T_c appears almost independent of hole concentration from $p\simeq 0.13$ to 0.24, but above this value, T_c begins to fall toward zero. Over the entire range above $p\sim 0.15$, the conductivity is always metallic, with a constant resitivity ratio of $\rho(300K)/\rho(50K)\sim 4.5$. These clear behaviors in the simplest system provide a key test of theories for High T_c.

3. Electronic phase diagram in the p (hole concentration) vs y (oxygen content) space in $YBa_2Cu_3O_y$-like system

The superconducting properties in $YBa_2Cu_3O_y$-system are critically dependent on both of the experimental parameters; the oxygen content(y) and the average hole concentration (p_{av}, or effective Cu valency, $p_{av}+2$). To decouple these aspects and better understand their respective roles in superconductivity, we have synthesized a series of new $YBa_2Cu_3O_7$-like compounds, each having a different total charge, Q, on the non-copper cations. Using the guideline of one small ion (on Y-site) and two large ions (on Ba-site) per unit cell in the so-called 1-2-3 structure, we could prepare the site-selectively doped systems; $Y(Ba_{1-x}La_x)_2Cu_3O_y$ and $(Y_{1-x}Ca_x)Ba_2Cu_3O_y$.[3] By substituting the small Ca^{2+} ion for Y^{3+}, the electronic charge (Q) on the non-copper cation sites has been reduced from 7 down to 6.5. Substitution of the large La^{3+} for the large Ba^{2+} yields compounds with larger values of Q, up to Q=7.8.

The average charge, p, per $[Cu\text{-}O]^{+p}$ unit, can be measured by using an iodometric titration technique[7]. From Q and this measurement of p, the oxygen content, y, is determined using the condition of electrical neutrality: $Q+3(p+2)=2y$. In other words, the experimental results in the Q-y space, which corresponds to the procedure of changing y in the specific sample having Q, can be immediately transformed to the data points in the p-y space, as shown in Fig.3 [3]. Large differences are observed in the conductivity behaviour of samples in these series and are included in Fig.3 by classifying these data into four types of behavior: (1) metals showing T_c>65K (open ellipses); (2) metals or semi-metals showing 65>T_c>20K (open rectangles); (3) semiconductors, but having 20K>T_c (half open/half solid rectangles); and (4) insulators (with fairly large activation energies ~0.1eV) showing no superconductivity (solid rectangles). The superconducting transition measured by resistivity had width of 1-10K and T_c was defined as the transition temperature of zero rsistance (included inside the ellipse or rectangle in Fig.3).

In the p-y space shown, there is a large region of compounds with high T_c(>65K), but more surprisingly a large region of compounds where the samples are insulating. In between those High T_c and insulating phases is a small region in which the samples show intermediate T_c (65>T_c>20K). The $YBa_2Cu_3O_7$-system with various y corresponds to the trajectory indicated by a dashed line Y1:2:3 in Fig.3 and its behavior is consistent with that observed for this larger series. The data for the samples in Fig.3 may be used to study the effects of y and p separately. For example, measured T_c in the samples with constant oxygen constant y~7.0 is plotted as a function

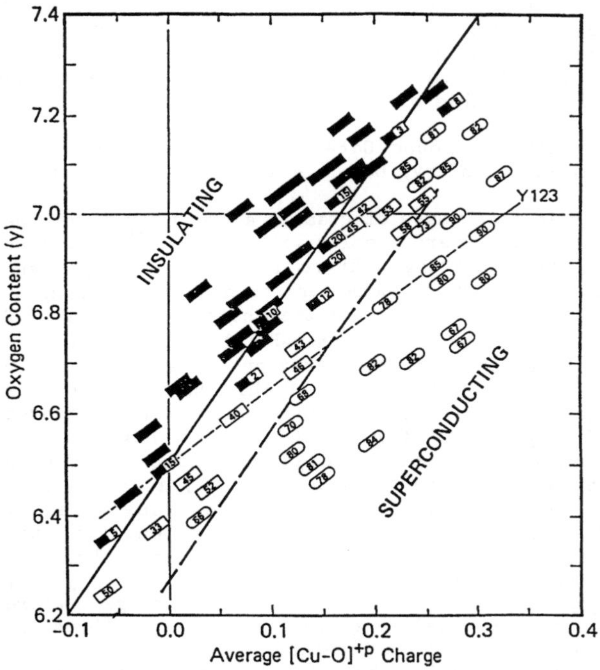

Fig.3. The electronic phase diagram in the p-y space in $YBa_2Cu_3O_y$-like compounds.

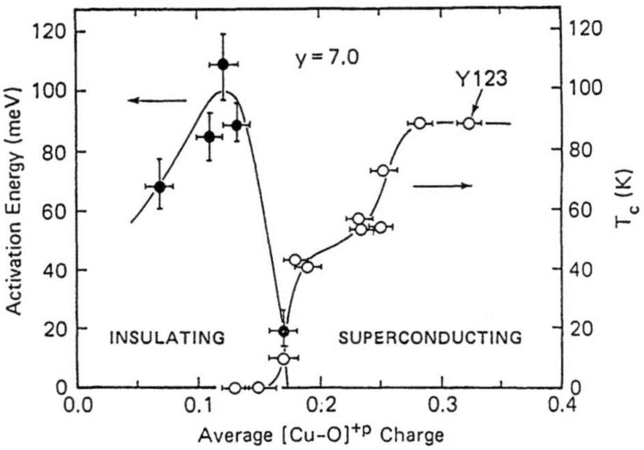

Fig.4. Activation energy and T_c for series of samples with y~7.0 as a function of the average $[Cu\text{-}O]^{+p}$ charge.

of the average hole concentration p in Fig.4, together with the thermal activation energy of conductivity in the insulating phase. The boundary between insulating and superconducting phases is evident near p=+0.17 and T_c exhibits two plateaux behavior as seen in $YBa_2Cu_3O_y$ observed by varying y. Such a boundary in the whole y-p space (fig.3) is determined neither by y nor p alone, but by a combination of y and p. It is to be noted here that the orthorhombic/tetragonal phase boundary [10] does not follow the supercondoctor/insulator boundary, indicating that the Cu-O chain-like order is not the key to understand this electronic phase diagram.

(Furthermore, the new 123-like compound $CaBa_{1-x}La_{1+x}Cu_3O_y$ has a very similar superconductor/insulator boundary, but is always tetragonal.[11])

The most striking feature in this phase diagram is a broad region of the insulating phase. For example, there are samples with a $[Cu\text{-}O]^{+p}$ charge as high as p=+0.25, yet they are insulating. There is only one simple and reasonable explanation for the pattern of behavior shown in Fig.3: samples with p>0 must be conducting if the holes are in the sheets. Then, all of the $[Cu\text{-}O]^{+p}$ charge (or holes) must be in the inter-Ba plane (or on the so-called chain site). On this assumption, the behavior for y=7.0 in Fig.5 is described as follows: for small p(<0.17), all of this charge resides on the chains and these holes are localized. As the charge p is increased beyond the threshold value p_c(=0.17 for y=7.0), holes begin to be introduced into the sheets, giving rise to metallic conductivity (and superconductivity). The measured $[Cu\text{-}O]^{+p}$ charge in Fig.3 is an average over the charge in the sheets and that in the inter-Ba plane (chain): $p=(2p_{sh}+p_{Ba})/3$. Below the threshold value p_c, or equivalently in the insulating phase, $p_{sh}=0$ and then $p_{Ba}=3p$. The experimental phase boundary in Fig.3 is approximately given by $3p_c=y-6.5=p_{Ba}$. At threshold for y=7.0, $p_{Ba}=+0.5$, meaning that every other [Cu-O] in the chains is charged before holes are put into the sheets. For smaller oxygen content, the threshold charge p_c decreases, as expected. In particular, $p_c=3p_{Ba}=0$ for y=6.5. This means the remarkably simple situation where the one extra oxygen added to the y=6.5 chain has the capacity to trap one hole. Taking account of

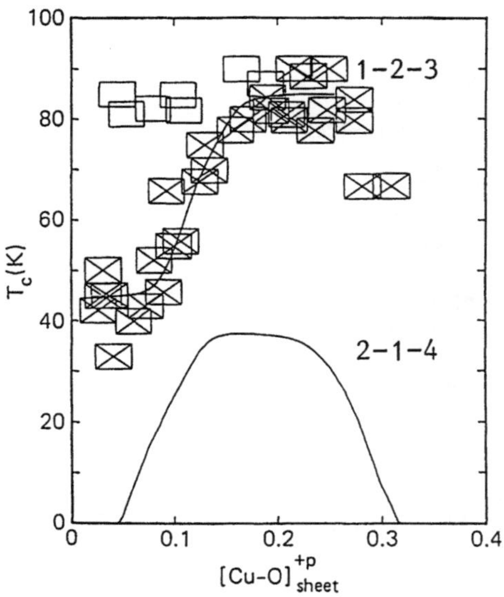

Fig.5. T_c for all samples in Fig.3 vs the $[Cu-O]^{+p}$ charge on the sheets. The open squares are for samples with y>7. The lower curve represents the T_c vs p relation in $La_{2-x}Sr_xCuO_4$ (see Fig.1), for comparison.

inter-site (Madelung-type) Coulombic interaction, Kondo [12] has recently presented the explanation for the presently observed pattern of distribution of doped holes in the sheets and chains.

The above-mentioned analysis can be continued beyond threshold into the region of superconducting samples if we assume that the charge in the inter-Ba plane cannot be increased beyond the threshold value ($p_{Ba}=3p_c$), so that all of the increased charge comes from the sheets. Then $p_{sh}=3(p-p_c)/2$. Using this analysis to determine p_{sh} for the superconducting sample, T_c is plotted versus p_{sh} in Fig.5 [3], including all samples in Fig.3, for comparison with the results on $La_{2-x}Sr_xCuO_4$ described in the previous section. The results in these $YBa_2Cu_3O_y$-like compounds show plateax of T_c near 40K and 80K. As clearly seen in this plot, the same well defined behavior is obtained for samples with oxygen contents ranging from 6.25 to 7.00 and with Q from 6.5 to 7.6. This demonstrates that the principal variable correlated with T_c in this system is the hole concentration on the sheets like in the case of $La_{2-x}Sr_xCuO_4$. (A few samples do not have the same behavior for small p_{sh}, perhaps because they have y>7).

4. New superconductor $(Nd,Ce,Sr)_2CuO_4$ with single layer sheet of Cu-O pyramids: charge compensation effect

High T_c cuprates exhibit four types of Cu-O sheets; networks of (a) octahedra, (b) pyramids, (c) isolated squares, and (d) chains. For example the strucure of $La_{2-x}Sr_xCuO_4$ (Fig. 6(a) shows the sheet of octahedra, whereas $YBa_2Cu_3O_7$ consists of the Cu-O pyramids and chains. Furthermore, triple- or quadruple-layer Bi(Tl)-Ca-Sr(Ba)-Cu-O compounds with T_c above 100K show the both of the pyramids and squares. All High T_c cuprates so

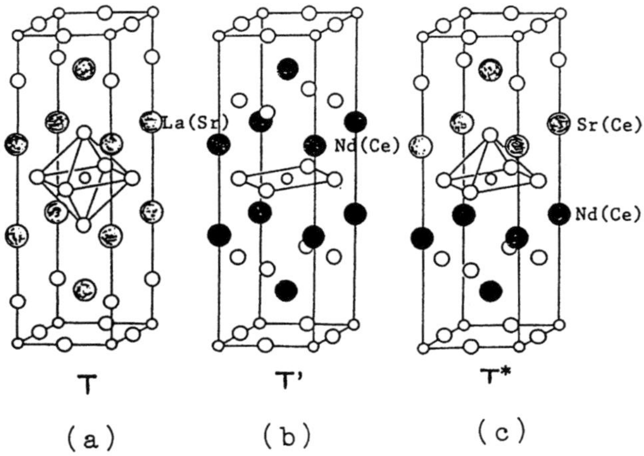

Fig.6. Crystal structures of (a) $La_{2-x}Sr_xCuO_4$ (T-phase), (b) $Nd_{2-x}Ce_xCuO_4$ (T'-phase), and (c) $Nd_{2-x-y}Ce_xSr_yCuO_4$ (T*-phase).

far discoverd have at least either of octahedron- or pyramid-type Cu-O sheets, which are doped with holes. At present, however, it is still controversial where the chemically doped holes are dominantly positioned in the Cu-O sheet, or more specifically whether the apical oxygen above and/or below the CuO_2 sheet are necessary or nct for High T_c. To approach this problem from an experimental point of view, compounds with single-layer Cu-O sheet may be useful as simple model systems. Fig.6 shows candidates for such compounds; (a) $La_{2-x}Sr_xCuO_4$ (T-phase), (b) $Nd_{2-x}Ce_xCuO_4$ (T'-phase), and (c)$Nd_{2-x-y}Ce_xSr_yCuO_4$ (T*-phase), with the Cu-O network of octahedra, squares and pyramids, respectively. Among them, superconductivity in $(Nd,Ce,Sr)_2CuO_4$ has been discovered by Akimitsu et al.[13] and its structure (shown in Fig.6(c)) has been solved recently by means of electron and neutron diffraction experiments[14,15]: it is essentially the hybrid of the T-and T'-phase, as shown in the figure, due to the partial ordering of Nd(Ce) and Sr along the c-axis. Systematic studies on attempts to introduce extra electrons or holes into Cu-O sheets in these Nd-based cuprates have been recently carried out by the present authors and their coworkers[4].

Let us first describe the attempts to dope electrons or holes into Nd_2CuO_4 of the T'-phase. In the T'-phase (Fig.6(b)), additional oxygens are located above and below the oxygen sites on CuO_2 sheet and hence the Cu-O network forms the sheet of isolated squares. If the Nd_2CuO_4 type cuprates could be made to be one of High T_c compounds, it would be quite decisive of the position of doped holes. However, all attempts to introduce an appreciable density of hole to the T'-phase have been so far unsuccessful: for example, Sr-doping into Nd_2CuO_4 causes rapid phase separation, where a dominant second phase is double-layer perovskites (e.g. Nd_2SrCuO_6) with p=0.

On the other hand, Ce ions can be dissolved in the T'-lattice to some extent (up to ca.10%). Fig.7 [4] shows temperature-dependence of resistivity in $Nd_{2-x}Ce_xCuO_{4-\delta}$. In the figure is also plotted the result for the Ce-undoped, but highly oxygen-deficient ($\delta\sim0.04$) sample, which was obtained by rapid quenching from air atmosphere at 1150°C. The Ce-doped or

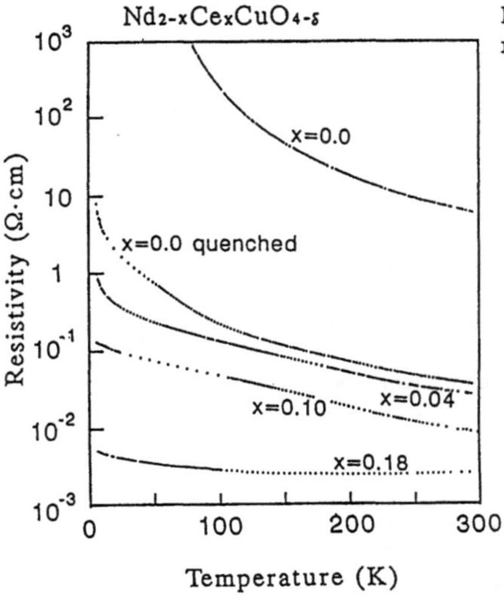

Fig.7. Temperature-dependence of resistivity in $Nd_{2-x}Ce_xCuO_4$.

oxygen-deficient samples show much reduced resistivity and semi-metallic, but not superconducting above 10K, behavior. Preliminary meausrements on Hall coefficient and thermopower in these samples both indicate that the charge carriers are electron-type. This clearly shows that the tetravalent Ce ion embedded in T'-Nd_2CuO_4 or oxygen defficiency introduces itinerant electrons into the compound. This is in contrast with the case of T-phase La_2CuO_4, where holes, not electrons, can be introduced by chemical doping of divalent ions or La deficiency.

Contrary to the Ce-undoped T'-phase compound, $Nd_{2-x}Ce_xCuO_4$ with x=0.1-0.2 can be doped with Sr ions without any trace of phase separation. However, this procedure makes the sample less conducting, indicating the doped divalent Sr ions partly compensate the charge and hence reduce the number of itinerant electrons. When the concentration of Sr is further increased, in particular beyond that of Ce, the lattice of $Nd_{2-x-y}Ce_xSr_yCuO_4$ begins to undergo the distinct transformation from the T'-phase (Fig.6(b)) to the T^*-phase (Fig.6(c)). In this case, Sr^{2+} ion over-compenstates the charge initially increased by Ce^{4+} in Nd^{3+} lattice, so that the the observation implies again that the T'-lattice with no apical oxygens cannot support doped holes, but the T^*-lattice can.

The T^*-phase, which forms single-layer sheets of Cu-O pyramids, can be superconducting, as shown for the compound $Nd_{1.4}Sr_{0.4}Ce_{0.2}CuO_{4-\delta}$ in Fig.8[4]. Properties of the T^*-phase compounds have been found to be quite senstive to annealing treatments in oxygen atmosphere. Fig.8 shows the Meissner data on the samples prepared by different thermal treatments. The sample quenched from air atmosphere at 1150°C down to room temperature does not show superconductivity as shown in Fig.8 and the temperature dependence of resistivity is semiconducting (not shown). Annealing the same sample in oxygen atmosphere much reduced resistivity and gave rise to superconductivity. In particular, extensive annealing in the oxygen atmosphere of 80bars appears to well fill oxygen vacancies and much improve the superconducting property: by this procedure the onset T_c in resitivity (shown in the right part of Fig.8) is increased up to 31K. Hole

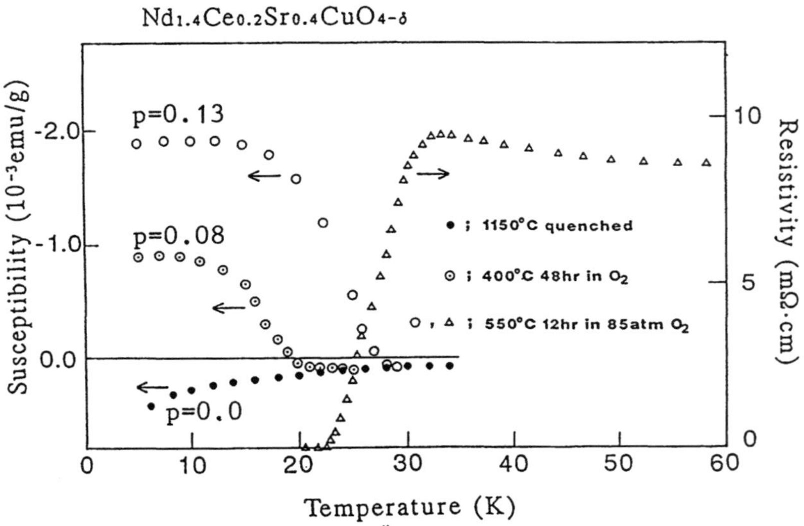

Fig.8. Meissner effect in the T^*-phase compounds of $Nd_{1.6}Ce_{0.2}Sr_{0.4}CuO_{4-\delta}$.

concentration (p) measured on each sample is shown in the figure. It appears that p is again the major variable correlated with T_c, although in this case it is difficult to separate contribution of oxygen vacancies from that of hole concentration.

In the T'-phase, superconducting carriers are hole-type, as is evident from the remarkable dependence on oxygen deficiencies. Measurements on Hall coefficients were also carried out on several superconducting samples, which always showed positive values of $(1\text{-}3)\times10^{-3}cm^3/C$, indicating again the hole-type carriers. It is to be noted here that the superconducting T^*-phase is stabilized by subtle balance of electronic charge on cation sites. Crystallographycally, the concentration of Sr included in the T^*-phase must be high enough, because the sequential ordering of Nd and Sr is essential for the T^*-phase[14,15]. With too high Sr-concentration, however, the sample would have an amount of oxygen vacancies, which would rather decrease the hole concentration, like in the case of highly Sr-doped La_2CuO_4 samples, or even unstabilize the T^*-structure. Tetravalent Ce ions dissolved in the Nd-lattice compensate the over-decreased charge on the non-copper cation sites and hence serves to prevent oxygen vacancies or phase separation. The tetravalency of Ce ion in the superconducting T^*-phase was confirmed by Ce core-level X-ray photoemission spectroscopy by Fujimori et al.[4].

From the above results on Nd-based cuprates, it appears that holes can be chemically doped into the sheets of Cu-O pyramids or octahedra, but perhaps not into the CuO_2 sheets of squares. However, it does not necessarily mean that the doped hole is positioned on the apical oxygen site. We will have to further pursue a real role of the apical oxygen: how does it change electronic structures of the ground state and of lower-lying excited states in High T_c cuprates?

We would like to gratefully acknowledge collaboration and discussion with T.C. Huang, S.S.P.Parkin, A.Bezinge, Y.Sakai, S.Okuda, K.Ikeda, H.Matsubara. H.Watabe and A.Fujimori. Y.T. also wishes to thank IBM Japan and IBM Almaden Research Center for generously supporting his stay at Almaden.

References

1. G.Bednorz and K.A.Muller : Z.Phys.B 64, 189(1980).
2. J.B.Torrance, Y.Tokura, A.I.Nazzal, A.Bezingl, T.C.Huang and S.S.P.Parkin : Phys.Rev.Lett. 61, 1127(1988).
3. Y.Tokura, J.B.Torrance, T.C.Huang and A.I.Nazzal : Phys.Rev.B 38, 7156(1988).
4. Y.Tokura et al.: to be sumitted.
5. M.W.Shafer, T.Penny and B.L.Olson: Phys. Rev. B36, 4047 (1987).
6. N.Nguyen, F.Suder and B.Raveau: J.Phys.Chem.Solids: 44, 389(1983).
7. A.I.Nazzal et al.: Physica 153-156C, 1367(1988).
8. The samples prepared in ref.[2] have p above the dashed line (p=x), which may be attributed to the presence of La vacancies ($\sim$1.5%). It has been later noticed that these La vacancies can be much reduced by the improved method to prepare the homogeneous samples of $La_{2-x}Sr_xCuO_4$. See the paper by Takagi et al. in the Proceedings.
9. R.B.van Dover, R.J.Cava, B.batlogg and E.A.Reitman: Phys.Rev.B35, 5337(1987).
10. T.C.Huang, Y.Tokura, J.B.Torrance, A.I.Nazzal and R.Kamiri: Appl.Phys. Lett. 52, 1901(1988).
11. Y.Tokura et al.: to be submitted.
12. J.Kondo: J.Phys.Soc.Jpn.(to be published).
13. J.Akimitsu, S.Suzuki, M.Watanabe and H.Sawa: Jpn.J.Appl.Phys.27, L1859(1988).
14. E.Takayama-Muromachi et al.: Jpn.J.Appl.Phys. (to be published).
15. H.Sawa et al.: to be published.

Electron Microscopic Study on Ti-Ba-Cu-O Superconductor Oxides

S. Iijima

NEC Corporation, Fundamental Research Laboratories, Miyazaki, Miyamae-ku, Kawasaki 213, Japan

Electron microscopic observations on Tl-Ba-CA-Cu-O (TBCCO) complex oxide systems reveal new, ordered-intergrowth phases, solid-state reaction between the 2212 and 2223 phases, and superstructure modulation in the 2201 phase. No Ca-containing 2201 crystals, which show significant variation in Tc values from non-superconductor to 80K, are examined in terms of super-lattice modulation. The superlattice, having a lattice vector of $Q = \pm[3/14, \pm 1/14, \pm 1]$, occurs on {310} planes. The disordered superlattice results in diffuse rings appearing in the electron diffraction pattern.

1. INTRODUCTION

Since the TBCCO oxides were found to be superconductors [1], the oxides were intensively studied by many researchers. The oxides form a homologous series of compounds whose chemical formulae are expressed by $Tl_mBa_2Ca_{n-1}Cu_nO_{2n+6+x}$, which is denoted as (m,2,n-1,n). Their crystal structures are composed of perovskite-like slabs of CuO_2-Ca-CuO_2 and also single slabs of TlO or double slabs of TlO-TlO, denoted by m = 1 or 2. The thickness of the CuO_2-Ca slabs is described by n, corresponding to the number of the CuO_2 layers. The Tc value has been thought to rise with the increase in the number of n. This idea, however, was found to hold only for structures up to n = 3 [2].

Among a series of the compounds common phases are the 2212 and 2223. They correspond to n=2 and 3, and their Tc values are around 105 and 120K, respectively. These Tc values have been recognized by many researchers. On the contrary, the Tc value for the 2201 phase showed a wide range of variations between nonsuperconductor and 80K [3,4,5]. This phase is particularly of interest because it contains no Ca atoms at all. Therefore, ambiguity in Ca ion substitution for other metal ions in the 2201 structure can be avoided. In earlier reports the Tc variation was thought to be caused by a trace of Ca atoms [5]. The role of the Ca additives, however, was denied by the experiment where the 2201 crystals prepared without Ca showed superconductivity [4].

X-ray crystal structures for the 2201 crystals, which were prepared by sintering the raw powder materials at a temperature of around 870°C, are mostly tetragonal. Some crystals showed superconductivity, but others did not. During the preparation of this paper, Hewat and co-researchers [6] reported that superconductive 2201 crystals were tetragonal, while non-superconductive ones were orthorhombic. Shimakawa et al.[7], however, found that the superconductivity of the 2201 phase occurred equally in both tetragonal and orthorhombic crystals. According to the electron microscope observation of the TBCCO oxides, these crystals contained superlattice modulation [3]. The lattice modulations similar to that of the TBCCO oxides have been reported on the BSCCO oxides [8]. We have been looking for a possible correlation between the lattice modulation and variation in the Tc values.

Springer Series in Materials Science, Vol. 11
Mechanisms of High Temperature Superconductivity
Editors: H. Kamimura and A. Oshiyama
Springer-Verlag Berlin Heidelberg © 1989

In the present report, we shall show some microstructures on TBCCO oxide crystals revealed by an electron microscope. In particular, high resolution electron microscope images of TBCCO, "ordered intergrowth phases", solid state reaction between the 2212 and 2223 phases, and superlattice modula tions in the 2201 phase are presented.

2. SPECIMENS AND OBSERVATION

TBCCO oxides examined in the present experiment were prepared by the conventional sintering of the powdered materials. For specimen 2201, powder materials, Tl_2O_3, BaO and CuO, with a 2:2:1 mixture ratio were pressed into pellet forms. Then they were heated at 900°C for 3hrs, in a 760 Torr oxygen atmosphere. After heating, the pellets were oven-cooled, or some of them were quenched.

For electron microscope observation, the sintered specimens were crushed in an agate mortar and pestle in ethylalcohol. A drop of the suspension was taken on an electron microscope carbon-microgrid. In this way specimen flakes were suitably distributed on the grid. Unlike BSCCO crystals, the TBCCO crystals do not have a dominant (001) plane cleavage habit. This makes it easier to find crystal orientations perpendicular to the c-axis. High resolution electron microscope images (HREM), selected area diff-raction (SAD) patterns, convergent beam electron diffraction (CBED) patterns, and microbeam electron diffraction (MBED) patterns, were observed by an ABT-002B electron microscope operated at 200keV.

3. "ORDERED-INTERGROWTH" PHASES

Difficulty in preparing a single phase specimen for the 2212 and 2223 phases has been experienced in the TBCCO and also the BSCCO systems. This means that a conventional X-ray or neutron diffraction structure analysis on these oxides may not be entirely reliable. Furthermore, a lattice modulation in these crystals, which causes superlattice spots and also diffuse scattering, may introduce some ambiguity in the conventional diffraction analysis. For these reasons, an electron optical method using electron microscopes seemed appropriate to analyze local structures as well as local chemical compositions.

Figure 1 shows a high resolution electron micrograph taken from a disordered region of the TBCCO crystal. This is oriented with its [110] axis parallel to the

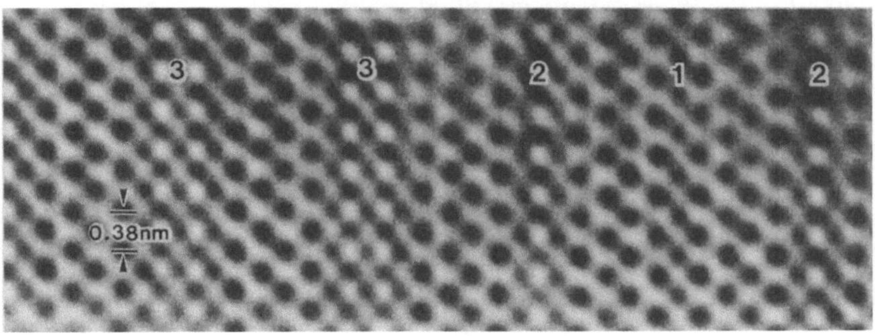

Fig. 1: A HREM image for a TBCCO crystal showing arrangements for the metal atoms. The figures indicate positions of the CuO_2 slabs and their numbers.

electron beam. The image was recorded under the condition so called "optimum focus" of the microscope, where positions of metal atoms appear as dark blobs [9]. Therefore, large and dark blobs are interpreted in terms of positions of Tl and Ba atoms, and the small blobs are those of Cu and Ca atoms. The micrograph contains three different structures, 2201, 2212 and 2223 structures. The figures shown in the image indicate the locations and the numbers of the CuO_2 slabs. The double TlO slabs are a common structural element in all these structures. Therefore, the three structures can be stacked coherently in their c-axis direction. In other words, intergrowth structures are easily introduced in these oxide systems as was reported. A frequent intergrowth causes a difficulty in attaining a single phase material.

Apart from discussion on the TBCCO oxide, it is worthwhile to mention that high quality HREM images such as shown in Fig. 1 could provide metal atom coordinations with accuracy, which is close to those obtained by a X-ray diffraction analysis [10,11]. The HREM technique demonstrates a techno-logical achievement in the HREM structure analysis. It is also mentioned that lighter atoms such as that of oxygen, could not be detected straight-forwardly as those in heavier atoms. However, the effect of oxygen atoms on the HREM image contrast has been examined in terms of the computer image simulation.

As mentioned above, intergrowth structures take place as randomly intermingled structures in the TBCCO oxides. Among them, however, we found occasionally, regular arrangements consisting of more than two structures in a homologous series of the compounds. We call them "ordered-intergrowths" [3]. An example for such ordered intergrowths consists in the 2212 and the 2223 phase. Its low magnification image and electron diffraction patter are shown in Figs. 2a and 2b, respectively. A region on the left of Fig. 2a shows a lattice fringe image of c/2 (= 1.70nm) of the 2223 structure, and its SAD pattern is given in the upper half of Fig. 2b. Referring to the 2223 structure, the crystal structure of the right hand side of the Fig. 2a was found to be described by a unitcell [2212+2223+2223] which is stacked in the c-axis direction. Its c-axis length becomes a summation of the three unitcells, namely, 10.06nm. Similarly, we found another ordered-intergrowth which is expressed as [2212+2223]. These ordered intergrowths are found in some limited regions of the specimen, and the largest ordered region was an order of submicron in the c-axis direction.

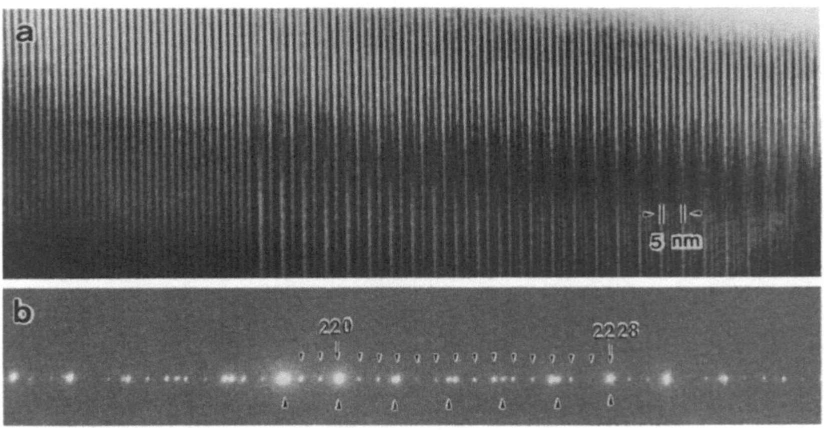

Fig. 2: A lattice image and a SAD pattern showing an "ordered-intergrowth" structure, $[2212+(2223)_2]$,(right), and the 2223 phase (left).

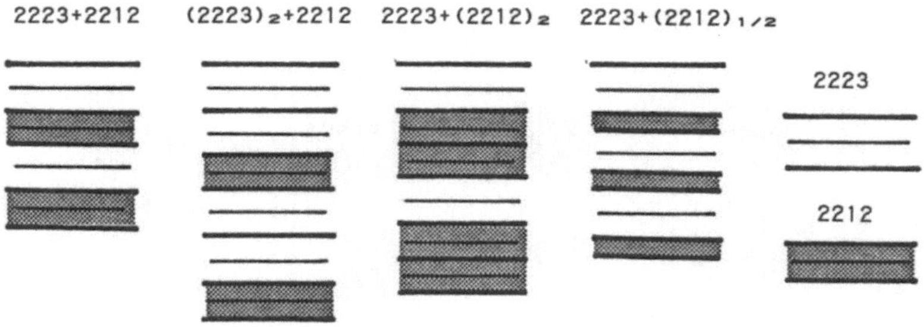

Fig. 3: Some examples for the "ordered-intergrowth" phases.

The ordered-intergrowth structures mentioned above immediately suggest a family of ordered-intergrowth structures which are represented by combi-ation of some of the homologous series of the compound. Some of the examples for the compounds with double TlO slabs are illustrated in Fig. 3. The variations in ordered-intergrowths can be diversified by taking a stacking unit as a c/2 axis length which results in the c-glide symmetry. Similarly, a combination of the single TlO slabs and double TlO slabs of the compounds causes more stacking variations.

From a superconductivity point of view, how the Tc values of the source of charge carriers in the ordered-intergrowths are affected by being a sandwiched intergrowth structure is of interest. If a separation between neighboring superconducting CuO_2 layers is large enough to avoid inter-erence, two-dimensional superconductor layers can be obtained. In the case of the TBCCO crystals, it is not clear, at the moment, where the charge carriers are applied to the CuO_2 layers. The fact that each compound in the homologous series has a different Tc value suggests that if the charges come from the TlO slabs, a sandwiched intergrowth structure may have a different charge amount from those of a single phase structure alone.

4. SOLID-STATE REACTION

It has been recognized by many researchers that growing single phase crystals for TBCCO oxides is extremely difficult. As was described in the previous section, TBCCO crystals prepared by a conventional sintering method contain inevitably intergrowth structures. Removal of the intergrown 2212 phase from the dominant 2223 phase, for instance, has been carried out experimentally to some extent by annealing the sample for a longer period of time, or adding some additives such as Pb. A mechanism for the annealing process seems to be important for understanding a solid-state reaction in the mixed TBCCO phases.

Consider a solid-state reaction between the 2212 and the 2223 phases. The most common phase boundary for these two phases is on (001) planes which are parallel to the BaO-TlO-TlO-BaO slabs. In order to eliminate the 2212 structure by moving the phase boundary, it should move along a direction perpendicular to the (001) plane.Such a solid-state reaction obviously will involve the rearrangement of whole atoms on the boundary, and, thus, will take considerable energy. Another type of phase boundary between the 2212 and 2223 was occasionally observed [12]. The boundary lies on a (110) plane as indicated by the arrows in Fig. 4. In this case, a 2212 - 2223 solid state reaction can be achieved by moving the boundary parallel to the (001) plane.

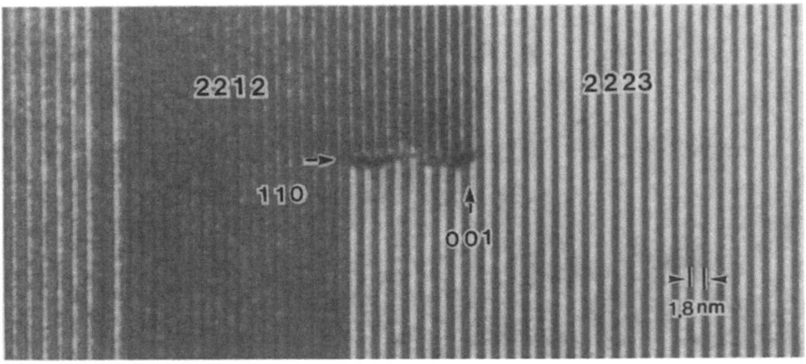

Fig. 4: Two kinds of phase boundaries between the 2212 and 2223 phases. On (001) and (110) planes (arrows).

It was found that there is a geometrical relationship between the widths of the (110) boundaries between the 2212 and 2223 phases. The (110) boundary widths are characterized by a 4:5 ratio for the number of c/2 axis length along the boundary. This relation can be explained by the fact that every fifth BaO-TlO-TlO-BaO slab in the 2212 phase corresponds to that in the 2223 phase within a 2.9% lattice mismatching ratio. Every fifth BaO-TlO-TlO-Ba slab, therefore, can be preserved unchanged across the boundary. When a (110) boundary width does not meet with the condition of the 4:5 ratio, the boundary displayed strain contrast around it. Such boundaries will have much lower free energy than those with a 4:5 lattice matching boundary. As a result, the boundaries can be removed easily during the heat treatment of the specimen. Although the matching BaO-TlO-TlO-BaO slabs are continuous across the boundaries, other mismatching BaO-TlO-TlO-BaO slabs and CuO_2-Ca-CuO_2 slabs will be terminated partially or completely near the boundary regions. Since the 2212 and 2223 phases are different in compo-sition, the above-mentioned reaction should be accompanied with material transport. The transport will occur through the specimen surfaces where the boundaries come out.

5. SUPERLATTICE MODULATION IN 2201 CRYSTALS

Although some of the 2201 phase materials were reported to show superconductivity, the materials that will be described below are non-superconductor [3]. The exact [001] zone axis SAD pattern of the 2201 crystal shows a square net pattern similar to a [001] perovskite-like pattern. A difference, however, is that when the crystal is rotated by a few degrees on any axis parallel to the (001) plane, diffuse ring patterns appear around the fundamental diffraction spots. The maximum intensities of the diffuse rings are obtained by a 7.5 degree rotation about [010] axis (Fig. 5a). Similar ring patterns have been reported for the 2212 crystals, and the diffuse ring patterns are common observation for the TBCCO oxide system [13].

Another [001] zone axis SAD pattern is shown in Fig. 5b. In this pattern, two sets of satellite spots are seen, but no second order spots. Their reflection vectors are approximately $\pm[3/14, \pm 1/14, 0]$. It is mentioned that some crystals show second and third order reflections. The satellite spot intensities change systematically as the scattering angle increases, and the reflection vectors in the

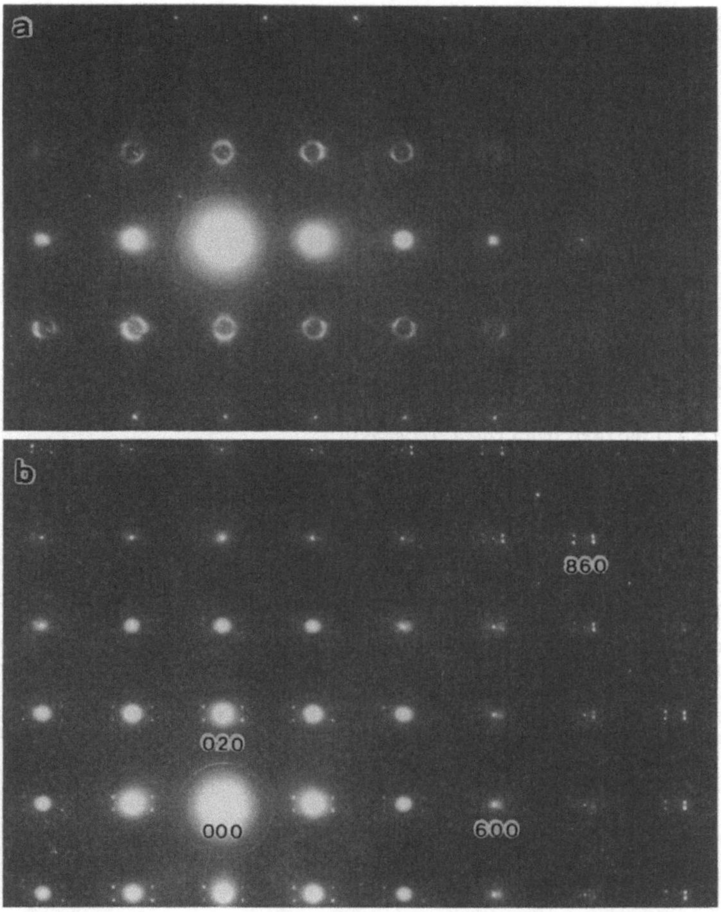

Fig. 5: The [001] zone axis electron diffraction patterns from 2201 crystals.
(a)Diffuse ring patterns appearing after tilting about [100].
(b)Superlattice reflections in the [3/7,1/7,0] direction.

higher scattering region become half of those of the lower angle region. These satellite spot characteristics are explained nicely in terms of the Ewald sphere effect.

As expected from the SAD pattern shown in Fig. 5b, a reciprocal lattice section perpendicular to the [-1,3,0] direction will show strong satellite spots as shown in Fig. 6. An electron micrograph taken from this direction is shown in Fig. 7a. Bright and dark alternating bands are found to be due to the twins which are caused by a superlattice taking place in two different orientations. The superlattice gives rise to the satellite spots mentioned above. The twin planes are probably on the (110) plane.

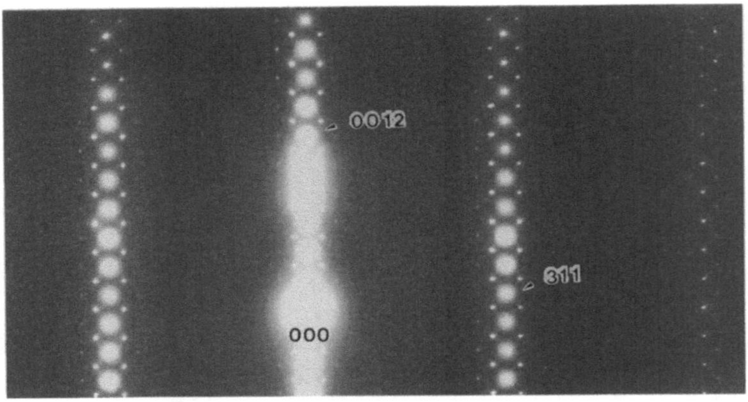

Fig. 6: A [-1,3,0] SAD pattern from the 2201 crystal.

The region indicated by a rectangle in Fig. 7a is enlarged, and reproduced in Fig. 7b. Dark band regions display one-dimensional lattice images, but bright band regions show two-dimensional arrays of dark blobs in a face-centered arrangement. The SAD pattern corresponding to the electron micrograph shown in Fig. 7b does show higher order satellite spots. This means that the two-dimensional array of the dark blobs are extended to large areas. On the contrary, the dark band regions do not show the superlattice, but its occurrence has been confirmed by observing CBED patterns. The result indicates that the superlattice is formed on the (3,-1,0) plane. This plane is one of the four {310} plane variants. It is not supprising to have the superlattice on all four {310} lattice planes. This was confirmed by observing a specimen prepared by quenching. The SAD diffraction patterns show four sets of {310} satellite spots. The spots appearing

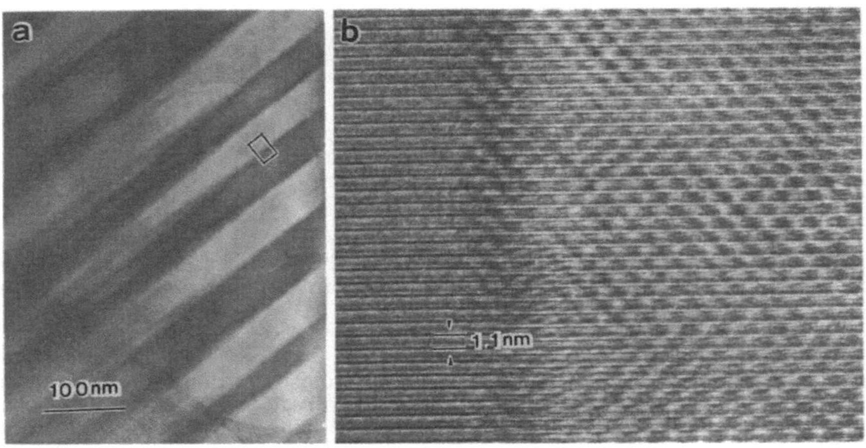

Fig. 7: Electron micrographs. (a)Repeating twin bands due to the superlattice observed on the crystal shown in Fig. 6. (b)Its enlarged lattice image showing a lattice modulation.

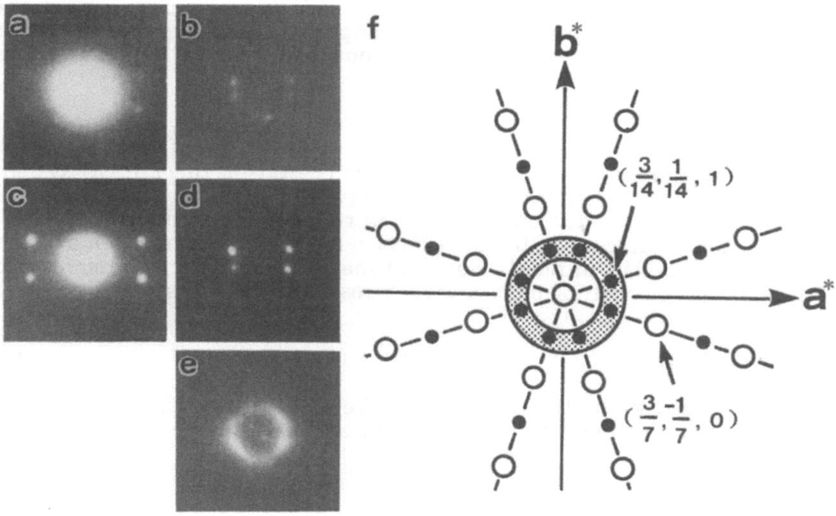

Fig. 8: Some superlattice reflections from the 2201 crystal. (a)Four sets of superlattice spots around (220) and (b) around (860). (c)Two sets of spots around (220) and (d) around (860). (e) Diffuse ring pattern obtained after tilting. (f) Explanation of superlattice spots and diffuse ring. Open circles represent spots with $l=0$, and solid ones with $l=1$. The superlattice vector is Q= $\pm[3/14.\pm1/14.\pm1]$.

around (220) and (860) which are fundamental reflections are shown in Figs. 8a and 8b, respectively. They are compared with those patterns accompanied by two sets of satellite spots as shown in Figs. 8c and 8d. The diffuse ring pattern shown in Fig. 5a is also included.

Three dimensional distribution of the satellite spots, which is deduced from the observation described above, is illustrated in Fig. 8f. The satellite spots appearing around the higher order fundamental reflections such as the (880) spot, correspond to a reciprocal lattice vector with $l=1$. The satellite spots around the (220) fundamental spots are those with $l=0$ (indicated by open circles). From these observations, the superlattice vector is found to be Q= $\pm[3/14,\pm1/14,\pm1]$. A (001)* plane projection of the superlattice vector is coincident with the radius of the diffuse scattering rings (Fig. 5a) as illustrated by a hatched ring in Fig. 8f.

For the specimens which show the diffuse ring patterns, no higher order satellite spots were observed. This can be explained by an imperfect ordering, and also a small size of the superlattice domains which are formed on the {310} planes. The sizes of the superlattice domains vary from specimen to specimen. The domains, which give the SAD pattern shown in Fig. 5b, was in order of several nm in size. The diffuse ring patterns which have been observed in the 2212 and 2223 specimens can be explained by the similar domain structure to those mentioned above.

For the origin of the superlattice, a vacancy model for oxygen and Tl atoms which takes place in the TlO slabs, has been proposed [6]. The nature of the superlattice awaits further study since a supply of the charge carriers in the structure could cause some kind of crystal structure modification. On the correlation between the superstructure and superconductivity, those crystals which do not show strong superlattice modulations tend to have a higher Tc value.

6. CONCLUSION

Under the circumstances where large-size single crystals of the TBCCO system are unavailable, and also intergrowths are unavoidable, neither single crystal analysis nor Rietveld powder analysis are unreliable for structural studies. Furthermore, the occurrence of the superlattice modulation seems to make the crystal structure analysis difficult. For the time being, the electron microscope will provide useful information on local structures of the TBCCO system.

ACKNOWLEDGMENT

The author would like to thank his co-workers, Y. Kubo, T. Ichihashi, Y. Shimakawa and T. Manako, for collaboration on the high Tc oxide research project.

REFERENCES

1. Z.Z. Sheng and A.M. Hermann: Nature 332, 136(1986).
2. H. Ihara, R. Sukise, M. Hirabayashi, N. Terada, M. Jo, K. Hayashi, A. Negishi, M. Tokumoto, Y. Kimura and T. shimomura: Nature 334, 510(1988).
3. S. Iijima, T. Ichihashi, Y. Shimakawa, T. Manako and Y. Kubo: Jpn. J. Appl. Phys. 27, L1061(1988).
4. R. Beyers, S.S. Parkin, V.Y. Lee, A.L. Nazzal, R. Savoy, G. Gorman, T.C. Huang, and A. La Placa: Appl. Phys. Phys. Lett. 53, 432(1988).
5. C.C. Torardi, M.A. Subramanian, J.C. Calabrese, J. Gopalakrishnan, K.J. Morrissey, T.R. Askew, R.B. Flippen, V. Chowdhry, and A.W. Sleight: Science 240, 631(1988).
6. E.A. Hewat, P. Bordet, J.J. Capponi, C. Chaillout, J. Chenavas, M. Godinho, A.W.Hewat, J.L.Hodeau and M.Marezio: Physica C, 156, 375(1988).
7. Y. Shimakawa, Y. Kubo, T. Manako, T. Satoh, S. Iijima, T. Ichihashi and H. Igarashi: submitted to Physics C.
8. Y. Matsui, H. Maeda, Y. Tanaka, and S. Horiuchi: Jpn. J. Appl. Phys. 27, L361(1988).
9. S. Iijima: J. Appl. Phys. 42, 5891(1971).
10. S. Iijima, T. Ichihashi and Y. Kubo: Jpn. J. Appl. Phys. 27, L1054(1988).
11. M.A. Subramanian, J.C. Calabrese, C.C. Torardi, J. Gopalakrishnan, T.R. Askew, R.B. Flippen, K.J. Morrissey, V. Chowdhry, and S.W. Sleight: Nature 332, 420(1988).
12. S.Iijima, T. Ichihashi and Y. Kubo: Jpn. J. Appl. Phys. 27, L1168(1988).
13. J.D. Fitz Gerald, R.L.Withers, J.G. Thompson, L.R. Wallenberg, J.S. Anderson, and B.G. Hyde: Phys. Rev.Lett. 60, 2797(1988).

Orthorhombic-II Superstructure and Significance of Oxygen Ordering for Superconductivity in $YBa_2Cu_3O_{7-\delta}$

Y. Kubo and H. Igarashi

Fundamental Research Laboratories, NEC Corporation,
4-1-1 Miyazaki Miyamae-ku, Kawasaki 213, Japan

There are two bulk superconducting phases, orthorhombic-I ($\delta \sim 0, T_c \sim 90$ K) and orthorhombic-II ($\delta \sim 0.5, T_c \sim 60$ K), in $YBa_2Cu_3O_{7-\delta}$ system. The former has fully-lined-up Cu-O chains on the Cu1 plane, while the latter has alternately-lined-up Cu-O chains. It should be emphasized that the superconductivity becomes percolative (inhomogeneous superconductivity) for other δ values. These facts mean that the T_c is not a simple function of the carrier density but that it should be related to the local oxygen ordering on the Cu1 plane. In order to explain this point, a percolation model is proposed and the importance of local atomic ordering is discussed in connection with the very short coherence length of superconductivity.

1. INTRODUCTION

In the study of the 90-K superconductor $YBa_2Cu_3O_{7-\delta}$, the real role of the Cu-O chain structure on the Cu1 plane is still an open question. In the initial stage of the study, many people believed that the Cu-O chain structure was essential because superconductivity is strongly suppressed by introducing oxygen defects into the chain sites. However, this idea was inconsistent with the results revealed by Fe- or Co- substitution studies [1], where the substituted samples showed rather high T_c values of ~80 K in spite of their crystal structural change into tetragonal, suggesting the absence of Cu-O chain structure. This led the researchers to the viewpoint that the CuO_2 plane (Cu2 plane) is most important for the high-T_c mechanism and that the role of the Cu1 plane is only a subsidiary one, such as controlling the hole concentration of the system [2].

However, recent studies [3,4] using high-resolution electron microscopy on Fe- or Co- substituted Y system oxides with tetragonal symmetry have revealed the existence of orthorhombic microtwin domains, 20-30 Å in size, due to short-range ordering of Cu-O chain structure (Fig. 1). This fact is to be related to the incomprehensive phenomenon that both transition temperature and Meissner volume fraction decrease with increasing substitution [5], because magnetic flux may penetrate into the microdomain boundaries, whose density increases as the domain size decreases with substitution. Moreover, recent study [6] on those samples has shown that the ac susceptibility largely depends on the measuring field for values as small as ~1 Oe. These facts strongly suggest that the superconductivity for the Fe- or Co- substituted systems can be explained in terms of an inhomogeneous superconductivity or a percolative superconductivity consisting of weakly coupled orthorhombic microdomains, whose transition temperature is essentially 90 K. Therefore, the possible importance of the Cu-O chain structure, even if a short-range ordering, for high-T_c superconductivity can not be excluded yet.

Recently discovered $CaBaLaCu_3O_{7-\delta}$ [7,8] has a tetragonal symmetry with no twinning and shows a bulk (not percolative) superconductivity of ~80 K.

Springer Series in Materials Science, Vol. 11
Mechanisms of High Temperature Superconductivity
Editors: H. Kamimura and A. Oshiyama
Springer-Verlag Berlin Heidelberg © 1989

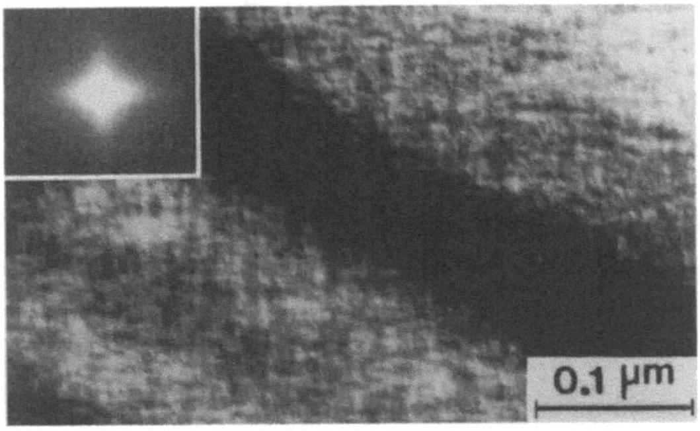

Fig. 1 Electron micrograph and the (220) diffraction spot of $YBa_2(Cu_{0.95}Co_{0.05})_3O_y$ with an incident beam along the [001] direction. Fine texture image and spot elongation to [110] and [110] directions indicate the existence of orthorhombic microtwins.

However, electron diffraction studies have revealed a superstructure of $2a \times 2b \times 2c$, which suggests oxygen ordering on the Cu1 plane induced by cation ordering on the Ba plane [7-9]. In addition, the importance of oxygen-ordering in this system has been suggested by the fact that the superconducting volume fraction largely depends on both sintering temperature and cooling rate, although the oxygen content remains the same [8].

In this paper, we present other evidence for the significance of the oxygen ordering on the Cu1 plane; the existence of orthorhombic-II superstructure phase with T_c of ~60 K. Also, we propose a percolation model for superconductivity in the oxygen deficient Y system. Finally, we discuss the importance of local ordering and the influence of local irregularity for superconductivity in connection with weakened superconductivity (small volume fraction) which is often observed in samples with reduced T_c's prepared by substitution or in other ways. It is considered that such phenomena should be related to the very short coherence length of the high-T_c materials.

2. ORTHORHOMBIC-II SUPERSTRUCTURE PHASE

Samples with various δ values were prepared by quenching sintered samples onto a metal block from high temperatures. This metal-block-quenching method gives a relatively slow quenching rate of about a few hundred degrees per second, which is slow enough to rearrange oxygen atoms in a unit cell toward equilibrium state (short distance diffusion) but is still fast enough to prevent oxygen desorption from the sample (long distance diffusion). Therefore, the obtained samples seems to be very close to the equilibrium state and showed no essential change in their superconducting transitions after low temperature long annealing in a quartz capsule [9, 10]. Details of the preparation and results were described elsewhere [9].

The most important result is that bulk superconductivity appears only at $\delta \sim 0$ (orthorhombic-I, $T_c \sim 90$ K) and $\delta \sim 0.5$ (orthorhombic-II, $T_c \sim 60$ K), and for other δ values, superconductivity becomes percolative. Figure 2 shows transition temperatures and their field sensitivities plotted against oxygen deficiency δ. The field sensitivity clearly shows that the quality of superconductivity changes as

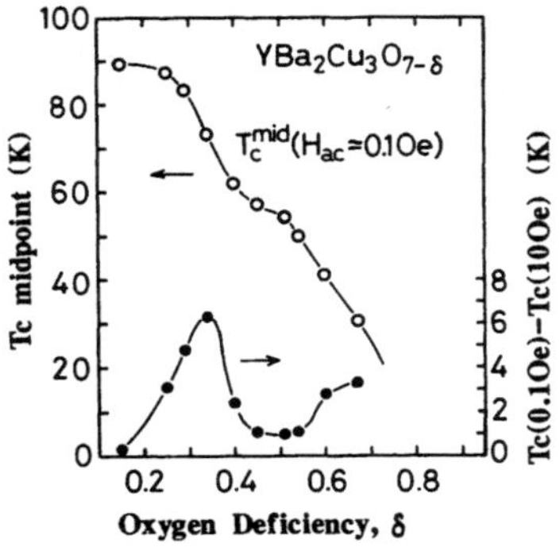

Fig. 2 Transition temperatures and field sensitivities for $YBa_2Cu_3O_{7-\delta}$ samples measured by ac susceptibility in the field of 0.1-10 Oe.

strong ($\delta \sim 0$), weak ($0 < \delta < 0.5$), strong ($\delta \sim 0.5$) and weak ($\delta > 0.5$). Obviously, such a complicated behavior is not explained by the change in the average carrier concentration determined by the total oxygen content. It should be attributed to further oxygen ordering at $\delta \sim 0.5$ (orthorhombic-II). The 60-K phase at $\delta \sim 0.5$ was studied using electron microscopy and diffraction, and superstructure with a doubled a-axis was found as shown in Fig. 3. Consequently, the most probable structure for the 60-K phase is that every other Cu-O chain is missing on the Cu1 plane.

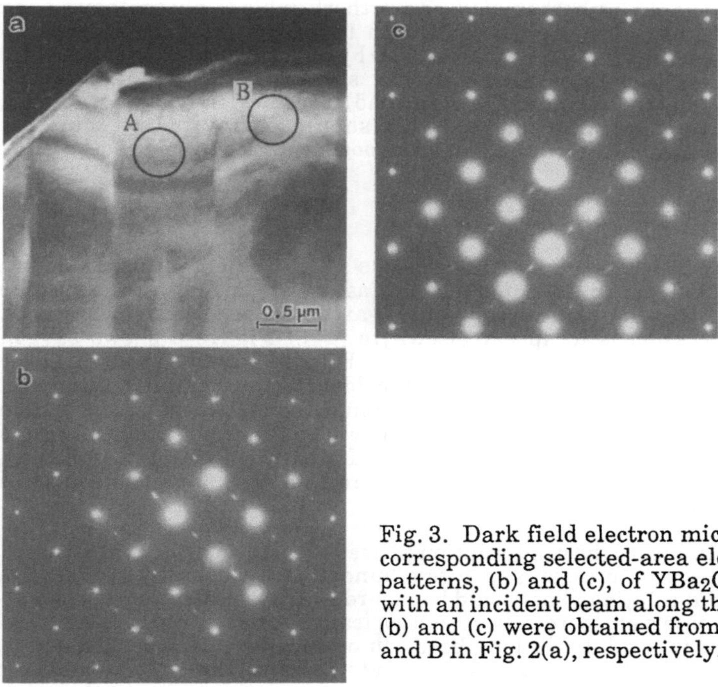

Fig. 3. Dark field electron micrograph, (a), and corresponding selected-area electron diffraction patterns, (b) and (c), of $YBa_2Cu_3O_{7-\delta}$ ($\delta = 0.56$) with an incident beam along the [001] direction. (b) and (c) were obtained from the regions of A and B in Fig. 2(a), respectively.

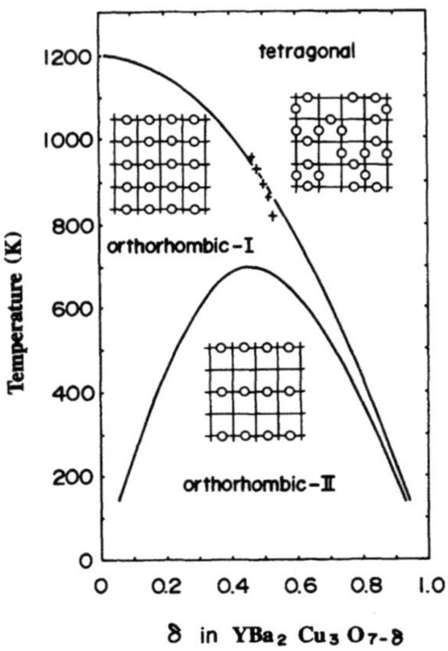

Fig. 4. Calculated δ-T phase diagram for $YBa_2Cu_3O_{7-\delta}$. Cross marks show the orthorhombic-tetragonal phase boundary determined by thermogravimetry (Ref. 11)

Figure 4 shows the δ-T phase diagram, calculated by considering order-disorder transition of 2D oxygen square lattice on the Cu1 plane, where pairlike interactions between nearest and next-nearest neighbor oxygen atoms were assumed [11-13]. Both orthorhombic-I and orthorhombic-II phases have wide solid-solution regions. This agrees well with the experimental fact that no evidence for phase separation was detected by electron microscopy study. Therefore, the question to be solved is why superconductivity for the off-stoichiometric δ values (not near $\delta = 0$ and 0.5) becomes so weak, even in a "homogeneous" solid-solution phase. This should be related to the local irregularity (local oxygen disordering) in a "homogeneous" solid solution.

3. PERCOLATION MODEL

To clarify the inhomogeneity in a homogeneous solid solution, let us consider a site-percolation problem in 2D square lattice as shown in Fig. 5, because the oxygen atoms on the Cu1 plane occupy only b-axis sites at lower temperatures. The detailed discussion is to appear elsewhere [14]. Figure 6 shows oxygen arrangements simulated for various δ values. Figures 6(a)-(f) are obtained by removing oxygen atoms randomly from the ideal orthorhombic-I structure. Figures 6(a')-(f') are obtained by attaching or removing oxygen atoms to or from the ideal orthorhombic-II structure. Therefore, Figs. 6(a)-(f) and 6(a')-(f') show a series of solid-solution phases for orthorhombic-I and II, respectively. Actually, the oxygen-deficient samples at room temperature seem to change as a-b-c'-d'-e'-f as the δ value increases.

As indicated in Fig. 6, the solid-solution phase is "homogeneous", only when it is considered in a certain scale, say, a few hundred Å. Generally speaking, a homogeneous solid-solution phase could be regarded as an inhomogeneous phase-mixture if it is observed extremely microscopically. That is, large statistical fluctuation exists in a solid-solution phase. In ordinary metal superconductors, such fluctuation does not produce weak links because the superconducting coherence length is large enough to average out such randomness in the solid-

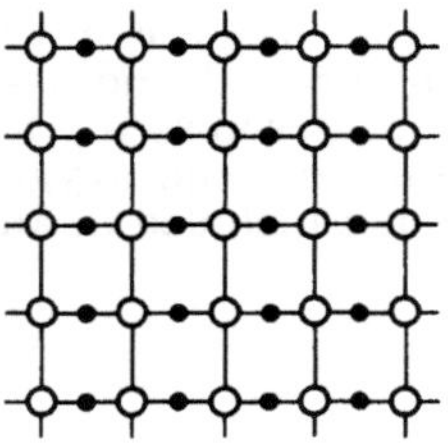

Fig. 5 Site-percolation model of the 2D oxygen square lattice on the Cu1 plane

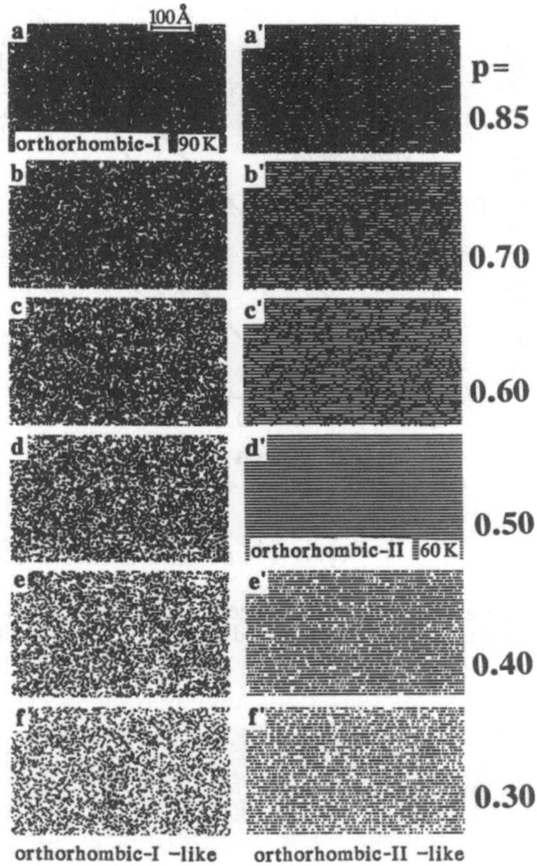

Fig. 6. Simulated oxygen arrangement on the Cu1 plane of the $YBa_2Cu_3O_{7-\delta}$. Site occupancies are represented by $p=1-\delta$. (a)-(f) and (a')-(f') show the arrangement in the category of orthorhombic-I and II, respectively.

solution phase. In the $YBa_2Cu_3O_{7-\delta}$, however, relatively oxygen-deficient microdomains due to the fluctuation of oxygen defects may behave as weak links because the coherence length of 20-30 Å in the a-b plane [15, 16] is too small to overcome the fluctuation. Roughly speaking, inhomogeneous superconductivity occurs if the superconducting coherence length is smaller than the structural coherence length of the solid solution.

As shown in Fig. 2, the percolative superconductivity sets in at $\delta \sim 0.25$ (site occupancy $p \sim 0.75$). However, this occupancy is much higher than the critical occupancy for the ordinary site-percolation problem, about 0.59 for 2D square lattice [17, 18]. Thus, the critical occupancy should be pushed up to ~ 0.75 in order to explain the phenomenon. One possible method is to stress the oxygen-deficient portions by eliminating the oxygen atoms which do not participate in the ordering. If we assume that a 2×2 clusterlike ordering is crucial for the superconducting path, we can ignore the oxygen atoms which are not included in any 2×2 cluster at all, as shown in Fig. 7. The site occupancy of the oxygen atom which is included in some 2×2 cluster, $Q_2(p)$, is expressed by

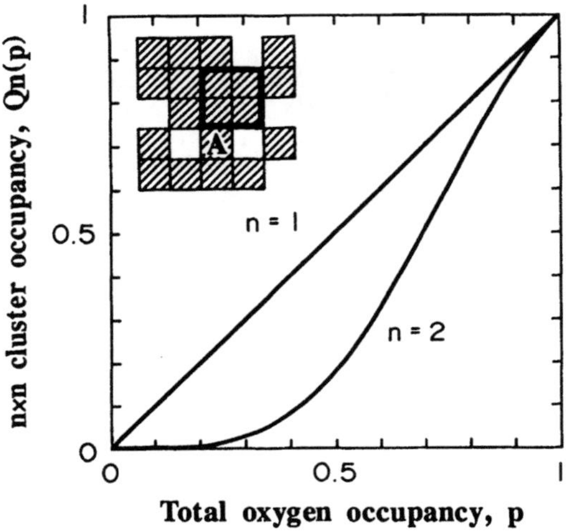

Fig. 7. Calculated n×n cluster occupancies $Q_n(p)$ for n=1 and n=2 [Eq. (1)]. The inset shows a 2×2 cluster (thick line) and the oxygen site (A) which is not included in any 2×2 clusters.

$$Q_2(p) = p^4 [p^5 + 8p^4q + 26p^3q^2 + 36p^2q^3 + 20pq^4 + 4q^5] \qquad (1)$$

where p is the total site occupancy of oxygen atom, and q=1−p. Equation (1) is shown in Fig. 7. Simulated oxygen distribution patterns are shown in Fig. 8.

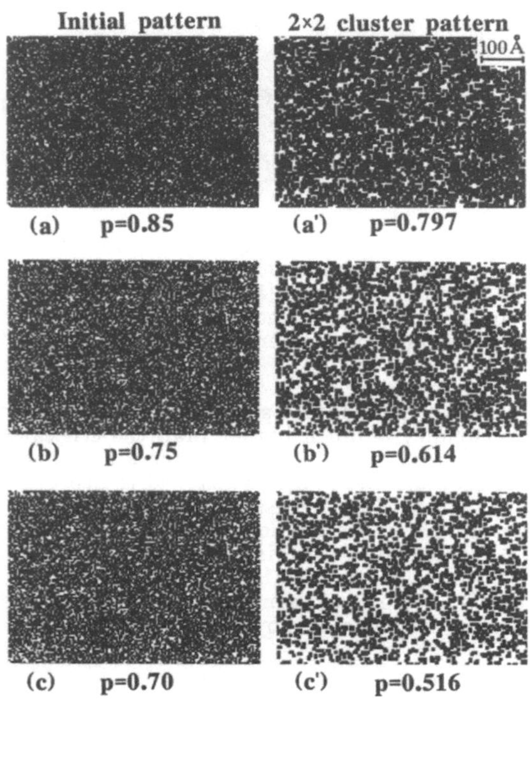

Fig. 8. Initial random-distribution patterns (left) and corresponding 2×2 cluster patterns (right). The p values for (a')-(c') are the resultant site occupancies $Q_2(p)$ obtained by the simulation for a 150×180 lattice model

The percolation threshold for this model will be obtained by some simulations such as Monte Carlo methods. However, for the present purpose, only a rough estimate is sufficient. If we simply assume that the percolation threshold for this model exists near the ordinary critical occupancy of ~0.59 [17, 18], that is $Q_2(p)$~0.59, we obtain a critical total oxygen occupancy p_c of ~0.74 (δ~0.26). This simple assumption seems plausible as demonstrated in Fig. 8, where pattern (b') for $Q_2(p)=0.614$ is percolated while pattern (c') for $Q_2(p)=0.516$ is not. The true critical value is considered to be somewhat higher than 0.59 because of the clustering tendency in the 2×2 model. Anyway, the critical total oxygen occupancy p_c of ~0.74 (δ~0.26) corresponds well to the oxygen deficiency at which the sample begins to show a weak, percolative behavior. For an oxygen content larger than p_c, the ordered 90-K domains are directly connected to each other and the sample shows a normal bulklike superconductivity. For a less oxygen content than p_c, however, the ordered 90-K domains are isolated and connected each other through oxygen-deficient weak links, which should show a percolative superconductivity as is often observed in inhomogeneous superconductors. Further decrease of oxygen content (p~0.6) will destroy the short-range ordering itself which is necessary for the appearance of the high-T_c superconductivity. However, further oxygen-ordering toward the double-spaced chain structure (orthorhombic-II) will occur at the oxygen content of p=0.6-0.4, and the sample shows a strong bulklike superconductivity again.

4. DISCUSSION

The above-mentioned model means that at least the 2×2 oxygen cluster is necessary for the strong connection between the superconducting microdomains. The importance of local oxygen ordering on the Cu1 plane might suggest the importance of interplane interaction between CuO_2 planes via the Cu1 plane, even if the superconducting current mainly flows on the CuO_2 planes.

It should be noted that the T_c suppression caused by oxygen defects or impurities is always accompanied by weakened supercnductivity, like percolative or inhomogeneous superconductivity. In the YBCO system, either Ca substitution for Y [19] or La substitution for Ba [20] cause broadening of transition and reduction of Meissner volume fraction, although the onset T_c value decreases rather slowly. In the Bi system, the suprconducting volume fraction is reduced by Y substitution for Ca, while the T_c onset remains the same [21]. A similar behavior is also observed in the Tl system [22]. In the $(La, Sr)_2CuO_4$ system, the Meissner volume fraction seems to decrease more rapidly than the onset T_c value does [23]. These facts strongly suggest that the superconductivity is not uniformly suppressed, but that it is only locally suppressed. This may be related to the random potential induced by the substitution.

Generally speaking, it is very difficult to change the carrier density without disturbing the local lattice order. All previous experiments attempting to change only the carrier density have simultaneously changed the local lattice order, because they were carried out by introducing other elements with different valences or vacancies. The case of orthorhombic-II seems to be a rare exception, because the randomly introduced oxygen vacancies tend to rearrange into a new ordered structure.

Considering the very short coherence length of these materials, such a random potential may act as a barrier for superconductivity. This may be the reason why the volume fraction decreases with increasing substitution. The relatively small decrease in onset temperature strongly suggests that the local state of superconductivity remains unchanged. In other words, the local transition temperature appears to be fixed at a particular value which is characteristic of the particular crystal structure. And, the random potential seems to destroy the superconductivity only locally, leaving the superconducting quality in other area unchanged. Therefore, it is suggested that superconductivity tends to be localized in a random potential, in order to keep the local transition temperature at an intrinsic value of the crystal structure. That is, the local transition temperature could remain unchanged, if the local carrier density is kept at an optimum value by localizing the superconducting carriers. In other area, where the local carrier density is not optimum, too low or too high, superconductivity would be largely depressed, and weak links would be formed. In conclusion, superconductivity is very sensitive to the local lattice order, probably because the superconducting coherence length is very small. Thus, the T_c dependence on the average carrier density should be interpreted by taking account of the quality of superconductivity, which is largely affected by the local lattice order.

We would like to thank T. Ichihashi and S. Iijima for the electron microscopy and diffraction.

REFERENCES

1. Y. Maeno, T. Tomita, M. Kyogoku, S. Awaji, Y. Aoki, K. Hoshino, A. Minami, and T. Fujita: Nature (London) 328, 512(1987)
2. Y. Tokura, J.B. Torrance, T.C. Huang, and A.I. Nazzal: Phys. Rev. B38, 7156(1988)
3. T. Ichihashi, S. Iijima, Y. Kubo, Y. Shimakawa, and J. Tabuchi: Jpn. J. Appl. Phys. 27, L594(1988)
4. G. Roth, G. Heger, B. Renker, J. Pannetier, V. Caignaert, M. Hervieu, and B. Raveau: Z. Phys. B71, 43(1988)
5. S. Noguchi, J. Inoue, K. Okuda, Y. Maeno, and T. Fujita: Jpn. J. Appl. Phys. 27, L390(1988)
6. Y. Shimakawa, Y. Kubo, K. Utsumi, Y. Takeda, and M. Takano: Jpn. J. Appl. Phys. 27, L1071(1988)
7. W.T. Fu, H.W. Zandbergen, C.J. van der Beek, and L.J. de Jongh: Physica C156, 133(1988)
8. D.M. de Leeuw, C.A.H.A. Mutsaers, H.A.M. van Hal, H. Verweij, A.H. Carim, and H.C.A. Smoorenburg: Physica C156, 126(1988)
9. Y. Kubo, T. Ichihashi, T. Manako, K. Baba, J. Tabuchi, and H. Igarashi: Phys. Rev. B37, 7858(1988)
10. H. Igarashi, Y. Kubo, T. Manako, T. Ichihashi, K. Baba, and J. Tabuchi: Physica C153-155, 854(1988)
11. Y. Kubo, Y. Nakabayashi, J. Tabuchi, T. Yoshitake, A. Ochi, K. Utsumi, H. Igarashi, and M. Yonezawa: Jpn. J. Appl. Phys. 26, 1888(1987)
12. Y. Kubo and H. Igarashi: Jpn. J. Appl. Phys. 26, 1988(1987)
13. Y. Kubo, Y. Nakabayashi, J. Tabuchi, T. Yoshitake, T. Manako, A. Ochi, K. Utsumi, H. Igarashi, and M. Yonezawa: In High-Temperature Superconductors, edited by M.B. Brodsky, R.C. Dynes, K. Kitazawa and H.L. Tuller, Materials Research Symposia Series, Vol. 99 (Materials Research Society, Pittsburgh 1988) p.89
14. Y. Kubo and H. Igarashi: Phys. Rev. B (in press)
15. T. Sakakibara, T. Goto, Y. Iye, N. Miura, H. Takeya, and H. Takei: Japn. J. Appl. Phys. 26, L1892(1987)
16. T.K. Worthington, W.J. Gallagher, D.L. Kaiser, F.H. Hortzberg, and T.R. Dinger: Physica C 153-155, 32(1988)
17. H.L. Frisch, E. Sonnenblick, and V.A. Vyssotsky: Phys. Rev. 124, 1021(1961)
18. M.F. Sykes and J.W. Essam: Phys. Rev. 133, A310(1964)

19. A. Manthiram, S.-J. Lee, and J.B. Goodenough: J. Solid State Chem. 73, 278(1988)
20. R.J. Cava, B. Batlogg, R.M. Fleming, S.A. Sunshine, A. Ramirez, E.A. Rietman, S.M. Zahurak, and R.B. van Dover: Phys. Rev. B37, 5912(1988)
21. T. Fujita: private communication
22. T. Manako: private communication
23. J.B. Torrance, Y. Tokura, A.I. Nazzal, A. Bezinge, T.C. Huang, and S.S.P. Parkin: Phys. Rev. Lett. 61, 1127(1988)

Part IV

Superconductivity of Ba-K-Bi-O Compound

A Comparison Between Bi-O and Cu-O Based Superconductors

B. Batlogg

AT&T Bell Laboratories, Murray Hill, NJ 07974, USA

The two classes of cuprate[1] and bismuthate[2] superconductors are special because their transition temperatures T_c are higher than in all conventional superconductors. It appears worthwhile, therefore, to discuss those properties that set them apart from other superconductors, particularly from other superconducting oxides. Their basic crystal-chemical structures have much in common and their T_c's are not only the highest ones on an absolute scale, but especially high when compared to the electronic density of states at the Fermi level. Many of their physical properties, however, are different, and are extensively documented in the literature.[3-8]

1. Electronic Structure and Bonding

A first insight into the electronic structure comes from band structure calculations.[9-11] Accordingly, the bands both in the cuprates and bismuthates are characterized, in a simple view, by strong hybridization between the oxygen and the metal atom states (Bi6s, Cu3d), forming a broad band. The Fermi level is located in the antibonding part of the $s{-}p_{\parallel}$ ($d{-}p_{\parallel}$) derived band. Other states, including the nonbonding $p_{\perp}$, lie several eV below the calculated Fermi level. A rough representation of the total density of states is given in Fig. 1. Most notable is the significant O2p character of the states at E_F. An extensive discussion of the electronic

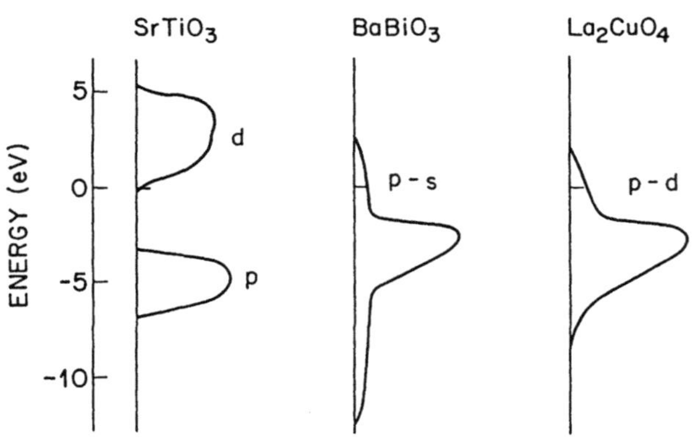

Fig. 1 Band structure density of states for various oxides (after Refs. 9-11, 13). Following the trend in atomic energy level position, the metal states (3d, 6s) move towards the oxygen 2p states when proceeding to the right of the periodic table, closing the band gap. Strong interaction between the metal and oxygen states (covalent bonding) is characteristic for the bismuthates and the cuprates, leading to significant oxygen 2p character at E_F.

Springer Series in Materials Science, Vol. 11
Mechanisms of High Temperature Superconductivity
Editors: H. Kamimura and A. Oshiyama
Springer-Verlag Berlin Heidelberg © 1989

properties of the cuprates, particularly of correlation energies and tight binding fitting parameters to the band structure calculation results can be found in Ref. 12, which also includes discussion of the relevant electron spectroscopy results.

The electronic structure of the bismuthates and cuprates with broad bands involving the oxygen 2pσ states is to be contrasted with those of the oxides of the early transition metals.[13] In $SrTiO_3$, for example, an energy gap of ~3eV separates the predominately p-valence band from the predominantly d-conduction band. Similarly, the 7th valence electron of Re in ReO_3 is located in a conduction band of mainly d character. (ReO_3 is a good metal and is not superconducting). The band structures mentioned here reflect trends in energy levels of the atoms.[14] On proceeding across the transition metal series, the d levels drop in energy and are lower by ~8eV in Cu than in Ti. The bonding is of ionic character in the titanate, and becomes covalent in the cuprates and bismuthates. Thus, the two groups of high T_c materials share common aspects of their electronic structure which are ultimately rooted in the atomic properties.

Superconducting oxides are known and have been studied for 25 years and examples of each group are given in Table I.[15]

TABLE I Superconducting Oxides

NbO,TiO	<2K	Hulm et. al, (1965)
$SrTiO_3$	0.7K	Schooley, et. al (1964)
M_xWO_3	6K	Raub, et. al (1964)
$Ag_7O_8HF_2$	1K	Robin et. al (1966).
$LiTi_2O_4$	13K	Johnston et. al (1973)
$Ba(Pb,Bi)O_3$	13K	Sleight, et. al (1975).
$(La,Ba)_2CuO_4$	35K	Bednorz and Müller (1986)

The states near the Fermi level in these superconductors are predominantly of d character and lack the dominant oxygen p character present in the Bi-O and Cu-O compounds. In more appropriate chemical terms the conduction states in the low T_c oxides result from metal-oxygen π^* interaction, whereas in the bismuthates and cuprates they derive from metal-oxygen σ^* interaction. These differences are discussed in more detail in Ref. 16.

2. Relative and Absolute Scales for T_c

There is little doubt that transition temperatures in excess of 100K are properly called "high". Not usually recognized, however, is that a T_c of 10K can also be "high". The scale on which T_c is measured is not necessarily the absolute one, as we will discuss in the following.

To illustrate the idea, we recall a basic result of the BCS theory: $T_c = \omega \exp[-1/(N(0)V)]$ which relates the transition temperature to a typical Boson (phonon) energy ω, an effective pairing potential V between the electrons, and the electron density of states N(0) at the Fermi level E_F. Accordingly, a high T_c is favored by a large value of either one of the parameters ω, N(0) or V.

To compare superconductors with various structures and elemental compositions, it is instructive to plot their T_c as function of density of states N(0). Equivalently, one may choose to plot T_c vs the Sommerfeld parameter γ, which is the experimental quantity proportional to the density of states. On such a T_c–γ plot, (Fig. 2) the superconductors with similar γ have a high T_c either when the prefactor ω or the interaction potential V is large. As indicated by the broken line in Fig. 2, there exists a more or less well-defined maximum T_c for a given γ. The A15 compounds with the highest T_c among the conventional superconductors also have the largest density of states. Improving T_c within this class would require even larger γ's and the resulting limitations due to structural instabilities have been extensively discussed in the past.

Unusual superconductors can be easily identified on the T_c–γ plot. Heavy fermion superconductors, e.g., fall in the lower right corner (T_c<1K, γ > 100). In the other

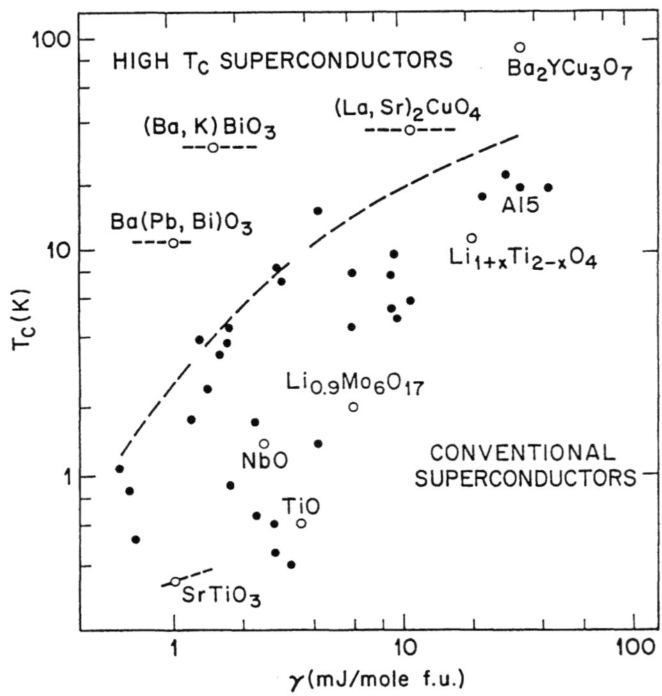

Fig. 2 T_c vs the Sommerfeld constant γ for various superconductors. Bismuthates and cuprates have the highest T_c's on an absolute scale and particularly for their density of states at E_F, measured in γ. The extraordinary nature of these two groups with respect to all other superconductors, including oxides, is highlighted in the T_c–γ plot.

extreme of the parameter range, the bismuthates and cuprates have T_c's which are several times higher than in conventional compounds with similar γ's. (The experimental estimates of γ values for the Bi-O will be discussed below). $Ba(Pb,Bi)O_3$ had been recognized earlier as an exceptional case because its T_c of ~12K is "enhanced" by a factor of 3-5. Together with $(Ba,K)BiO_3$,[17,18] it forms a group of high T_c superconductors.

The other oxides have rather low T_c's for their density of states. This becomes clear when $LiTi_2O_4$ is compared to $Ba(Pb,Bi)O_3$. While their T_c's are the same, the density of states is ~20 times larger in $LiTi_2O_4$. The Bi-O and Cu-O superconductors are clearly separated from all others, including also other oxides. This points to the electronic structure, characterized by metal-oxygen σ^* interaction, as one of the essential components giving rise to high T_c superconductivity in these compounds.

3. Density of States in the Bi-O Superconductors

Usually, the Sommerfeld parameter in a superconductor is derived from measurements of the specific heat anomaly at T_c and/or the specific heat at $T \to 0$ in magnetic fields large enough to suppress superconductivity. The anomaly at T_c can also be estimated from measurements of the thermodynamic critical field H_c in the vicinity of T_c. This alternative method involves the lower and upper critical fields H_{c1} and H_{c2}, respectively, which are intensive quantities, and therefore puts less stringent requirements on the sample quality than specific heat measurements. Experimental details and a critical discussion of the data are given in Refs. 19 and 20. The main conclusion is reflected in the T_c–γ plot: bismuthates are high T_c superconductors. Furthermore, a comparison can be made with band structure calculations and the density of states enhancement can be estimated. A summary of these results is shown in Fig. 3, where the calculated[9, 21] density of states N(0) and the measured one, N*(0), are plotted as function of x in $Ba(Pb_{1-x}Bi_x)O_3$ and $(K_{1-x}Ba_x)BiO_3$. Variation in the K/Ba or Pb/Bi ratio has essentially the same effect in shifting the Fermi level.[9] In $BaBiO_3$, the anti-bonding $s\text{–}p_{\|}$ band is half filled and the material is a CDW semiconductor. The abscissa in Fig. 3 reflects, therefore, the filling factor ranging from 0 to 1/2. The density of states increases when the Fermi level is raised, implying that T_c could be significantly enhanced if the metallic state could be maintained closer to 1/2 filling. A generalized phase diagram will be discussed below.

The density of states enhancement $(1 + \lambda)$ can be estimated from comparisons of N(0) with $N^*(0) = (1+\lambda)N(0)$. It is quite clear that λ is not very large: $\lesssim 0.8$ in $Ba(Pb,Bi)O_3$ and ~0.5 in $(Ba,K)BiO_3$. (The data base in the latter compounds is not sufficient to give a more precise value for λ.) Earlier estimates gave a somewhat larger value for λ[6,22].

A schematic phase diagram for various degrees of band filling is shown for compound series based $BaBiO_3$ and La_2CuO_4. At half filling, the ground state is either a magnetic insulator or a CDW semiconductor. When the Fermi level is lowered, high T_c superconductivity is observed over a narrow range of band filling, determined in detail by crystal-chemical properties. If guided only by arguments that the behavior be symmetric around 1/2 filling, one expects superconductivity to occur also when the Fermi level is raised enough as to destroy the non-metallic ground. Indeed, most recently, superconductivity was discovered in $(Nd,Ce)_2CuO_4$ with n-type conduction[23], whereas all other cuprate superconductors are of p-type. (The crystal

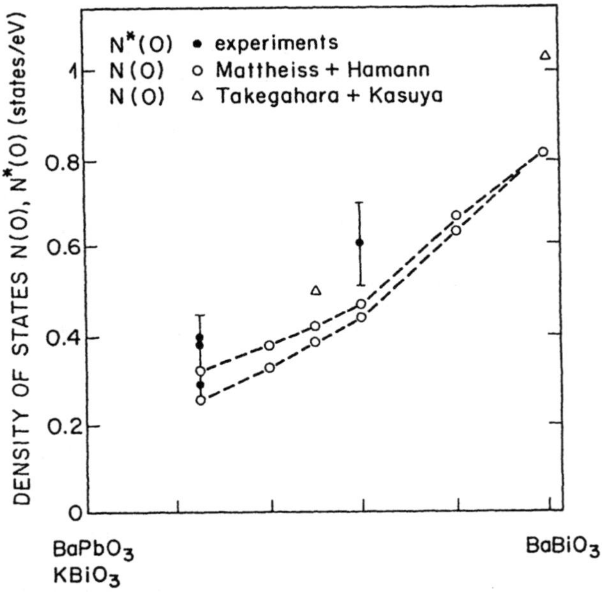

Fig. 3 Density of states in $(Ba,K)BiO_3$ and $Ba(Pb,Bi)O_3$. Open symbols represent band structure calculation results for N(0) [9,22] and the closed circles show the measured values for $N^*(0) = (1+\lambda)N(0)$. The enhancement factor λ can be estimated from this data. (The two sets of open circles correspond to different methods to calculate N(0)).

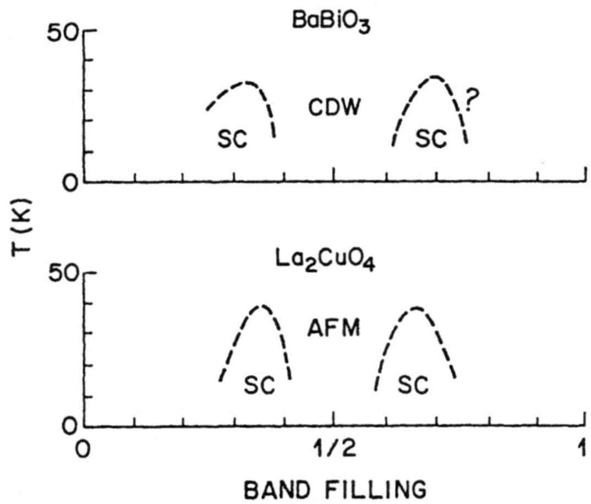

Fig. 4 A schematic phase diagram as function of band filling for compounds based on La_2CuO_4 and $BaBiO_3$.

structure of Nd_2CuO_4 is not identical, but very similar to the one of La_2CuO_4). Fig. 4 reflects this new regime of superconductivity qualitatively and one might speculate that adding electrons to $BaBiO_3$ would also lead to superconductivity, after the CDW state is destroyed.

The CDW state in $BaBiO_3$ is commonly interpreted as a charge disproportionation $Bi^{IV} \rightarrow Bi^{III} + Bi^{V}$, supported by the observation that the oxygen atoms are statically displaced towards or away from the Bi atoms in a "breathing" type distortion. The symmetry of the oxygen arrangement changes when only a small amount of K replaces Ba[24], still maintaining a non-metallic ground state. Each oxygen octahedron around Bi is compressed along one direction and elongated along the two others (or vice versa). Each Bi has 2(4) short and 4(2) long bonds to oxygen with the same bond lengths as in $BaBiO_3$. The inequivalent Bi sites are therefore a result of bond charge $(s\text{–}p_{\|}, \sigma^*)$ disproportion, which can be visualized as a 3-dim array of three 1-dim bond-CDW's. This points again to the strong $s\text{–}p_{\|}$ hybridization.

4. Differences Between Bi-O and Cu-O Superconductors

After having emphasized similarities between these two groups of compounds, the most striking differences need to be pointed out. For simplicity, they are represented in Table II. Numerical values are best estimates and might change somewhat as more complete experimental results become available. The difference of most significance might be the number of Cooper pairs in a coherence volume. While large in the bismuthates, it is rather small in the cuprates.

TABLE II: A comparison between cuprate and bismuthate superconductors.

	Bismuthates	Cuprates
Structure	3-dim	2-dim
Magnetism in parent compound	no	yes
T_c (K)	>30	>125
Effective mass m*/m	~0.5	3-5
Coherence length $\sqrt{\xi . 1}$, ξ_{ab}(Å)	70-100	~15
$2\Delta/kT_c$	3.5±0.1	3.5-8
Cooper pairs/coherence volume	10^3–10^4	5-10

5. Summary

The bismuthate and cuprate superconductors have the highest T_c's on an absolute, and particularly on a relative, scale which is based on the density of electronic states at E_F. Their underlying metal-oxygen bonding is unique among the superconducting oxides and is characterized by strong hybridization between the metal (6s,3d) and oxygen ($2p_{\|}$) states. The Fermi level is located in a wide band with relatively low density of states. Beyond these essential similarities, which might suggest a common mechanism for high T_c superconductivity, several other properties are distinctly different. Worth noting is that in the cuprates only few (~5–10) Cooper pairs exist within a coherence volume, whereas in bismuthates this number is large (10^3–10^4) as in usual superconductors.

Acknowledgement

It is a pleasure to acknowledge collaborations with R. J. Cava, A. S. Cooper, R. C. Dynes, G. P. Espinosa, J. P. Remeika, L. W. Rupp, Jr. and L. F. Schneemeyer, as well as stimulating discussions with C. M. Varma.

References

[1] J. G. Bednorz and K. A. Müller, *Z. Phys.* **64,** 189 (1986).

[2] A. W. Sleight, J. L. Gillson and P. E. Bierstedt, *Solid State Commun.* **17,** 27 (1975)

[3] For cuprates, see for instance, proceedings of recent conferences: "LT 18 Kyoto", *Japn. J. Appl. Phys.* **26** (1987), Suppl. 26-1, 2, 3;

[4] "Int'l. Conf. on High-Temperature Superconductors and Materials and Mechanisms of Superconductivity - Interlaken, 1988", J. Müller and J. L. Olsen, eds., *Physica C* **153-155** (1988).

[5] Progress in High Temperature Superconductivity, Vol. 9, eds. R. Nicholsky, R. A. Barrio, O. F. de Lima, R. Escudero, World Scientific (1989).

[6] Summaries for bismuthates can be found in B. Batlogg, *Physica* **126B** 275 (1984) and in [7] and [8].

[7] K. Kitazawa, S. Uchida and S. Tanaka, *Physica* **135B,** 505 (1985).

[8] R. J. Cava and B. Batlogg, *MRS Bulletin,* Vol. XIV/1, 49 (1989).

[9] L. F. Mattheiss and D. Hamann, *Phys. Rev. Lett.* **60,** 2681 (1988).

[10] L. F. Mattheiss, *Phys. Rev. Lett.* **58,** 1028 (1987).

[11] Jaejun Yu, A. J. Freeman and J.-H. Xu, *Phys. Rev. Lett.* **58,** 1035 (1987).

[12] M. S. Hybertsen, M. Schlüter and N. E. Christensen, *Phys. Rev. B (1989).*

[13] L. F. Mattheiss, *Phys. Rev.* **B6,** 4718 (1972). See also: W. A. Harrison, "Electronic Structure and the Properties of Solids", p. 438 ff.; W. H. Freeman & Co., San Francisco (1980).

[14] F. Herman and S. Skillman, "Atomic Structure Calculations", Prentice Hall, Englewood Cliffs, NJ (1963).

[15] J. K. Hulm, C. K. Jones, R. Mazelsky, R. A. Hein and J. W. Gibson, "Proc. 9th Int'l. Conf. on Low Temperature Physics", J. G. Daunt, D. O. Edwards, F. J. Milford, M. Yacub, eds., p. 600 (1965).
J. F. Schooley, W. R. Hosler and M. L. Cohen, *Phys. Rev. Lett.* **23,** 474 (1964).
Ch. J. Raub, A. R. Sweedler, M. A. Jensen, S. Broadston and B. T. Matthias, *Phys. Rev. Lett.* **13,** 746 (1964).
M. B. Robin, K. Andres, T. H. Geballe, N. A. Kuebler and D. B. McWhan, *Phys. Rev. Lett.* **17,** 917 (1966).

D. C. Johnston, H. Prakash, W. H. Zachariasen and R. Viswanathan, *Mat. Res. Bull.* **8,** 777 (1973).

[16] A. W. Sleight, in "Superconductivity: Synthesis, Properties and Processing", ed. W.Hatfield, Marcel Dekker, Inc. (1988).

[17] L. F. Mattheiss, E. M. Gyorgy and D. W. Johnson, Jr., *Phys. Rev.* **B37**, 3745 (1988).

[18] R. J. Cava, B. Batlogg, J. J. Krajewski, R. C. Farrow, L. W. Rupp, Jr., A. E. White, K. T. Short, W. F. Peck, Jr. and T. Y. Kometani, *Nature* (London), **332,** 814 (1988).

[19] B. Batlogg, R. J. Cava, L. W. Rupp, Jr., A. M. Mujsce, J. J. Krajewski, J. P. Remeika, W. F. Peck, Jr., A. S. Cooper and G. P. Espinosa, *Phys. Rev. Lett.* **61,** 1670 (1988).

[20] B. Batlogg, R. J. Cava, L. F. Schneemeyer and G. P. Espinosa, *IBM J. Research and Development* (1989).

[21] K. Takegahara and T. Kasuya, *J. Phys. Soc. Japn.* **56,** 1478 (1987).

[22] K. Kitazawa, M. Naito and S. Tanaka, *J. Phys. Soc. Japn.* **54,** 2682 (1985).

[23] Y. Tokura, H. Takagi and S. Uchida, *Nature* (London), February 1989.

[24] L. F. Schneemeyer, J. K. Thomas, T. Siegrist, B. Batlogg, L. W. Rupp, Jr., L. Opila, R. J. Cava and D. W. Murphy, *Nature,* **335,** 421 (1988).

Part V

Summary of Discussion Session

Notes on Hakone Conference

G.A. Sawatzky

Department of Applied and Solid State Physics, Materials Science Center, University of Groningen, Nijenborgh 18, 9717 AG Groningen, The Netherlands

General Remarks:

I guess we all agree that the Cu containing compounds should be viewed as Mott-Hubbard or charge transfer antiferromagnetic insulators into which charge carriers have been introduced by chemical substitution or non-stoichiometry. The substitutions done in all cases require the introduction of holes for charge neutrality which will enter the top of the valence band or to be more accurate they will correspond to the first ionization states of the antiferromagnetic insulators. In view of this most of the symposium dealt with some very basic aspects of Mott-Hubbard/charge-transfer antiferromagnetic insulators using virtually the whole arsenal of experimental and theoretical tools developed in the past 4 decades of solid state research. The questions one tried to answer concerned:

A) The influence of electron correlation and the magnitude of electron-electron interactions.
B) Magnetism.
C) The symmetry, character and charge distribution of the holes introduced by substitution/stoichiometry.
D) The influence of the holes on the magnetism, transport and optical properties.
E) Are these Fermi liquid systems?
F) Dirt
G) $2\Delta/kT_c$
H) How important is the Cu?
I) Mechanisms for superconductivity.

A) The influence of electron correlation and the magnitude of the electron-electron interactions:

The general consensus of this group of scientists was that the understanding of these materials asks for a theory which explicitly includes electron correlation effects. One particle band theory cannot, in the opinion of most, describe some of the most relevant properties of these materials like the insulating antiferromagnetic nature of the "unsubstituted" compounds, the large local magnetic moment even in the conducting materials, the non-conventional temperature dependent transport properties, the fact that the holes introduced by doping are primarily in O 2p bands, etc. Perhaps the most important basic discrepancy with band theory is that the lowest energy hole states are of primarily O 2p character and the Cu^{3+} (d^8) like states are very high in energy.

There was a lot of discussion on how to include the correlation effects and on the relative importance of various interactions. From the detailed analysis of photoelectron-inverse photoelectron, Auger spectroscopy, electron energy loss, and X-ray absorption it is quite clear that the Cu d-d Coulomb interaction U_{dd} is very large (≈ 8 eV for the d^8 singlet states). From model system studies like Cu_2O

Springer Series in Materials Science, Vol. 11
Mechanisms of High Temperature Superconductivity
Editors: H. Kamimura and A. Oshiyama
Springer-Verlag Berlin Heidelberg © 1989

and CuO the O p-p Coulomb interaction was also large ($\approx 4-5$ eV) and, there are indications that the interatomic Cu-O Coulomb interactions are small < 1 eV.

On the other hand the one electron like interactions yielding the oxygen band width (≈ 5 eV) and the Cu d ligand field splittings etc. are also large. The large Cu-O hybridization leads to very strong O-Cu exchange and Cu-O-Cu superexchange interactions so that it is not simple to justify limiting approximations often made in various model Hamiltonian approaches. Is it for example justified to take the $U_{dd} \rightarrow \infty$ limit? Perhaps it is as far as the electronic (low energy) charge degrees of freedom are concerned but certainly virtual excitations involving the d^8 states are of extreme importance for describing the spin fluctuations since they contribute strongly to both the superexchange and the O-Cu exchange.

It was also pointed out in several talks that one cannot suffice by neglecting the Cu(d^8) multiplet splitting as one usually implies by using only one U value. The multiplet spread due to the Racah R and B parameters is large $\approx$ 6 eV for d^8 and cannot be neglected.

To take these correlation effects into account numerous different model Hamiltonians were proposed of which most of them included a large U_{dd}. These were almost all 2 band Hubbard-like models including an O 2p band and a Cu 3d band for which in most cases the orbital degrees of freedom were neglected. Upon the introduction of approximations the original two band Hamiltonian was usually replaced by a single band Hubbard model, a Kondo lattice model, a Kondo lattice model with (large) superexchange, or an Anderson impurity model. It is still not clear which one (if any) of these approaches closely resembles the true system.

Quite a number of scientists stressed that the exchange and superexchange interactions are probably of the same order of magnitude as the effective band widths or Fermi energies resulting in a strong coupling of spin and charge fluctuations. This is quite a different situation from that usually encountered in Kondo or heavy fermion like systems.

B) <u>Magnetism:</u>

There were a lot of very interesting results reported on the magnetic properties. These include the antiferromagnetic order in La_2CuO_4, $YBa_2Cu_3O_6$ and the magnetic insulating analogue of the Bi compound. In each case we have a magnetic insulating analogue for which the long range 3d antiferromagnetic order disappears at small hole concentrations. In each case the in-plane superexchange is exceptionally large compared to normal antiferromagnetic 3d TM compounds other than Cu oxides. J values of 0.12 eV were found from both neutron diffraction and 2 magnon Raman spectroscopy. It was also reported that the superexchange for CuO is also very large <u>$J \approx 0.1$ ev</u>. In addition neutron studies displayed low energy spin fluctuations and a kind of incommensurate 2d spin-spin correlation for larger hole concentration in $La_{2-x}Sr_xCuO_4$. Of great importance also is the observation that the local magnetic moment is independent of x. This is quite contrary to the predictions of band theory.

There was considerable discussion concerning the sign and magnitude of the exchange interaction between the oxygen hole and the Cu spins. It was generally agreed upon that no matter what the sign, it would in any case tend to either weaken the antiferromagnetic superexchange or if large and antiferromagnetic would have the influence of a magnetic dilution. In either case a 2D spin 1/2 antiferromagnet is expected to be strongly influenced either by strongly increasing (locally) the already large quantum spin fluctuations or by introducing non colinear (spiral) spin correlations - if the CuO exchange is considerably larger than the Cu-O-Cu superexchange. In several talks it was pointed out that the Cu-O exchange is expected to be antiferromagnetic if the hole is the in-plane oxygen $2p_\sigma$ orbital (σ relative to the Cu-O band) and ferromagnetic if the holes are in π orbitals (p_y or p_z) or if the holes are in the out-of-plane oxygen. This can he concluded from symmetry and Hunds rule arguments. It was also pointed out that only for the O in plane p_σ orbital do we expect a magnitude much larger than the superexchange. In all other cases the O-Cu exchange is expected to be considerably smaller. This is quite important for deciding between various proposed models.

C) Charge distribution, character and symmetry of substitution induced holes:

Convincing evidence was presented for the assertion that the holes are in primarily O2p states, First this is consistent with a large Cu U found from photoemission, also it is consistent with the trends expected for late 3d transition metal oxides. Also the analysis of photoemission data in terms of many body theory clearly shows this. The most direct evidence is from high energy electron energy loss and X-ray absorption spectroscopy involving the O1s orbital. In addition one would have expected dramatic changes in the Cu NQR since the electric field gradient would dramatically increase if the extra hole went into a $d_{x^2-y^2}$ orbital or dramatically decrease if the extra hole went into a $d_{3z^2-r^2}$ orbital. Polarization dependent XAS and angle dependent ELS on the Bi compound both show that the addition holes are p_x or p_y like i.e. in-plane polarized. Although the measurements cannot distinguish between σ and π like in-plane orbitals, they do virtually eliminate models based on substantial p_z hole character be it from the in-plane oxygens or the out-of-plane oxygens. For the $Y_1Ba_2Cu_3O_7$ compounds things are less clear. The O_{1s} measurements show that about 2/3 of the holes are in x,y orbitals but 1/3 is in p_z like states. These could be holes in the CuO chains or BaO planes in the Apex oxygen 4.

As mentioned above theory tells us that if the holes are in p_σ like orbitals a very strong hybridization with Cu d states is expected leading to a very strong antiferromagnetic exchange whereas if the holes are in p_π like orbitals a weak ferromagnetic exchange is expected.

These results spell trouble for certain models and theories!

D) Influence of holes on magnetism, optical and transport properties:

As mentioned above the introduction of holes profoundly affects the magnetic properties in $La_{2-x}Sr_xCuO_4$ a few percent of holes is sufficient to destroy the long range antiferromagnetic order although very strong short range spin-spin correlations remain and the magnetic momentum per Cu ion hardly changes. The spin-spin correlation length seems to be close to the average hole separation. A detailed study of the temperature dependent magnetic susceptibility shows the typical antiferromagnetic behaviour for the insulators with a weak ferromagnetic component in La_2CuO_4 due to the Dzialoshinski-Moriya antisymmetric exchange. As x increases the susceptibility gets a paramagnetic like 1/T component which grows with x initially but finally decreases again for larger x.

As far as the optical properties are concerned, I guess we are quite sure that structure seen at 1.5 - 2 eV in the $La_{2-x}Sr_xCuO_4$ is due to an O2p-Cu3d charge transfer transition. This is consistent with the predictions of models assuming large Cu d-d Coulomb interactions since for $U > \Delta$ where Δ is the charge transfer energy the first ionization states are O 2p states and the first electron affinity states are Cu (3d) (($3d^{10}$)) states. This feature is only visible for polarization $\perp$ to the c axis consistent with the assertions that the holes in La_2CuO_4 are in x^2-y^2 (b_{1g} symmetry in D_{4h} point group) states. Upon doping this feature rapidly loses oscillator strength and broad almost structureless absorption appears below 2 eV. A rather interesting observation for the low energy optical reflection for $La_{2-x}Sr_xCuO_4$ is that if a simple Drude fit with an energy independent relaxation time (τ) is done then $\hbar\omega_p$ is independent of x and the change in the optical absorption would be due to a variation in t. This interpretation was pointed out to be ridiculous since the integrated absorption yields the expected variation of the carrier concentration with x. This all shows that the low energy optical properties are not simple Drude like.

Of great interest also were the optical properties of high quality single crystals of YBaCuO. A Drude like structure with a reflectivity of 1 appeared for $\hbar\omega < 0.1$ eV followed by a broad structure extending up to the charge transfer edge. A possible interpretation of this is that the low energy part corresponds to coherent motion of the quasi particles with their dressing (magnetic excitonic of phononic), followed by a structure due to the incoherent motion in which a wake of magnetic or excitonic excitations are left behind.

The most dramatic influence of the holes on the transport properties is of course the onset of superconductivity for $x > 0.06$ in $La_{2-x}Sr_xCuO_4$ and $y < 0.6$ in $YBa_2Cu_3O_{7-y}$. There was some skepticism concerning the reported decrease of Tc in $La_{2-x}Sr_xCuO_4$ for $x > 0.25$. The apparent decrease could be due to an observed strong increase in the width of the onset and a strong decrease in the bulk superconducting fraction as observed from the Meissner effect. Another possible conclusion could be that T_c saturates at 40 K and abruptly drops to zero for $x \geqq 0.3$. Of interest also was to report that if we take into account that a considerable fraction of the holes go into the chain CuO layers in YBaCuO then the behaviour of T_c with the concentration of holes in the planes is similar (except for a higher maximum T_c) to that of $La_{2-x}Sr_xCuO_4$.

Considerable attention was also placed on the very non-conventional behaviour of the conductivity in the non-superconducting state. The conductivity is highly anisotropic and shows also a different temperature dependent behaviour. The normal state conductivity has a temperature dependence which is not that of a normal metal or that of variable range hopping. Zettl reported that quite a number of materials had a behaviour given by T^{α} exp (Eg/kT) with $0.5 < \alpha < 1$ for the conductivity parallel to the c axis. It is probably of quite some importance to try to understand this.

E) Fermi liquid or not?

I believe that we reached a consensus that there is no really reliable evidence for or against a Fermi liquid. I guess everyone agrees that the NMR T_1 behaviour certainly is not conventional Korringa like extrapolating to zero at $T=0$ and being linear in T over at least a considerable temperature range. This does not exclude Fermi liquid behaviour however, since we know of many examples for which the ordinary Korringa behaviour is not observed. That the T_1 behaviour in Cu and O is strongly different is also not unique since it is also observed in for example Al impurities in βMn. Transport measurements like Hall effect thermo power conductivity and specific heat often behave more like that expected from a very dirty semiconductcr than a Fermi liquid metal. On the other hand, this does not exclude Fermi liquid like behaviour since the very low carrier density as well as the disorder due to non-stoichiometry and random (?) distribution of substituents could cause strong deviations in the transport properties for a dirty Fermi liquid.

Recent photoelectron spectroscopy results do (be it weakly) finally exhibit a Fermi cut-off indicating a low density of states in the vicinity of the Fermi level. The interpretation of these measurements are however hampered by the fact that the photoelectron escape depth is smaller than a unit cell in materials like the Bi-Cu high T_c's. This material I believe most scientists agree is the only one known so far for which a good surface can he exposed by cleaning. But then again we also heard that scanning tunneling microscopy showed a surface "buckling" and superstructure which must strongly influence the surface electronic structure.

What about the magnetic susceptibility? Well, it certainly is not Pauli like! I think most scientists agreed that the temperature independent susceptibility observed above T_c for some of the oxidic Cu compounds is due to strong (short range) antiferromagnetic correlations in a 2D (localized) spin system rather than being due to a Fermi liquid behaviour although a contribution from the latter cannot be excluded.

In connection with the photoelectron spectroscopy observed Fermi edge, it was pointed out that the linear and symmetric behaviour in dI/dV curve in a tunneling measurement for $eV > E_{gap}$ certainly does not agree with the rather constant density of states found in photoelectron spectroscopy.

In fact the tunneling data in numerous systems for $T > T_c$ shows a structure suggesting a gap which could be evidence against Fermi liquid. It was however pointed out repeatedly that the tunneling measurements are difficult to reproduce in most cases.

Band structure results which basically assume Fermi liquid like behaviour were found to be quite successful in describing a few results. However, it was also made clear that they were quite contradictory to many of the experimental results found.

F) Dirt

The experimentalists were continually bugged by a rather loud voice which in no uncertain terms suggested that the results obtained were a (garbage, smutz, dirt or rubbish) effect. At first, one was annoyed but in the course of time the experimentalists were confronted with more and more conflicting results and were somewhat less annoyed at the comments of the garbage man. The bad quality of samples and the influence this has especially on transport properties (of which perhaps the thermoelectric power is the least sensitive to garbage), surface techniques like photoemission, tunneling results especially concerning small structures, optical properties near the insulating phases prevent (or at least should prevent) us from drawing decisive conclusions.

G) $2\Delta/kT_c$

At this symposium there was a strong lobby for $2\Delta/kT_c$ considerably larger than BCS (i.e. $\simeq 6$). This conclusion was reached primarily from tunneling measurements of various kinds. Although optical data of the YBaCuO single crystals indicate BCS relation it also exhibits structure at energies corresponding to $2\Delta/kT_c \approx 6$. Very interesting and important is the report of a strong anisotropy in the gap being much smaller along C($2\Delta/kT_c \approx 3.6$) than $\perp$ to C ($2\Delta/kT_c \approx 6$)

H) How important is the Cu?

An important factor in this discussion is the high T_c of some no Cu containing oxide like $Ba_{0.6}K_{0.4}BiO_3$ as well as $BaPb_{0.75}Bi_{0.25}O_3$.

In comparing these to the Cu containing compounds there are some similarities and some differences. From band structure calculations the bands crossing the Fermi level in the non Cu materials have substantial O 2p character as do the states close to E_F for the Cu compounds. In both cases we are dealing with insulating or semiconducting materials which become superconducting upon chemical substitution of such a kind as to introduce holes (low density) in the valence bands. In both cases a long range order is destroyed by substitution. For the BiO samples this is the charge density wave and for the Cu based compounds it is the antiferromagnetic (spin density wave) order. Some scientists also claim a similarity in possible charge disproportionation $Bi^{4+} \rightarrow Bi^{3+} + Bi^{5+}$ and $Cu^{2+} \rightarrow Cu^{1+} + Cu^{3+}$. Although this is well known for Bi, supported by the fact that Bi^{4+} compounds are not stable, it is less evident for Cu since a large number of stable Cu^{2+} compounds like the dihalides and even CuO exist. In fact it is known now that the extra holes go primarily into the O 2p band. Perhaps it is interesting to note that Au behaves quite differently from Cu in that Au^{2+} compounds are not stable and disproportionation is expected. $Au^{2+} \rightarrow Au^{1+} Au^{3+}$ so Au is similar in this sense to Bi although I don't know of any high T_c Au compounds.

Other important differences between the BiO and Cu based compounds are: BiO compounds have a substantial isotope effect whereas the high T_c Cu based compound do not. The gap to T_c ratio in BiO is BCS like ($2\Delta/kT_c \approx 3.5$) whereas it might be considerably larger for the Cu based compound. There is nothing magnetic about the BiO based compounds whilst the Cu based compounds have very dominating magnetic properties.

From all of this we can say that those scientists who like magnetic mechanisms usually highlight the differences whereas those who like non-magnetic mechanisms often highlight the simultaneities between these two classes. By looking at the emphasis placed on magnetism by this author, it is obvious that he probably prefers something magnetic.

I) Mechanisms:

The mechanisms discussed can broadly be put into 3 classes. Magnetic, charge fluctuations and exotic. The possibility of a magnetic mechanism was mainly deduced from the fact that all of the Cu based superconducting classes are magnetically ordered with strong antiferromagnetic superexchange for the insulating phases and the breakdown of long range order for small hole concentrations indicating a strong coupling between the hole and the magnetic excitations. The magnetic mechanism proposed all involved an exchange coupling between an oxygen hole and the Cu spins. In some cases this was assumed to be ferromagnetic as for the magnetic polaron or antiferromagnetic as in the Kondo like models or the local singlet models.

Although these models differ considerably they all have in common a "magnetically polarized cloud" around the hole which is an attractive region for a second hole. In all of the cases except that of the magnetic polaron one requires a very strong exchange interaction between the hole and the Cu spins. It was pointed out that this exchange interaction would be antiferromagnetic and very large ($> J_{superexchange}$) for a b_1 symmetry σ bonding hole and ferromagnetic but small ($< J_{superexchange}$) for all the other possibilities taking parameters as obtained from various spectroscopies. Most of the talks on magnetic mechanisms stressed the importance of a large or comparable Cu-O hole exchange as compared to the superexchange and the O hole effective bandwidth. This is basically the difference between these systems and heavy Fermion or Kondo lattice systems.

In the charge fluctuation mechanisms various modes of polarizabilities were suggested to provide the polarization cloud around a hole which would be an attractive place for a second hole. Some stressed the importance of the Cu-O interatomic Coulomb interaction predicting that the hole on a Cu would move to an oxygen site if a second hole approached another nearest neighbor oxygen site. The polarizable medium is a charge transfer involving only in-plane ions. An argument against this is that the interatomic Coulomb interaction required is probably much too large. This model predicts an increase in the Cu^{1+} in plane content with hole concentrating rather than the conventionally assumed decrease. Others stressed the importance of charge transfer involving the out of plane oxygens in providing the polarizable medium, and the importance of $d_{x^2-y^2}$ to $d_{3z^2-r^2}$ excitations on the same ion as a polarizable medium, although in general one was rather skeptical about the required size of the interaction.

All of the above mechanisms have been proposed basically with Fermi liquid ideas in mind. They basically assume one has mobile charged Fermions as charge carriers and the aim is to find ways to get an attractive interaction upon which one assumes that a superconducting state will be the ground state. More exotic theories discard the Fermi liquid approach and suggest the existence of effective excitations involving non-integral charge, spinless particles (holons) and chargeless spins (spinons). Since there is at the moment no really good evidence for or against Fermi liquid behaviour such exotic mechanisms are serious candidates.

Index of Contributors

The Second NEC Symposium

Mechanisms of High-Temperature Superconductivity

October 24–27, 1988, Hakone

1. H. Kamimura
2. G. A. Sawatzki
3. A. Zettl
4. Mrs. Sawatzky
5. W.P. Weber
6. Y. Kubo
7. E. Burstein
8. Mrs. Zettl
9. I. Takeuchi
10. Mrs. Burstein
11. H. Aoki
12. H. Shiba
13. H. Takagi
14. R.J. Birgeneau
15. M. Imada
16. A. Fujimori
17. J. Mizuki
18. M.R. Beasley
19. A.J. Freeman
20. R.B. Laughlin
21. R.E. Walstedt
22. Y. Endoh
23. N. Shima
24. S. Matsuno
25. S. Ohnishi
26. Y. Tokura
27. S. Iijima
28. T. Nakayama
29. N. Hamada
30. Mrs. Kamimura
31. Mrs. Müller
32. Mrs. Weber
33. Mrs. Birgeneau
34. S. Sugano
35. S. Sugai
36. Y. Nishina
37. H. Katayama-Yoshida
38. *sweet-heart* Shohko Yoshida
39. Mrs. Yoshida
40. Y. Iye
41. J.E. Hirsch
42. T. Fujita
43. J.S. Tsai
44. B. Batlogg
45. S. Uchida
46. K.A. Müller
47. T. Shinjo
48. S. Nakajima
49. B.G. Kotliar
50. S.Maekawa
51. T. Manako
52. D. Shinoda
53. I. Hirosawa
54. A. Yoshimori
55. H. Yasuoka
56. K. Asayama
57. H. Abe
58. A. Oshiyama

MIX
Papier aus verantwortungsvollen Quellen
Paper from responsible sources
FSC® C105338

If you have any concerns about our products,
you can contact us on
ProductSafety@springernature.com

In case Publisher is established outside the EU,
the EU authorized representative is:
Springer Nature Customer Service Center GmbH
Europaplatz 3, 69115 Heidelberg, Germany

Printed by Libri Plureos GmbH
in Hamburg, Germany